T0271922

Optical Fiber Current and Voltage Sensors

Optical Fiber Current and Voltage Sensors is the first book to provide a complete, comprehensive, and up-to-date treatment of the domain of fiber optic and polarimetric sensors, covering fundamental operating principles, characteristics, and construction as well as their use for high voltage current and voltage sensing.

Written by one of the most recognized experts in polarimetric sensing, Optical Fiber Current and Voltage Sensors begins by covering the fundamentals of polarized light, as well as essential sensor components. The author goes on to outline various sensor types and their applications, with a focus on sensors for electric phenomena. The chapters then lay out the demands that sensors need to meet, the technical obstacles and limitations that need to be considered. The book also covers comparisons to corresponding traditional instruments, as well as covering alternative non-conventional sensors.

This book will be of interest to a broad audience of prospective readers ranging from graduate research students, to researchers in physics and engineering fields, to industry professionals active in the field who wish to learn about the technology and/or interested in the development of new commercial solutions based on polarimetric-type fiber sensing.

Klaus Bohnert has been a Senior Principal Scientist at the ABB Corporate Research Center in Baden, Switzerland. Fields of research included nonlinear optics, ultrafast phenomena, and photorefractive effects in semiconductor materials. In 1986, he joined the Corporate Research Center of Brown Boveri Ltd (ABB Ltd since 1988) in Baden, Switzerland. There, his focus was on research and development of optical sensors for applications in the electric power industry, oil and gas industry, and chemical process industries. He has been an author or co-author of more than 130 scientific publications, about 130 conference presentations, and an inventor on over 90 patents.

Series in Fiber Optic Sensors
Series Editor: Dr. Alexis Mendez, MCH Engineering, LLC

This series of practical, concise, and modern guidebooks encompasses all types of fiber optic sensors, including fiber Bragg grating sensors, Fabry-Pérot sensors, interferometric sensors, distributed sensors, and biomedical sensors. The aim of the series is to give a broadly approachable, essential overview of the fundamental science, core technologies, design principles, and key implementation challenges in applications, such as oil, gas, and mining; renewable energy; defense/security; biomedical sciences; civil and structural engineering; and industrial process monitoring. Scientists, engineers, technicians, and students in any relevant field of practice or research will benefit from these unique titles.

Titles in the Series

Fiber-Optic Fabry-Pérot Sensors: An Introduction
Yun-Jiang Rao, Zeng-Ling Ran, Yuan Gong

An Introduction to Distributed Optical Fibre Sensors
Arthur H. Hartog

Plastic Optical Fiber Sensors: Science, Technology and Applications
Edited by Marcelo M. Werneck, Regina Célia S. B. Allil

Optical Fiber Current and Voltage Sensors
Klaus Bohnert

For more information about this series, please visit https://www.crcpress.com/Series-in-Fiber-Optic-Sensors/book-series/CRCSERINFIB

Optical Fiber Current and Voltage Sensors

Klaus Bohnert

CRC Press is an imprint of the
Taylor & Francis Group, an **informa** business

Designed cover image: Shutterstock_2222103149

First edition published 2024
by CRC Press
2385 NW Executive Center Drive, Suite 320, Boca Raton FL 33431

and by CRC Press
4 Park Square, Milton Park, Abingdon, Oxon, OX14 4RN

CRC Press is an imprint of Taylor & Francis Group, LLC

ISBN: 978-0-367-55584-9 (hbk)
ISBN: 978-0-367-57007-1 (pbk)
ISBN: 978-1-003-10032-4 (ebk)

DOI: 10.1201/9781003100324

Typeset in Minion
by SPi Technologies India Pvt Ltd (Straive)

Contents

Series Preface

Optical fibers are considered among the top innovations of the twentieth century, and Sir Charles Kao, a visionary proponent who championed their use as a medium for communication, received the 2009 Nobel Prize in Physics. Optical fiber communications have become an essential backbone of today's digital world and internet infrastructure, making it possible to transmit vast amounts of data over long distances with high integrity and low loss. In effect, most of the world's data flows nowadays as light photons in a global mesh of optical fiber conduits. As the optical fiber industry turned fifty in 2016, the field might be middle-aged, but many more advances and societal benefits are expected of it.

What has made optical fibers and fiber-based telecommunications so effective and pervasive in the modern world? Its intrinsic features and capabilities make it so versatile and very powerful as an enabling and transformative technology. Among their characteristics we have their electromagnetic (EM) immunity, intrinsic safety, small size and weight, capability to perform multi-point and multi-parameter sensing remotely, and so on. Optical fiber sensors stem from these same characteristics. Initially, fiber sensors were lab curiosities and simple proof-of-concept demonstrations. Nowadays, however, optical fiber sensors are making an impact and serious commercial inroads in industrial sensing, biomedical applications, as well as in military and defense systems, and have spanned applications as diverse as oil well downhole pressure sensors to intra-aortic catheters.

This transition has taken the better part of thirty years and has now reached the point where fiber sensor operation and instrumentation are well understood and developed, and a diverse variety of commercial sensors and instruments are readily available. However, fiber sensor technology is not as widely known or deeply understood today as other more conventional sensors and sensing technologies such as electronic, piezoelectric, and MEMS devices. In part this is due to the broad set of different types of fiber sensors and techniques available. On the other hand, although there are several excellent textbooks reviewing optical fiber sensors, their coverage tends to be limited and does not provide sufficiently in-depth review of each sensor technology type. Our book series aims to remedy this by providing a collection of individual tomes, each focused exclusively on a specific type of optical fiber sensor.

The goal of this series has been, from the onset, to develop a set of titles that feature an important type of sensor, offering up-to-date advances as well as practical and concise information. The series encompasses the most relevant and popular fiber sensor types in common use in the field, including fiber Bragg grating sensors, Fabry-Pérot sensors, interferometric sensors, distributed fiber sensors, polarimetric sensors, polymer fiber sensors, structural health monitoring (SHM) using fiber sensors, biomedical fiber sensors, and several others.

This series is directed at a broad readership of scientists, engineers, technicians, and students involved in relevant areas of research and study of fiber sensors, specialty optical fibers, instrumentation, optics, and photonics. Together, these titles will fill the need for concise, widely accessible introductory overviews of the core technologies, fundamental design principles, and challenges to implementation of optical fiber-based sensors and sensing techniques.

This series was originally made possible by the vision and enthusiasm of the first series manager, Luna Han, to whom I owe a debt of gratitude for his passion, encouragement, and strong support — from the initial formulation of the book project and throughout the initial series development.

I also wish to thank Carolina Antunes, the current Series Editor and manager at CRC Press, who has been a tremendous resource and facilitator, and a delight to work with in the various stages of development for this and the other series' volumes.

Information — as the saying goes — is knowledge. And thanks to the dedication and hard work of the individual volume authors as well as chapter co-authors, the readers have enriched their knowledge on the subject of fiber-optic sensors. I thank all the authors and extend my deep appreciation for their interest and support for this series and for all the time and effort they poured into its writing.

To the reader, I hope that this series is informative, fresh, and of aid in his/her ongoing research, and wish much enjoyment and success!

Alexis Méndez, Ph.D.
Series Editor
President
MCH Engineering, LLC
Alameda, CA

Preface

For more than half a century, optical current and voltage sensors have been a subject of research and development in academia and industry. The most important applications of the sensors are in electric power transmission and distribution. Further applications are in industry, e.g., in the electro-winning of metals in traction (railways), sciences, and other fields. In the electric power world, the lightweight sensors are alternatives to heavy traditional instrument transformers. Besides saving tons of material, optical current and voltage sensors provide a more accurate image of the primary current or voltage waveforms, offer high application flexibility, ease of installation, and are safer to operate. Moreover, the sensors perfectly fit into modern digital systems for substation communication and control.

This book aims to trace the history of optical current and voltage measurement, present the numerous sensing concepts that have been developed in the course of time, and explain the physics behind the technology. The presentation portrays important transducer materials and their properties, special sensing fibers and dedicated optical components, and depicts examples of theoretical and experimental performance data. The book also points out the considerable technical obstacles that had to be overcome on the path to viable sensors capable of meeting the stringent industry demands. Examples of practical applications illustrate the benefits of optical current and voltage sensors versus conventional solutions. With somewhat lesser emphasis, the book also considers optical magnetic and electric field sensors for other purposes such as field mapping, e.g., on antennas, high frequency electronic circuits, or in electro-magnetic compatibility studies. Another important objective is a comprehensive overview of the literature in the field. The citations mostly refer to archival journal articles. Articles in conference proceedings are cited in the absence of a corresponding journal paper or for reasons of priority.

The book will be of interest to a broad audience ranging from graduate research students, to researchers in physics and engineering fields, to industry professionals active in the field who wish to learn about the technology and its applications. The first part of the book on *Fundamentals* introduces readers less intimate with the field of optics to the basics of optical fibers, polarized light, the Jones matrix calculus for describing the propagation of polarized light through birefringent media, and polarimetric and interferometric sensing techniques. *Part II* and *Part III* then present the actual sensors, the underlying physics, and applications.

I am indebted to many former colleagues at ABB Ltd for fruitful collaborations, an inspiring and enjoyable work atmosphere, and countless discussions. They are too numerous to mention all by name. Special thanks are due to my former team colleagues at Corporate Research: Philippe Gabus, Andreas Frank, Berkan Gülenaltin, Lin Yang, Xun Gu, Georg Müller, Miklos Lenner, Stephan Wildermuth, Sergio Marchese, Hubert Brändle, and Jürgen Nehring. I would like to thank Alexis Méndez, the editor of the book series, for his encouragement to undertake this endeavor. I also thank several professional societies, particularly Optica, IEEE, and CIGRE, for granting permission to reprint various Figures as indicated in the corresponding captions. Not least, I would like to thank the staff at Taylor & Francis for their support in the preparation of the manuscript.

List of Abbreviations

AIS	Air insulated switchgear/substations
BS	Beam splitter
CT	Current transformer
CVD	Capacitive voltage divider
DFB	Distributed feedback
DGD	Differential group delay
DOP	Degree of polarization
EOVT	Electro-optic voltage transducer (transformer)
FOCS	Fiber-optic current sensor
FRM	Faraday rotator mirror
FSR	Free spectral range
FWHM	Full width at half maximum
GIS	Gas-insulated switchgear
HV	High voltage
HVDC	High voltage direct current
IEC	International Electrotechnical Commission
LC	Liquid crystals
LED	Light-emitting diode
LHC	Left-handed circular
LN	Lithium niobate, $LiNbO_3$
MEMS	Micro electromechanical system
MIOC	Multi-functional integrated optical chip
MMF	Multimode fiber
MNF	Microfibers and nanofibers
MOCT	Magneto-optic current transducer
MV	Medium voltage
NEC	Noise equivalent current
NLC	Nematic liquid crystals
OCS	Optical current sensor
OCT	Optical current transducer (transformer)
OTDR	Optical time domain reflectometry
OVS	Optical voltage sensor

OVT	Optical voltage transducer (transformer)
PCF	Photonic crystal fiber
PD	Photodiode
PDL	Polarization dependent loss
PER	Polarization extinction ratio
PM	Polarization maintaining
PMF	Polarization maintaining fiber
PMB	Polarization mode beat
POVT	Piezo-optic voltage transducer
PZT	Piezo-electric transducer
RFOCS	Reflective fiber-optic current sensor
RHC	Right-handed circular
RT	Room temperature
SLED	Superluminescent light-emitting diode
SMF	Single-mode fiber
SOP	State of polarization
VT	Voltage transformer
WDM	Wavelength division multiplexer

Introduction

Optical current and voltage sensors have been developed predominantly as alternatives to conventional instrument current and voltage transformers in alternating current (ac) electric power transmission and distribution. Further fields of application include high voltage direct current (HVDC) power transmission, railways, electro-winning of metals and further industrial uses, scientific applications, e.g., in plasma physics and particle accelerators, and others. Conventional instrument current and voltage transformers for high voltage (HV) substations, commonly referred to in the industry as CT and VT, respectively, are bulky, typically oil and paper insulated devices (SF_6 gas insulation is an alternative), and heavy – with weights as large as several tons. By contrast, the all-dielectric nature of optical sensors drastically reduces size and weight, inherently insulates secondary equipment from high voltage, and provides new options for the integration of current and voltage measurement in HV systems. The oil-free insulation eliminates any risk of explosive failure, e.g., during earthquakes. Furthermore, the sensors' large bandwidth and absence of phenomena like magnetic saturation or ferro-resonances provide a more accurate image of the primary current or voltage waveforms, especially during faults, and the sensors' digital electronics is in line with modern digital substation communication and control.

On the other hand, whereas CT and VT devices are well-established instruments that have been optimized over many years, not least in terms of cost, optical current and voltage sensors represent a new technology in a conservative industry and had to overcome considerable challenges. The electric power industry demands high measurement accuracy, often to within ±0.2% at outdoor conditions. Reliability of the sensors, especially at transmission voltages (voltages from 110 kV onwards), are of utmost importance. The lifetime of conventional measurement transformers is in excess of 30 years and often taken as a reference. Obstacles have also included non-conventional signal formats that complicated the sensors' interfacing to conventional substation control and protection equipment and unfamiliar handling and installation procedures. The cost of conventional transformers rapidly rises with increasing system voltage, while this is much less so for optical sensors. Hence, optical sensors may be very cost-attractive at high rated voltages yet find it more challenging to compete with traditional solutions at lower voltages.

DOI: 10.1201/9781003100324-1

CONVENTIONAL INSTRUMENT CURRENT AND VOLTAGE TRANSFORMERS

Figure I.1 depicts conventional current transformers in a HV substation and illustrates a typical environment of current and voltage measurement in ac electric power transmission. CT generate via magnetic induction from the primary current I_p on high voltage a secondary current I_s of typically 1 A at the rated primary current I_r (2 A and 5 A or further options), that serves as input to metering or relaying devices [1]. Usually, CTs have several ferromagnetic cores optimized for metering and protection functions. Metering cores typically cover currents up 1.2 I_r, whereas protection cores cover the much larger ranges of fault currents, e.g., up to 30 I_r. In a so-called, top-core CT as shown in Figure I.1, the cores reside in a tank at the CT top, whereas in hairpin style CTs, schematically illustrated in Figure I.2a, they reside inside a tank at the bottom.

FIGURE I.1 Conventional 380 kV current transformers in a high-voltage substation.

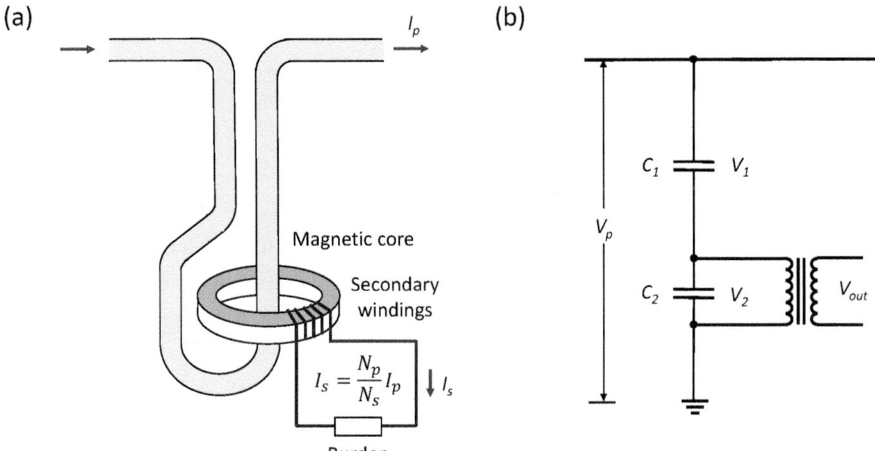

FIGURE I.2 (a) Schematic illustration of a hairpin-type current transformer; N_p and N_s are the number of primary and secondary windings, respectively ($N_p = 1$); (b) equivalent circuit diagram of a capacitive voltage divider.

In electric power transmission, typical rated currents are from 400 A up to 6 kA, and rated voltages range from 110 kV to 800 kV; even systems up to 1200 kV have been deployed. (Note: Throughout the book, the stated alternating voltages are root-mean-square (rms) values, if not specified otherwise.) Fault currents can well exceed 100 kA over a few current cycles until a protective relay initiates current interruption. A CT must comply with relevant industry standards like IEC60044-8 [2]. Among many other things, those standards specify allowed error limits for various accuracy classes, which must also be observed by corresponding optical sensors. The frequently applied IEC measuring class 0.2, for instance, requires accuracy to within ±0.2% between 100% and 120% of the rated current, ±0.35% at 0.2 I_r, and ±0.75% at 0.05 I_r. Phase error limits range from ±30 min at 0.05 I_r to ±10 min at I_r. Classes 0.2s and 0.1 specify even tighter limits (Table I.1). These limits are valid at the rated power frequency (50 Hz or 60 Hz); wider limits apply at harmonics. Larger errors are allowed for protective CT at fault currents (IEC class 5P requires ±5%). CTs in regional power distribution grids at voltages between 1 kV and 36 kV often obey accuracy classes

TABLE I.1 Limits of Current and Phase Errors at Rated Power Frequency for Measuring Current Transformers [2]

	± Percentage Current Error Limits *Current in % of Rated Current*					**± Phase Error Limits (minutes)** *Current in % of Rated Current*				
Accuracy Class	*1*	*5*	*20*	*100*	*120*	*1*	*5*	*20*	*100*	*120*
0.1	–	0.4	0.2	0.1	0.1	–	15	8	5	5
0.2s	0.75	0.35	0.2	0.2	0.2	30	15	10	10	10
0.2	–	0.75	0.35	0.2	0.2	–	30	15	10	10
0.5	–	1.5	0.75	0.5	0.5	–	90	45	30	30
1	–	3	1.5	1.0	1.0	–	180	90	60	60

TABLE I.2 Limits of Voltage and Phase Errors at Rated Power Frequency for Measuring Voltage Transformers [3]

	± Percentage Voltage Error Limits *Voltage in % of Rated Voltage*			± Phase Error Limits (minutes) *Voltage in % of Rated Voltage*		
Accuracy Class	*80*	*100*	*120*	*80*	*100*	*120*
0.1	0.1	0.1	0.1	5	5	5
0.2	0.2	0.2	0.2	10	10	10
0.5	0.5	0.5	0.5	20	20	20
1	1	1	1	40	40	40

0.5 or 1.0. Ideally, an optical current sensor should cover both metering and protection current ranges, if the sensor is to replace a multiple-core conventional CT.

Conventional voltage transformers for transmission voltages are either inductive transformers (VT) or capacitive voltage dividers (CVD). Externally, they look similar to CT. CVD are commonly the more economic option at voltages above 145 kV. Particularly at medium voltages (in the 1 kV to 36 kV range), resistive capacitive voltage dividers are another option. CVD comprise two connected series of capacitors C_1 and C_2 that divide the primary voltage V_p in voltages V_1 and V_2 (Figure I.2b) [1]. With $C_1 \ll C_2$, voltage V_2 across the bottom capacitor series C_2 is much lower than V (and V_1) and corresponds, e.g., to 22 kV/$\sqrt{3}$ at the rated voltage V_r. An auxiliary inductive transformer then steps down V_2 to a standard output voltage, which is typically is 110 V/$\sqrt{3}$ in Europe (ratings in other regions may differ but are similar). Again, V_2 serves as an input to a metering instrument or substation/grid control and protection. It should be noted that the rated primary voltage V_r refers to the rms voltage between two phases, whereas voltage transformers measure the phase-to-ground voltage, which is $V_r/\sqrt{3}$. Naturally, voltage transformers must meet similarly strict accuracy requirements as current transformers. Commonly, HV metering and measurement VT meet IEC accuracy class 0.2, which specifies voltage error limits of ±0.2% and phase error limits of ±10 min at voltages between 80% and 120% of V_r (Table I.2). Voltage error limits of protective VT according to class 3P are ±3% between $0.05V_r$ and F_v V_r, where the rated voltage factor F_v can be $F_v = 1.2$, 1.5, or 1.9. The maximum allowed phase error is ±120 min [3].

OPTICAL CURRENT AND VOLTAGE SENSORS

Initial exploratory research work on optical current sensors (OCS), also called optical current transformers or transducers (OCT), began in the 1960s and early 1970s both in academia and industry. From early on, the prospect of replacing heavy instrument transformers by a simple optical fiber around the current conductor fascinated the researchers. Electric power product manufactures with corresponding research projects included Siemens, Brown Boveri & Cie, Asea, AEG, and Alsthom in Europe, General Electric and Westinghouse in North America, and several companies in Japan including Toshiba, Hitachi, and Tokyo Electric Power Company (TEPCO). Even though there was no lack of field demonstrations, often those early sensors were still premature and far from viable products. Hence, for a long time, many in the industry considered the sensors as interesting but exotic, and it was not before the 2000s until optical sensors, in particular current sensors, started to become a more serious alternative to traditional instruments.

The vast majority of optical current sensors make use of the Faraday effect, i.e., the phenomenon that the plane of polarization of linearly polarized light rotates by a certain angle when the light traverses a transparent medium such as glass in the presence of a longitudinal magnetic field. Over several decades, researchers investigated numerous sensor designs. Magneto-optic sensing media included bulk pieces of glass, crystalline materials, and particularly optical fibers. The works benefitted from the advances in optical fiber communication and opto-electronics. Many communication system components such as fibers, fiber polarizers, directional couplers, connectors, phase modulators, polarization rotators, light sources, and photodetectors have also found uses in numerous optical sensors including current and voltage sensors. In addition, dedicated specialty fibers were developed.

Simple Faraday effect current sensors derive the current from a local magnetic field measurement in the vicinity of the current conductor but can be prone to field disturbances and crosstalk from neighbor currents. The combination of such sensors with a magnetic flux concentrating core around the conductor significantly improves the robustness of the current signal. True current sensors measure the current according to Ampere's law as a closed-loop integral $\oint \boldsymbol{H} \cdot d\boldsymbol{s}$ of the magnetic field around the conductor, and hence are largely insensitive to geometrical parameters, varying field distributions, and crosstalk. Corresponding glass block current sensors guide the light by multiple internal reflections around the conductor but offer little flexibility in form factor and sensitivity. By contrast, all-fiber-optic current sensors (FOCS) with a coil of sensing fiber around the conductor can be easily adapted to wide ranges of current and conductor cross-section. Early FOCS, however, were severely limited by mechanical and temperature-dependent stresses acting on the fiber. Particularly, the bending of the fiber to a coil, the fiber coating, and packaging were sources of stress. Also, mechanical shock and vibration could produce signal perturbations in excess of the current signal. Inadequate sensor components and manufacturing challenges represented further obstacles. But finally, specialty sensing fibers, sensor configurations that intrinsically cancelled mechanical disturbances and influences of temperature, and other measures resulted in robust FOCS with accuracy as high as ±0.1% in wide temperature ranges (Figure I.3).

Other optical magnetic field and current sensors make use of magnetostriction, field-induced refractive index changes in magnetic fluids, or Lorentz forces in micromachined devices. Hybrid HV current transducers combine traditional instrumentation with optical interrogation techniques and as a result, they too significantly reduce expenditures for electric insulation.

Optical voltage sensors (OVS) have been developed in parallel to optical current sensors. Commercial OVS were introduced as early as the 1990s but have not quite reached the appeal of fiber-optic current sensors. FOCS are attractive in that a simple optical fiber around the conductor measures a closed-loop integral $\oint \boldsymbol{H} \cdot d\boldsymbol{s}$, that is the current. An equivalent optical voltage sensor would need to measure the path integral of the electric field $\int \boldsymbol{E} \cdot d\boldsymbol{s}$ between ground and high voltage potentials. Conventional optical fibers are unsuited for this task. Researchers have made attempts towards electro-optic fibers, but the

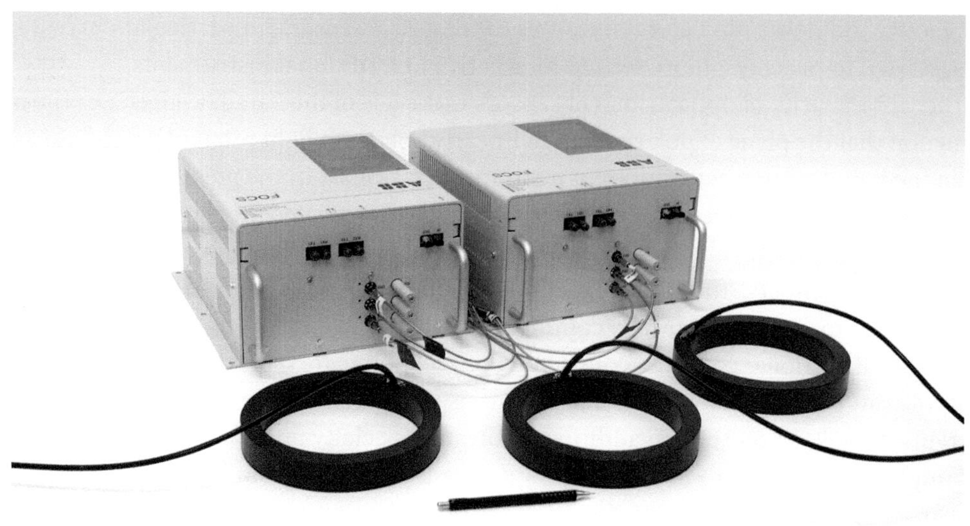

FIGURE 1.3 Three-phase fiber-optic current sensor system (redundant) for integration in HV circuit breakers. Each ring contains two optical fiber coils (Chapter 12) [4].

works did not proceed beyond exploratory stages. Instead, most optical voltage sensors employ electro-optic crystals or piezoelectric transducers with fiber-optic readout. Commonly, OVS for air-insulated HV substations require substantial efforts and expenditures for electric insulation and field steering, which to a certain degree outweigh the benefits of the optical technology.

Besides the linear electro-optic effect (Pockels effect) and converse piezoelectric effect, a variety of other phenomena have been employed for optical electric field and voltage sensing that include the Kerr effect, Stark effect, electro-absorption, electro-luminescence, electro-striction, electrostatic forces, and field-induced second harmonic generation. Further applications of electric field sensitive devices include mapping of electric field distributions, e.g., on antennas, high speed electronic circuits, electric power equipment, etc., electromagnetic compatibility studies, accelerators in high energy physics, or high power electromagnetic pulse detection to name a few.

REFERENCES

1. R. Minkner and J. Schmid, *The Technology of Instrument Transformers* (Springer Vieweg, 2021).
2. Standard of the International Electrotechnical Commission (IEC), IEC60044-1, Instrument transformers – Part 1: Current transformers (2003).
3. Standard of the International Electrotechnical Commission (IEC), IEC60044-7, Instrument transformers – Part 7: Electronic voltage transformers (1999).
4. M. Lenner, A. Frank, L. Yang, T. M. Roininen, and K. Bohnert, “Long-term reliability of fiber-optic current sensors,” *IEEE Sens. J.* 20(2), 823–832 (2020), doi: 10.1109/JSEN.2019.2944346

PART I

Fundamentals

CHAPTER 1

Optical Fibers

Fused silica optical fibers with loss below 20 dB/km became available in the early 1970s after breakthroughs in the fiber preform preparation and fiber drawing at Corning Glass Works [1, 2]. (Before 1970, typical fiber loss had been on the order of 1000 dB/km [3].) By the end of the 1970s, NTT in Japan had reduced losses to 0.2 dB/km at 1550 nm, close to the fundamental limit by light scattering [4]. The advent of low loss fibers initiated the era of optical fiber communication and in parallel opened up the field of optical fiber sensing. Fibers for sensors serve to guide light to and from a discrete sensing element, or the fiber itself—often a specialty fiber—acts as the sensing element. This chapter summarizes some basics of optical fibers. Readers can find more comprehensive information in several excellent books and reviews on the subject, e.g., [5–9].

1.1 LIGHT WAVE PROPAGATION

A standard step-index optical fiber consists of a cylindrical core with radius a and refractive index n_{co} and a cladding with radius b and index n_{cl}, where $n_{co} > n_{cl}$ (Figure 1.1) [8]. The light is guided in the core by total internal reflection at the core/cladding interface. In a simple geometrical optics picture, light rays impinging at the interface at angles θ larger than the critical angle θ_c, defined by $\sin\theta_c = n_{cl}/n_{co}$, remain confined to the core, whereas rays at $\theta < \theta_c$ gradually leak out. Accordingly, the acceptance angle θ_i for guided light and the numerical aperture NA in air ($n_0 = 1$) is defined by

$$NA = \sin\theta_i = \left(n_{co}^2 - n_{cl}^2\right)^{1/2}. \tag{1.1}$$

Typical optical fibers are weakly guiding waveguides ($n_{co} \approx n_{cl}$); the NA can then be approximated by

$$NA = n_{co}\left(2\Delta\right)^{1/2}, \tag{1.2}$$

with $\Delta = (n_{co} - n_{cl})/n_{co}$. The NA of single-mode fibers (see below) is typically between 0.1 and 0.3, whereas multimode fibers often have an NA as high as 0.4–0.5. Besides the

DOI: 10.1201/9781003100324-3

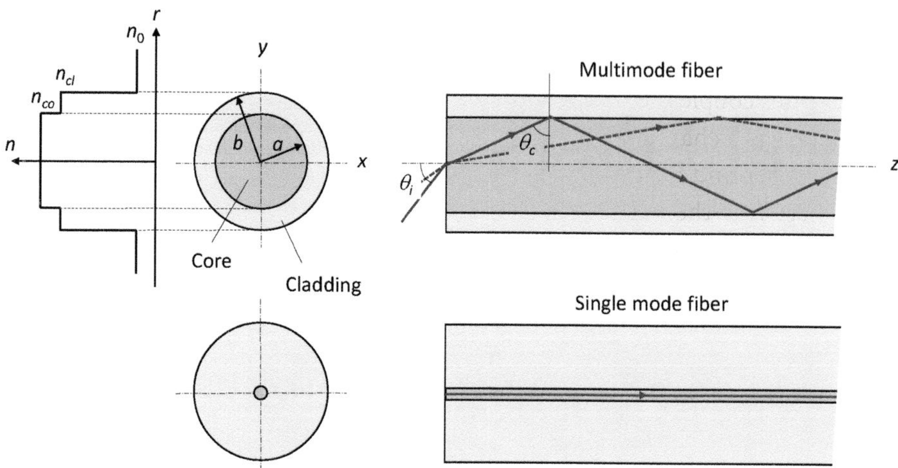

FIGURE 1.1 Multimode and single-mode optical fibers.

meridional rays depicted in Figure 1.1, these are rays that repeatably intersect the fiber axis; a fiber also supports so-called skew rays (not shown) not intersecting the axis and having an azimuthal path component.

Most optical fibers are made from fused silica. The fiber core commonly contains GeO_2 dopants that raise the refractive index, while the cladding is of pure silica (n_{cl} = 1.444 at a wavelength l of 1.55 mm). The index difference $n_{co} - n_{cl}$ is typically in the range 10^{-3} to 10^{-2} [8]. Modern long-haul communication systems employ fibers having a core of pure silica and a cladding with fluorine dopants that lower the refractive index. A pure silica core slightly reduces transmission loss.

Geometrical optics describes the light propagation reasonably well in multimode fibers with large cores ($a \gg l$). More accurately, the optical field in a fiber is derived from the wave equation, which follows from Maxwell's equations for electromagnetic waves. The solutions of the wave equation for the given boundary conditions, that is, the refractive index profile and wavelength, are specific spatial distributions of the electromagnetic field, called fiber modes [5–8]. There are three categories of modes: Guided or core modes, cladding modes, and a continuum of unconfined radiation modes. In the following, we will only consider guided core modes. (Cladding and radiation modes commonly terminate in the protective coating of a fiber (not shown in Figure 1.1).) The guided modes are denoted as HE_{lm} or EH_{lm} (l, m = 1, 2, 3, …) and are hybrid modes, because they have non-zero electric and magnetic field components, E_z and H_z, where z is propagation direction (fiber axis). In HE modes, H_z dominates over E_z, while the opposite is true for EH modes. The radial and azimuthal field components E_r, E_ϕ, and H_r, H_ϕ of the modes are not independent but can be expressed in terms of E_z and H_z. In addition to the hybrid HE and EH modes, the core modes include purely transverse modes; those are the electric modes TE_{0m} and the magnetic modes TM_{0m} with $E_z = 0$ and $H_z = 0$, respectively. The HE and EH modes correspond to the skew rays, and the TE and TM modes correspond to the meridional rays of the geometrical optics picture. Mathematically, the modes are orthogonal, i.e., there is no crosstalk between the modes of an ideal fiber, and their spatial field distributions remain constant as the light propagates along the fiber. (Perturbations such as fiber bends or fiber inhomogeneities couple the modes, however.)

Each mode has a characteristic propagation constant b that represents the mode's optical phase shift per unit length of fiber. Accordingly, the mode's effective refractive index is $n = b/k$, where k is the vacuum wavenumber: $k = 2p/l$; n lies in a range $n_{cl} < n < n_{co}$. Normalization of b in terms of the fiber parameters results in a normalized propagation constant b ($0 < b < 1$) [10]:

$$b = \frac{(\beta/k)^2 - n_{cl}^2}{\left(n_{co}^2 - n_{cl}^2\right)}. \tag{1.3}$$

For weakly guiding fibers ($n_{co} \approx n_{cl}$), b is approximately equal to

$$b \approx \frac{(\beta/k) - n_{cl}}{n_{co} - n_{cl}} = \frac{n - n_{cl}}{n_{co} - n_{cl}}. \tag{1.4}$$

Whereas n determines the phase velocity $v_p = c/n$, that is, the velocity of a strictly monochromatic light wave, the group refractive index n_g determines the group velocity $v_g = c/n_g$, that is, the velocity at which the envelope of a light pulse propagates through a medium [8]:

$$n_g = n + \varpi \frac{dn}{\varpi}, \tag{1.5}$$

where ϖ is the optical frequency. The group index also determines the optical path length L of a mode in a fiber of length l ($L = n_g l$), e.g., in fiber interferometers.

In weakly guiding fibers, both E_z and H_z are close to zero. The modes are then often approximated by linearly polarized modes, designated as LP_{lm} [5, 10]. A given LP-mode has $2l$ field maxima along an azimuthal line around the core center and m maxima along a radius vector. Figure 1.2 schematically depicts the electric field distributions of the three

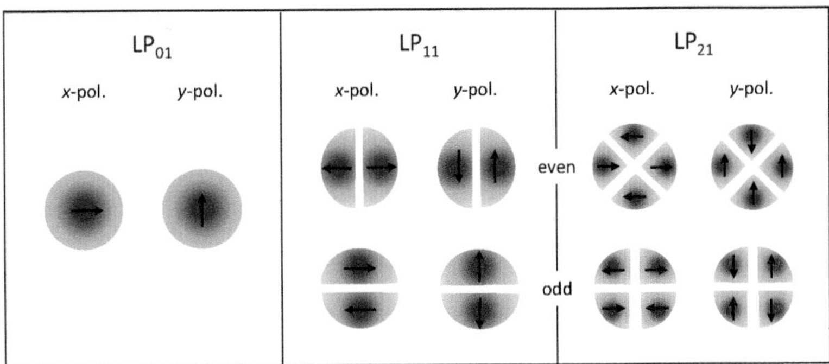

FIGURE 1.2 Transverse electric field distribution (schematically) of the three lowest order LP-modes. The arrows indicate the field directions (based on [3, 10, 12]).

lowest order LP modes. The modes with $l \geq 1$ can assume one of two spatial distributions (denoted as even and odd). Their fields along the azimuthal coordinate ϕ vary in proportion to $\cos(l\phi)$ and $\sin(l\phi)$, respectively. The modes are polarized either along the x coordinate or along the y coordinate. For a fiber with perfect circular symmetry, the spatial and polarization sub-states of a given mode are degenerate, i.e., all four states have the same propagation constants. Deviations from circular symmetry lift the degeneracy, for example, in fibers with an elliptical core that are used in certain types of optical fiber sensors [11]. It should be mentioned that each LP mode can be related to one or several exact fiber modes. For instance, the LP_{01} mode corresponds to the HE_{11} mode, while the LP_{11} mode corresponds to a superposition of the HE_{21}, TM_{01}, and TE_{01} modes, all of which have similar propagation constants [10]. Analytic formulas for the modal field distributions can be found in [10].

Another important parameter of a fiber is its normalized frequency V or V-number:

$$V = \frac{2\pi}{\lambda} a \left(n_{co}^2 - n_{cl}^2\right)^{1/2}. \tag{1.6}$$

Figure 1.3 shows the normalized propagation constants of some lower order modes as a function of V [10]. Step index fibers with $V < 2.405$ only support the fundamental mode LP_{01}. Such fibers are called single-mode fibers (SMF), as opposed to few mode fibers or multimode fibers with $V > 2.405$. The first higher order mode LP_{11} is cut off at the so-called cut-off frequency $V_c = 2.405$. At V_c, the LP_{11} effective refractive index equals n_{cl} and mode guiding ceases. It is obvious that the cut-off frequencies increase with increasing mode order (the corresponding cut-off wavelengths decrease). Note that at $V_c \to \infty$, n_e of all modes approaches n_{co}. The number of guided modes in a fiber with $V \gg 1$ is approximately $V^2/2$ [10].

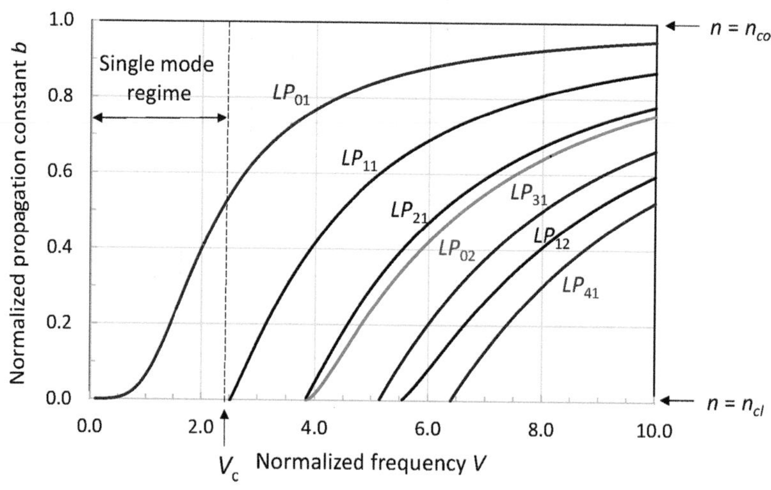

FIGURE 1.3 Normalized propagation constant b vs. normalized frequency V for some low order LP-modes (after Gloge [10], with permission from © The Optical Society).

The formulas for the modal field distributions are somewhat cumbersome. In practice, the fundamental mode LP_{01} is often approximated by a Gaussian function [8]:

$$E(r) = E(0)\exp\left(-\frac{r^2}{w^2}\right), \tag{1.7}$$

where w is the so-called mode field radius, that is, the radius where the mode field has dropped to $1/e$ and the intensity to $1/e^2$. It is important to note that the fields of the guided modes are not restricted to the fiber core but extend into the cladding. The power fraction of the fundamental mode propagating in the core, called confinement factor Γ, is the smaller the lower the V-number and given by [8]:

$$\Gamma = 1 - \exp\left(-\frac{2a^2}{w^2}\right). \tag{1.8}$$

Whereas at $V = 2$, about 75% of the mode power propagates in the core, that fraction reduces to 20% at $V = 1$. Therefore, telecommunications fibers commonly operate in the range $2 < V < 2.4$ [8]. Optical fiber sensors often combine different types of fibers. The transmission of a splice between two fibers having mode field radii w_1 and w_2 is [7]:

$$T = \left(\frac{2w_1w_2}{w_1^2 + w_2^2}\right)^2. \tag{1.9}$$

An important characteristic, in particular of telecommunication fibers, is dispersion. Contributions are material dispersion, waveguide dispersion, and intermodal dispersion (only in multimode fibers). Dispersion sets an upper limit to the bandwidth of the transmitted signal. However, dispersion plays hardly a role in the fiber sensors of the present text. We will consider dispersion when necessary, in particular polarization mode dispersion in specialty fibers, but otherwise refer the interested reader to the literature cited further above.

1.2 FIBER TYPES

As already noted, optical fibers are either single-mode or multimode fibers. Single-mode fibers only support the fundamental mode HE_{11} (LP_{01}). They also can be operated as few mode fibers by choosing a wavelength of operation below the cut-off wavelength. The most important application of SMF is long haul communication. Having no intermodal dispersion, SMF provide much higher bandwidths than MMF. Typical core diameters of SMF for telecom wavelengths (1.3–1.6 nm) are 9–10 μm. The standard outer diameter is 125 μm (not counting the protective buffer coating). Single-mode fiber doped with rare earth ions such as erbium ions are employed in fiber amplifiers, fiber lasers, and broadband fiber light sources [13].

Fibers designated as multimode fibers typically have core diameters ranging from 50 μm to about 1 μm and support hundreds or even thousands of fiber modes. Due to their large core, MMF can be combined with inexpensive LED sources; alignment tolerances, e.g., at fiber connectors, are less critical than with SMF. MMF are often employed for short haul communication. In sensors, they serve to send light to or return light from a bulk sensing element such as an electro-optic crystal. In certain hybrid current and voltage sensors, MMF provide optical power to electronic circuits on high voltage potential.

Graded index fibers represent a subcategory of MMF and have a parabolic index profile that reduces intermodal dispersion and thus enhances the bandwidth. Further MMF types include hard clad silica (HCS) fibers with a silica core and a polymer cladding. HCS fibers are particularly suited for harsh environments. Polymer optical fibers (POF) are low-cost and easy-to-install solutions for short distances. At this point, it is also worth mentioning that more recently, multicore fibers, each core supporting several modes, have become the subject of intense research. By combing wavelength-division multiplexing (WDM) and so-called space-division multiplexing (SDM), the transmission capacity of the fibers reaches into the Pbit/s regime [9].

The optical attenuation in fused silica fibers strongly increases beyond 1.8 nm [8]. For a review on fiber glasses and fiber types for the infrared, see, e.g., Tao et al. [14]. Another important class of fibers that has been developed since the 1990s are photonic crystal fibers (PCF), or more generally, micro-structured fibers [15]. PCFs have also found uses in optical fiber sensors [16]. Other than conventional fibers, PCFs do not guide the light via a radially varying material composition, but an array of air holes parallel to the fiber axis makes for the wave guiding. PCF properties including the dispersion, birefringence, or nonlinear optical characteristics can be custom-designed within wide limits.

An important component of any glass fiber is a protective buffer coating. Typically, fibers are provided with a UV cured dual acrylate coating consisting of a soft inner layer that prevents fiber micro-bending and a harder outer layer for mechanical robustness. A standard overall diameter of acylate coated 125-mm-diameter fibers is 245 μm. Acrylate coatings are suited for temperature from −55°C to 85°C. Special acrylates enable operation at somewhat higher temperatures. Fibers with polyimide coatings can be operated up to about 300°C [17]. Other coating options for harsh environments include metal coatings and carbon coatings (hermetically sealed fibers) [18].

1.3 POLARIZATION-MAINTAINING FIBERS

The two orthogonal polarization states of the fundamental mode, $HE_{11,x}$ and HE_{11y}, of an ideal single-mode fiber are degenerate, i.e., they have the same propagation constants ($b_x = b_y$). Hence, an ideal fiber will preserve the polarization of injected light. In practice, perturbations like a non-perfectly circular core, fiber bends, and others lift the degeneracy of the orthogonal states. As a result, the light assumes varying elliptical polarization states along the fiber. Polarization maintaining fibers (PMF) are highly birefringent fibers with well-defined birefringent axes x, y (see Figure 1.4 below). The propagation constants b_x, b_y differ significantly, so that coupling between the orthogonal polarization modes is strongly suppressed. Typically, the birefringence of PMF is due to an elliptical core or a stress field embedded into the fibers (Figure 1.4) [19]. Elliptical core fibers [11] are easier to manufacture, and the birefringence varies less with temperature, which can be beneficial in sensor

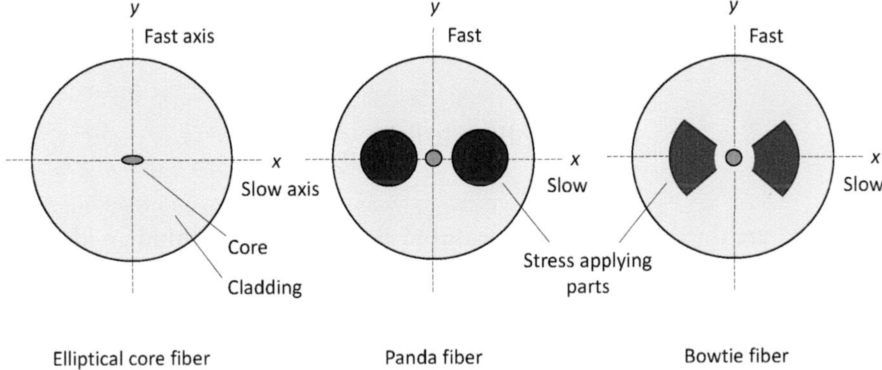

FIGURE 1.4 Cross-sections of some polarization-maintaining fibers.

applications. On the other hand, the elliptical mode field may result in higher loss at splices to circular core fibers.

Typical ratios of the major and minor core axis are between 1.5 and 4. The cladding of fibers with stress-induced birefringence comprises so-called stress applying parts (SAP). The SAP cross-sections may be circular (the fibers are then called Panda fibers), bowtie shaped (bowtie fiber), or the SAP may be designed as an inner elliptical cladding. Commonly, the SAP are doped with boron (B_2O_3) and have a larger thermal expansion coefficient than the pure silica cladding. Hence, a stress field builds up when the fiber cools down after drawing. The slow and fast birefringent axes are oriented as indicated in Figure 1.4. A linear state of polarization (SOP) injected with its polarization parallel to a principal axis (x or y) is preserved.

A PMF is characterized by its beat length L_B, that is, the length of fiber over which the two orthogonal polarization modes accumulate a phase difference of 2p [19]:

$$L_B = \frac{\lambda}{B} = \frac{2\pi}{\left|\beta_x - \beta_y\right|}. \tag{1.10}$$

The phase birefringence B is defined as

$$B = \left|n_x - n_y\right|, \tag{1.11}$$

where n_x and n_y are the effective refractive indices of the two orthogonal SOP (slow and fast indices). A linear SOP (or any other SOP, for that matter) injected at 45° to x passes as a function of z through a sequence of elliptical polarization states including circular that repeats itself with a period of L_B (see Figure 2.3 further below). Typical beat lengths at 1310 nm are between 2 mm and 4 mm, i.e., B is between 10^{-3} and 10^{-4}. Significantly smaller beat lengths ($L_B < 0.5$ mm) have been achieved with photonic crystal PMF [20]. The beat length is a measure for the resistance of a PMF against external perturbations; the smaller L_B, the more mechanical stress, e.g., due to bending, a PMF tolerates before modal cross-coupling becomes too large. Assuming that the light is injected into the fiber with its polarization parallel to x, the polarization crosstalk in units of dB is defined as [19]:

$$CT_x\left(\mathrm{dB}\right) = 10\log_{10}\frac{P_y}{P_x}, \tag{1.12}$$

where P_y and P_x are the optical powers measured at the fiber end behind an ideal polarizer aligned parallel to x and y, respectively. An analogous equation holds for CT_y. At fiber lengths l on the order of 10 m, polarization crosstalk in an undisturbed PMF is typically in the range of 35 to 40 dB [19]. The h-parameter of a PMF is a measure for the polarization cross-coupling per unit length. For sufficiently high CT-values, h can be approximated as [19]:

$$h \approx \left(\frac{1}{l}\right)10^{\frac{CT}{10}}. \tag{1.13}$$

It should be mentioned that there are fundamental limits to the polarization preserving properties of a PMF because the orthogonal modes of a PMF inherently lack perfect linear polarization [21]. Besides the phase birefringence B and corresponding beat length, another important parameter is the group birefringence B_g, which determines the optical path length difference of the orthogonal modes. It is worth noting that, depending on wavelength, B and B_g can significantly differ in elliptical core fibers, whereas B and B_g are similar in fibers with stress-induced birefringence [22–24].

PMF applications include coherent communication, polarimetric and interferometric fiber sensors, and others. PMF for applications like fiber-optic gyroscopes that require tight fiber bending often have a reduced diameter (80 μm instead of 125 μm) to limit bending-induced stress and birefringence. At a given bending radius, bending-induced birefringence decreases quadratically with decreasing fiber diameter [25]. Parts II and III will present current and voltage sensors that utilize PMF. We will also encounter spun PMF that preserve circular polarization states.

CHAPTER 2

Polarization

An electric or a magnetic field applied to a sensing medium, e.g., an electro-optic crystal or an optical fiber, alters the medium's indices of refraction via electro-optic or magneto-optic effects, that is, the fields induce linear or circular birefringence. The birefringence, in turn, alters the state of polarization (SOP) of light propagating through the medium. The change in the SOP then serves as a measure for the applied field. The following subsections explain the nature of polarized light and introduce the Jones matrix formalism that describes the propagation of polarized light through birefringent media.

In this context, it is necessary to define the handedness of circular and elliptical polarized light and the sense of polarization rotation in optically active media, as conventions differ, unfortunately. We will refer to circular or elliptically polarized light as right-handed, if, *at a given point* on the optical path, the electric field vector rotates clockwise as seen by an observer looking against the propagation direction, and left-handed, if the rotation sense is counterclockwise. This is the common convention in physics and chemistry; by contrast, engineers usually define the rotation sense conversely. Likewise, optics textbooks commonly define the angle of polarization rotation in an optically active medium (that is, a circularly or elliptically birefringent medium) as positive (right-rotatory) if the polarization rotation as seen by an observer looking towards the light source is clockwise, and negative (left-rotatory) if the polarization rotation is counterclockwise. This chapter will adhere to this convention. However, the literature on the Faraday effect and optical activity of twisted fiber and spun fiber defines the sense of polarization rotation opposite. The following chapters will follow this convention for consistency. Two articles by D. Clarke recall the history of the nomenclature of polarized light and point out the differences between the "snapshot picture" of a polarization state and the "angular momentum perspective" [26, 27].

DOI: 10.1201/9781003100324-4

2.1 POLARIZATION ELLIPSE

The electric field of a monochromatic plane light wave propagating through a medium with refractive n ($n = 1$ in vacuum) is characterized by its amplitude E_0, wave vector $\boldsymbol{k}$, and angular frequency w:

$$\boldsymbol{E} = E_0 \cos\left(wt - \boldsymbol{k}z\right)\hat{\boldsymbol{u}}. \tag{2.1}$$

The wave vector points in the propagation direction, that is, the positive z-axis of a Cartesian coordinate system. The electric field vector $\boldsymbol{E}$ at a given location z then lies in the x, y-plane; $\hat{\boldsymbol{u}}$ is a unit vector pointing the direction of the field. The absolute value of the wave vector, called wave number, is defined as $k = 2pn/l$ with l being the wavelength in vacuum. The magnetic field component of the light wave can be neglected for most practical purposes due to its weakness. An arbitrary state of polarization (SOP) of monochromatic light is commonly represented as the vector sum of two orthogonal linearly polarized field constituents $\boldsymbol{E}_x$, $\boldsymbol{E}_y$ having amplitudes E_{0x}, E_{0y} and additional phase terms d_x, d_y:

$$\boldsymbol{E}_x = E_{0x} \cos\left(\varpi t - kz + \delta_x\right)\hat{\boldsymbol{x}}, \tag{2.2}$$

$$\boldsymbol{E}_y = E_{0y} \cos\left(\varpi t - kz + \delta_y\right)\hat{\boldsymbol{y}}. \tag{2.3}$$

For $d_y \neq d_x$, the tip of the sum vector $\boldsymbol{E} = \boldsymbol{E}_x + \boldsymbol{E}_y$ follows a helical path around the propagation direction z. As an example, Figure 2.1 shows the special case of two waves $\boldsymbol{E}_x$, $\boldsymbol{E}_y$ with equal amplitudes and a phase difference of $d = d_y - d_x = p/2$ ($\boldsymbol{E}_y$ leads $\boldsymbol{E}_x$), which results in (right-handed) circular polarization. (Note: In modern literature, the phase term is often written as $[kz - wt + d]$ rather than $[wt - kz + d]$ as above. In that case, $\boldsymbol{E}_y$ lags behind $\boldsymbol{E}_x$ for

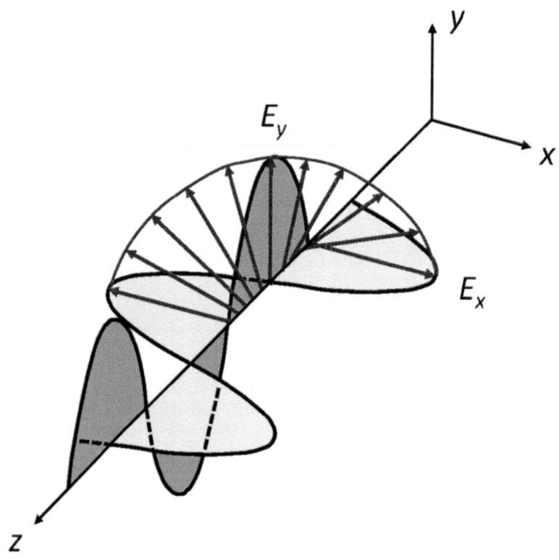

FIGURE 2.1 The vector sum of two orthogonal light fields E_x, E_y of equal amplitude with E_y leading E_x by $d = p/2$ results in right-handed circularly polarized light.

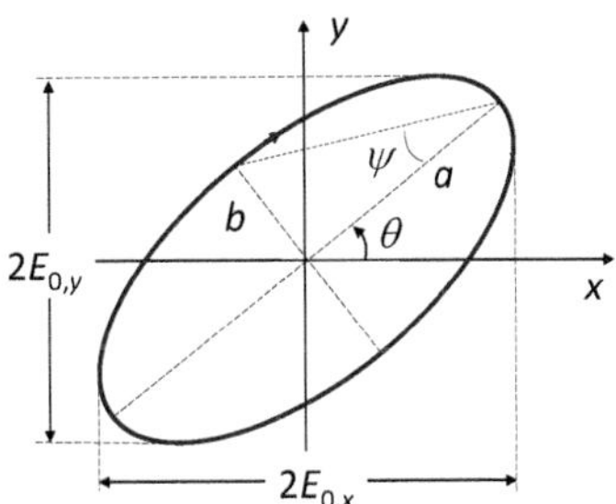

FIGURE 2.2 Polarization ellipse characterized by its azimuth θ and ellipticity angle ψ. The indicated state of polarization is right-handed ($\psi > 0$).

positive d; $d = +\mathrm{p}/2$ then results in left-handed circular polarization. Moreover, the polarization variable that we will introduce further below is then given by $\chi = E_x/E_y$ rather than by $\chi = E_y/E_x$; the alternative notation also leads to sign swaps in the Jones matrices below).

The projection of the helix into the x, y-plane is called polarization ellipse (Figure 2.2) [28–31]:

$$\left(\frac{E_x}{E_{0x}}\right)^2 + \left(\frac{E_y}{E_{0y}}\right)^2 - 2\left(\frac{E_x}{E_{0x}}\right)\left(\frac{E_y}{E_{0y}}\right)\cos\delta = \sin^2\delta. \tag{2.4}$$

The polarization ellipse is characterized by its ellipticity $e = b/a$ ($-1 \le e \le 1$) or corresponding ellipticity angle ψ, azimuth θ, and sense of rotation; a, b are the lengths of the semi-major and semi-minor axis of the ellipse, respectively. θ and ψ can be expressed in terms of E_{0x}, E_{0y}, and d as follows [30]:

$$\tan(2\theta) = \left[\tan(2a)\right]\cos d \qquad -\mathrm{p}/2 \le \theta \le \mathrm{p}/2,$$

$$\sin(2\psi) = \left[\sin(2a)\right]\sin d \qquad -\mathrm{p}/4 \le \psi \le \mathrm{p}/4,$$

$$\tan\alpha = \frac{E_{0y}}{E_{0x}} 0 \le a \le \mathrm{p}/2. \tag{2.5}$$

The ellipticity angle in terms of a and b is defined by [30]

$$\tan\psi = \pm\frac{b}{a} - \mathrm{p}/4 \le \psi \le \mathrm{p}/4 \tag{2.6}$$

The electric field $\boldsymbol{E}$ vibrates in a fixed plane, and the polarization is said to be linear ($e = 0$, $\psi = 0$) if $\boldsymbol{E}_x$ and $\boldsymbol{E}_y$ are in phase or anti-phase ($d = m\mathrm{p}$, $m = 0, \pm 1, \pm 2 \ldots$). All other d result in elliptical polarization. The special case of $E_{0x} = E_{0y}$ and $d = \mathrm{p}/2 + 2m\mathrm{p}$ ($\boldsymbol{E}_y$ leads $\boldsymbol{E}_x$ by p/2 modulo 2p) corresponds to right circularly polarized light, RHC ($e = 1$, $\psi = 45°$), and the special case of $E_{0x} = E_{0y}$ and $d = -\mathrm{p}/2 + 2m\mathrm{p}$ ($\boldsymbol{E}_y$ lags $\boldsymbol{E}_x$ by p/2 modulo 2p) corresponds to

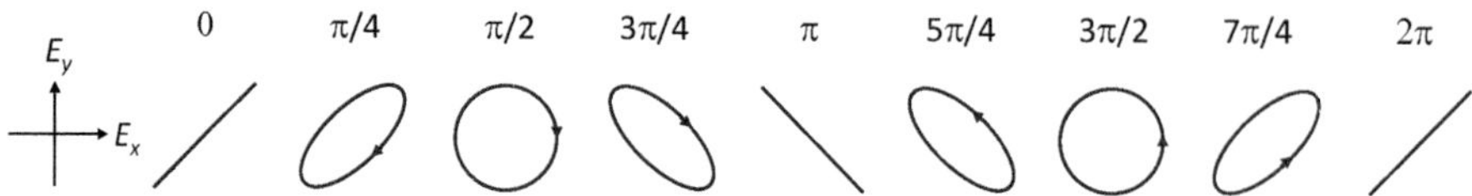

FIGURE 2.3 Polarization ellipses for equal amplitudes of the orthogonal constituents E_x, E_y and selected values of their phase difference *d*; E_y leads E_x by *d*.

left circularly polarized light, LHC ($e = -1, \psi = -45°$). As already noted in the introduction above, the polarization is referred to as right-handed circular (or right-handed elliptical) if the electric field vector at a given point *z* rotates clockwise and left-handed circular (left-handed elliptical) if the electric field vector rotates counterclockwise as seen by an observer looking against the propagation direction; $\psi > 0$ indicates clockwise rotation and $\psi < 0$ counterclockwise rotation.

Figure 2.3 shows for the special case of $E_{0x} = E_{0y}$ how the azimuth and ellipticity of the polarization ellipses vary at changing phase *d*. Independent of *d*, the SOP azimuth equals ±45° if $\mathbf{E}_x$, $\mathbf{E}_y$ have equal amplitudes. Note also that the superposition of left and right circular waves of equal amplitude (and wavelength) results in linear polarization. The linear polarization is parallel to *x* if the circular SOP is in phase and rotates counterclockwise if the LHC-state advances versus the RHC-state. The sign of the azimuth angle θ is commonly defined in the mathematical sense, i.e., in a right-handed coordinate system, θ is positive in the direction indicated in Figure 2.2 [28, 30].

Any elliptical polarization state, defined by its ellipticity angle ψ and azimuth θ, can be expressed in the form of a single complex number, that is, the ratio of the orthogonal field constituents (Cartesian complex-plane representation) [28]:

$$\chi = \frac{E_y}{E_x} = \frac{E_{0y}}{E_{0x}} e^{i(\delta_y - \delta_x)} = \frac{E_{0y}}{E_{0x}} e^{i\delta}. \tag{2.7}$$

The polarization variable χ contains the full information on the polarization ellipse and can be expressed in terms of ψ and θ as

$$\chi = \frac{\tan\theta + i\tan\psi}{1 - i\tan\theta\tan\psi}. \tag{2.8}$$

The real axis of the complex χ-plane represents all linear SOP (the origin represents *x*-polarized light ($\chi = 0$), infinity corresponds to *y*-polarized light); the upper half-plane represents all right-handed elliptical SOP, and the lower half-plane represents all left-handed elliptical SOP; $\chi = \pm i$ correspond to right and left circular light. Note that the polarization state orthogonal to χ is given by

$$\chi_{\text{orth}} = \frac{1}{\chi^*} = \frac{-\chi}{|\chi|^2} \tag{2.9}$$

with $\chi\,\chi^*_{orth} = \chi^*\chi_{orth} = -1$. The asterisk indicates the complex conjugate. Orthogonal polarization states have the same ellipticity, their azimuths are mutually orthogonal, and their polarization ellipses are traced in opposite senses. If a polarization state χ is known, its ellipticity and azimuth angles are given by

$$\sin(2\psi) = \frac{2Im(\chi)}{1+|\chi|^2}, \tag{2.10}$$

$$\tan(2\theta) = \frac{2\operatorname{Re}(\chi)}{1-|\chi|^2}. \tag{2.11}$$

As an alternative to the Cartesian complex-plane representation, a polarization state may also be represented as a point in the circular complex-plane with left and right circular SOP, $\boldsymbol{E}_L$ and $\boldsymbol{E}_R$, as basis states (Figure 2.4) [28]. The polarization variable χ_{circ} is given by

$$\chi_{\text{circ}} = \frac{E_R}{E_L} = \frac{E_{0R}}{E_{0L}} e^{i(\delta_R - \delta_L)} \tag{2.12}$$

where E_{0R}, E_{0L}, δ_R and δ_L are the amplitudes and phases of the circular SOP. χ_{circ} may also be expressed in terms of ψ and as

$$\chi_{\text{circ}} = \tan\left(\psi + \frac{\pi}{4}\right) e^{-i2\theta}. \tag{2.13}$$

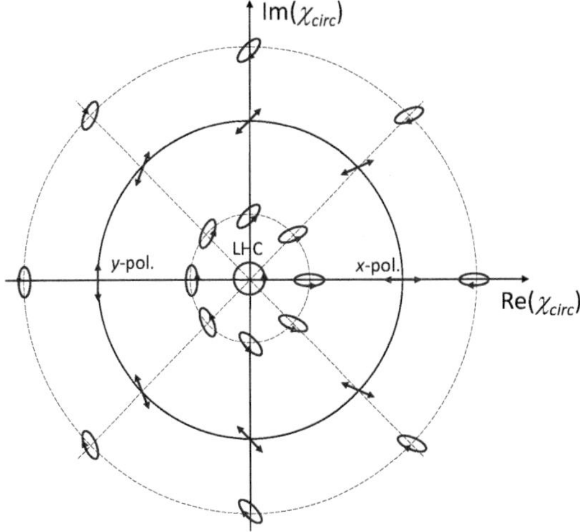

FIGURE 2.4 Circular complex-plane representation of polarization states with left and right circular SOP, E_L and E_R, as basis states; see text for details.

Conversely, ψ and θ are given in terms of χ_{circ} as

$$\tan(\psi_{\text{circ}}) = \frac{|\chi_{\text{circ}}| - 1}{|\chi_{\text{circ}}| + 1}, \tag{2.14}$$

$$\theta = -(1/2)\arg(\chi_{\text{circ}}). \tag{2.15}$$

The origin and infinity of the circular complex plane correspond to left and right circular polarizations, respectively; points on the unit circle (solid line) correspond to linear SOP ($E_L = E_R$). Points inside the unit circuit represent left-handed elliptical SOP ($E_L > E_R$), and points outside the unit circuit represent right-handed elliptical SOP ($E_L < E_R$). SOP of equal ellipticity (ψ = constant) are on concentric circles around the origin, while SOP of equal azimuth (θ = constant) are on radial rays from the origin to infinity.

2.2 POINCARÉ SPHERE

The Poincaré sphere representation of polarized light visualizes polarization states χ as points on the surface of a sphere and is essentially equivalent to the complex-plane representations [28]). (In fact, the Poincaré sphere follows from the Cartesian or circular complex-plane representation through corresponding stereographic projections.) The twofold azimuth and ellipticity angles, 2θ ($-p \leq 2\theta \leq p$) and 2ψ ($-p/2 \leq 2\psi \leq p/2$), represent the SOP's longitude and latitude, respectively, as indicated in Figure 2.5. Starting at horizontal linear polarization, θ is positive in the eastern hemisphere and negative in the western hemisphere; ψ is positive in the northern hemisphere and negative in the southern hemisphere.

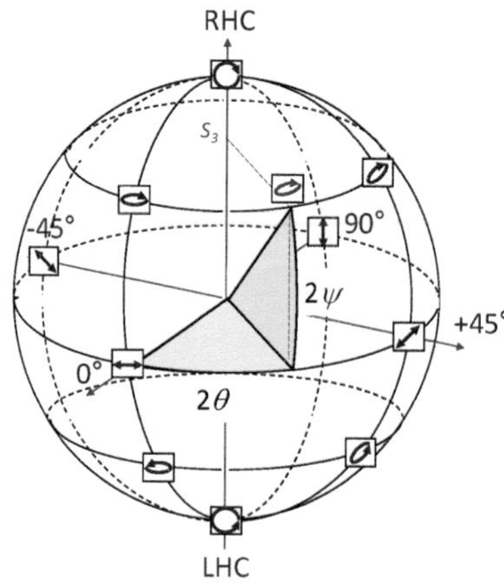

FIGURE 2.5 The Poincaré sphere visualizes polarization states according to the azimuth θ and ellipticity angle ψ of the corresponding polarization ellipse. Points on the equator correspond to linearly polarized light ($\psi = 0$); points in the northern hemisphere correspond to right-handed elliptical polarizations. The North pole ($2\psi = 90°$) corresponds to right-handed circular polarization (RHC). Analogously, points in southern hemisphere indicate left-handed senses of rotation.

Accordingly, points on the equator correspond to linearly polarized light ($\psi = 0$); points in the northern hemisphere correspond to right-handed elliptical polarizations. The North pole ($2\psi = 90°$) represents right-handed circular polarization (RHC). Analogously, points in the southern hemisphere correspond to left-handed elliptical polarizations; the south pole ($2\psi = -90°$) corresponds to left-handed circular polarization (LHC). Points on equal-latitude circles represent SOP of equal ellipticity. A full revolution around the equator is equivalent with a 180°-polarization rotation of linearly polarized light (corresponding to a 360°-change in the phase d). Any pair of antipodes on the sphere represents orthogonal polarization states. The corresponding polarization ellipses have equal ellipticity but with opposite handedness and their major axes are mutually orthogonal. The SOP evolution of light propagating through birefringent media is often visualized as a trajectory on the Poincaré sphere, as we will see in Chapter 7 for specialty optical fibers for current sensing.

2.3 LINEAR BIREFRINGENCE

In optically isotropic materials such as glass or crystals of cubic symmetry (examples are sodium chloride, silicon, or gallium arsenide), the speed of light is independent of the direction of propagation, i.e., these materials have a single index of refraction. Crystals with trigonal, hexagonal, or tetragonal symmetry are uniaxial crystals in which the light propagation is governed by an ordinary and extraordinary index of refraction (n_o, n_e). Such materials are said to be double-refractive or (linearly) birefringent, since light impinging on their surfaces is, apart from special orientations, refracted in two different directions [31]. Well-known examples are calcite ($CaCO_3$, n_o = 1.658, n_e = 1.486 at 589.3 nm) and quartz (n_o = 1.544, n_e = 1.553 at 589.3 nm). Corresponding materials are called positive uniaxial, if $n_e - n_o > 0$, and negative uniaxial, if $n_e - n_o < 0$. Amorphous materials like glasses or plastics also become birefringent when their anisotropy is perturbed by unidirectional pressure. Crystals of still lower symmetry (orthorhombic, monoclinic, and triclinic crystal classes) are biaxial and exhibit three different indices of refraction.

The index ellipsoid or indicatrix serves to illustrate the phenomenon. The indicatrix is defined by the equation

$$\frac{x^2}{n_x^2} + \frac{y^2}{n_y^2} + \frac{z^2}{n_z^2} = 1, \tag{2.16}$$

where $n_x = \sqrt{\varepsilon_{rx}}$, $n_y = \sqrt{\varepsilon_{ry}}$, $n_z = \sqrt{\varepsilon_{rxz}}$. $e_{r,x}$, $e_{r,y}$, $e_{r,z}$ are the principal relative permittivities (dielectric constants) at the optical frequency ϖ [32]. Figure 2.6 shows the index ellipsoid of a positive uniaxial crystal [32] with the z-coordinate axis coinciding with the optical axis of the crystal. The cross-section of the ellipsoid in the x, z and y, z planes is then an ellipse with semi-major and semi-minor axes given by n_e ($n_z = n_e$) and n_o ($n_x = n_y = n_o$), respectively, while the cross-section in the x, y-plane is a circle with radius n_o. The electric field vector of light traveling parallel to the optical axis (z) lies in the x, y-plane. The light then sees an index of refraction given by n_o, independent of the field direction (polarization direction) and is called ordinary wave (o-ray). Light traveling along y has its field in the x, z-plane and

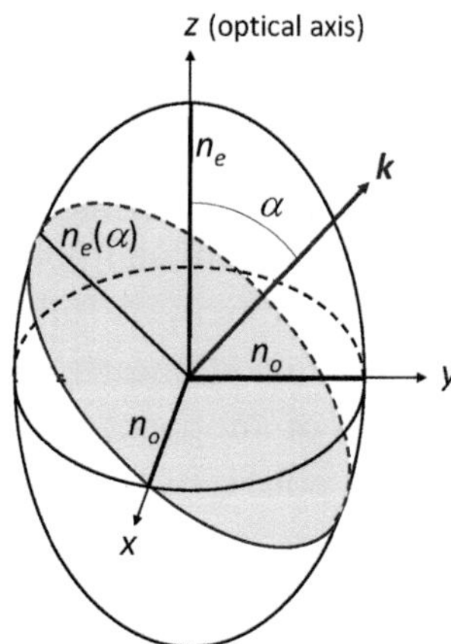

FIGURE 2.6 Index ellipsoid of a positive uniaxial crystal with ordinary and extraordinary indices of refraction n_o, n_e ($n_e - n_o > 0$). The wavevector $\boldsymbol{k}$ indicates the propagation direction of the light.

experiences a refractive index of n_o for the field component polarized along x and of n_e for the component polarized along z. The latter is called extra-ordinary wave (e-ray). Finally, the orthogonal polarization components of light traveling at an arbitrary angle α to the z-axis experience refractive indices of n_o and $n_e(\alpha) < n_e$, as illustrated in Figure 2.6. As this type of birefringence affects two orthogonal linearly polarized components of a light wave, it is commonly referred to as linear birefringence, in contrast to circular birefringence, which relates to orthogonal circular polarization states. We will return to index ellipsoids in the context of electro-optic voltage sensors (Chapter 13).

2.4 CIRCULAR BIREFRINGENCE

Media with chiral structure exhibit circular birefringence (optical activity), i.e., the plane of vibration of linearly polarized light rotates by an angle $a_r = A_r\, d$, when the light propagates through a medium of thickness d; the so-called rotary power A_r measures the polarization rotation per unit length. If the light is elliptically polarized, the *major axis* of the polarization ellipse rotates by a_r. The polarization rotation is due to different indices of refraction, n_L and n_R, and hence different phase velocities of the left and right circular constituents of the SOP. In terms of n_L and n_R, the rotary power is [28, 30, 31]:

$$A_r = -\frac{\pi\left(n_R - n_L\right)}{\lambda}. \tag{2.17}$$

In analogy to linear retarders, optically active media are called circular retarders. Examples are aqueous solutions of sugars such as glucose and fructose and crystals with chiral structure such as quartz ($A_r = 21.7°/\text{mm}$, $|n_R - n_L| = 7.1 \times 10^{-5}$ at 670.8 nm for light propagating along the optic axis [31]). The medium is called right-rotary (A_r is positive, $n_L > n_R$) if the polarization rotates clockwise (looking towards the source) and left-rotary for counterclockwise rotation (A_r is negative, $n_L < n_R$). Materials under torsional stress, for example, a twisted optical fiber, also exhibit optical activity (Chapter 7). Furthermore, a magnetic field applied to any medium induces circular birefringence as a result of the Faraday effect for light propagating parallel to the field. The Faraday effect is the basis of many optical magnetic field and current sensors. It is important to note that the Faraday effect is non-reciprocal, i.e., opposite beam directions with respect to the field direction result in polarization rotation of opposite sense. By contrast, optical activity is a reciprocal effect, i.e., the sense of rotation is the same for forward and backward propagating beams.

2.5 PRISM POLARIZERS

Birefringent prism polarizers split an incoming beam of polarized or unpolarized light into two spatially separated beams with orthogonal linear polarization states that correspond to ordinary and extra-ordinary rays. A variety of schemes is known, all of which make use of the difference between n_o and n_e [31]. Figure 2.7 shows two important examples. The Glan-Foucault polarizer (Figure 2.7a) consists of two calcite prisms separated by a narrow air gap. The optic axis of the prisms is perpendicular to the plane of projection as indicated. The light enters the prism at normal incidence. The electric field component vibrating perpendicular to the optic axis (arrows) is reflected at the air gap by total internal reflection (o-ray) and is commonly discarded, while the field component parallel to the optic axis (dots) is transmitted (e-ray). The condition for total internal reflection of the o-ray is $n_e < 1/\sin\theta < n_o$, where θ is the angle of incidence at the calcite/air interface. The prism angles must be chosen accordingly. Polarizers with the two prisms cemented together instead of having an air gap are known as Glan-Thompson polarizers.

Wollaston prisms (Figure 2.7b) serve as polarizing beam splitters and are of particular interest for applications that make use of both the o-ray and e-ray (Chapter 5). Other than in a Glan-Foucault polarizer, the optic axes of the two prisms (calcite or quartz) are perpendicular to each other as indicated. As a result, the o-ray and e-ray are refracted at the interface into two different directions. High-quality prism polarizers provide polarization extinction ratios (PER) in excess of 10^5 (50 dB) compared to PER of around of 10^3 (30 dB) or less of typical wire-grid-type polarizers. On a logarithmic scale, the polarization extinction ratio is defined as

$$\mathrm{PER}(\mathrm{dB}) = 10\log_{10}\frac{P_0(0^\circ)}{P_c(90^\circ)}, \tag{2.18}$$

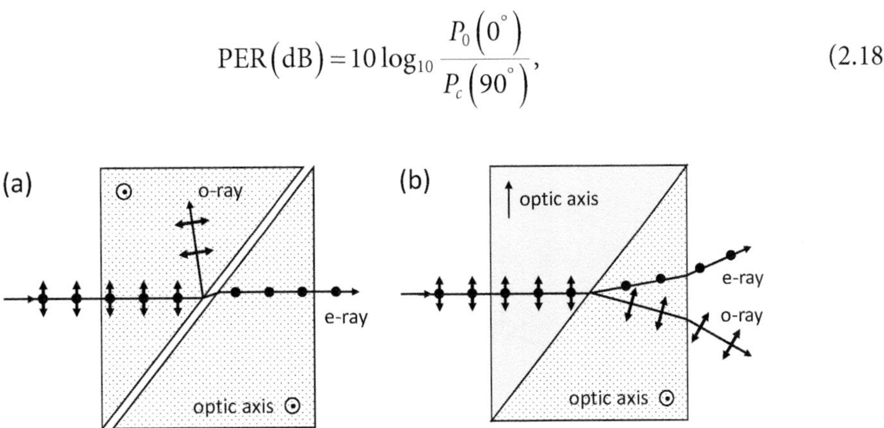

FIGURE 2.7 Glan-Foucault (a) and Wollaston prisms (b) [31].

where $P(0°)$ and $P(90°)$ are the light powers transmitted by the polarizer, when oriented at 0° and 90°, respectively, to the polarization direction of linearly polarized input light.

2.6 LINEAR RETARDERS

Linear retarders, also called wave plates, consist of a birefringent crystal material such as quartz with refractive indices n_o and n_e [31]. The optic axis lies in the plane of the plate. A linear polarization state parallel to the optic axis sees the refractive index n_e, while a polarization state orthogonal to the optic axis sees the index n_o. As a result, the two polarization states accumulate a relative path difference *DL* and corresponding differential phase delay δ:

$$\Delta L = \left(\left|n_0 - n_e\right|\right)d, \tag{2.19}$$

$$\delta = \left(2\pi/\lambda\right)\left(\left|n_0 - n_e\right|\right)d, \tag{2.20}$$

where d is the thickness of the plate. The most common retarders are quarter-wave retarders (QWR) and half-wave retarders (HWR) that generate path differences corresponding to a quarter of a wavelength and half of a wavelength, respectively (i.e., phase delays of 90° and 180°). Figure 2.8 schematically depicts a quarter-wave retarder. The slow (fast) axis corresponds to the polarization direction with the larger (smaller) of the two indices n_o, n_e. In case of quartz ($n_o < n_e$), the optic axis is the slow axis. It is obvious that a QWR converts incoming linear polarization states at ±45° to the slow axis to right circular (E_y leads E_x) or left circular polarization, respectively. Vice versa, incoming circular polarization is converted to linear polarization. A QWR is said to be of zero order, if $DL = l/4$. Higher order QWR have delays of $(3/4)l$, $(5/4)l$, etc. A quartz zero-order retarder for a wavelength of 1550 nm has a thickness of about 52 mm. Zero-order retarders are delicate to handle but less sensitive to deviations from the nominal operating temperature. It should be mentioned that the optical activity of quartz is not evident in a retarder, since the light propagates perpendicular to the optic axis.

Another common retarder material is the biaxial crystal mica, mostly in the form of muscovite [31]. Whereas quartz must be cut and polished, mica can be easily cleaved, and its principal axes (slow and fast axes) lie in the cleavage planes. At normal incidence, mica then behaves like a uniaxial crystal; $(|n_0 - n_e|) = 0.005$ (at 589.3 nm).

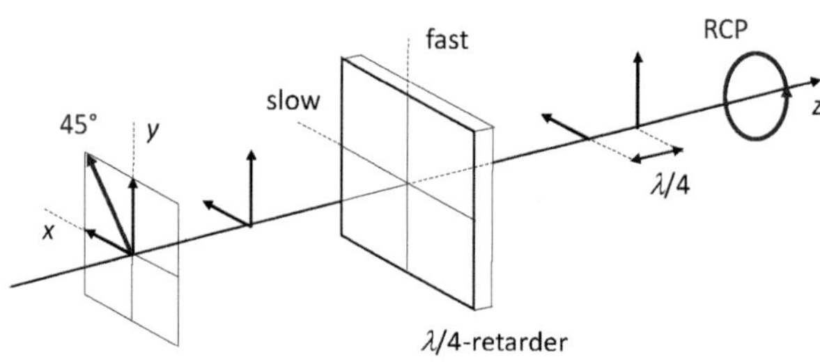

FIGURE 2.8 Quarter-wave retarder.

Half-wave retarders are commonly utilized as polarization rotators. A HWR rotates a linear polarization state oriented at an angle θ to the slow axis by an angle of -2θ. The same applies to the azimuth of elliptical polarization. In particular, a linear SOP at 45° is rotated by −90°, i.e., the transmitted linear SOP is orthogonal to the incident SOP. Furthermore, HWR convert left-handed elliptical SOP to right-handed elliptical SOP and vice visa; the SOP ellipticity remains unchanged.

2.7 JONES MATRIX CALCULUS

The Jones calculus, established by the American physicist R. C. Jones in 1941 [33], serves to describe the propagation of polarized light through (lossless) non-depolarizing optical elements such as polarizers, beam splitters, or linearly and circularly birefringent retarders. Often, such elements are referred to as polarization elements. In Chapters 7 and 8, we will apply the Jones calculus to the polarization evolution in optical fibers for current sensing and to derive the corresponding sensor signals.

2.7.1 Jones Vectors

Jones vectors describe the polarization of a light wave in vector form (Jones vector) and read in complex notation [28–31]:

$$\boldsymbol{E} = \begin{pmatrix} \boldsymbol{E}_x \\ \boldsymbol{E}_y \end{pmatrix} = \begin{pmatrix} E_{0x}\, e^{i\delta_x}\, \hat{\boldsymbol{x}} \\ E_{0y} e^{i\delta_y}\, \hat{\boldsymbol{y}} \end{pmatrix}. \tag{2.21}$$

The oscillatory terms $\exp[i(\varpi t - kz)]$ of the fields can be dropped in this context. Note that the ratio E_y/E_x of the two vector components is the polarization variable χ from (2.7). The product $\boldsymbol{E}^{\dagger}\boldsymbol{E}$ corresponds to the light wave intensity (or irradiance) I:

$$I = \boldsymbol{E}^{\dagger}\boldsymbol{E} = E_{0x}^2 + E_{0y}^2. \tag{2.22}$$

The dagger superscript † indicates the Hermitian adjoint (complex conjugate of the transpose matrix). (Note: The intensity in units of W/cm² is $(1/2)\varepsilon_0 c\ \boldsymbol{E}^{\dagger}\boldsymbol{E}$, but the pre-factor is commonly omitted, when only relative intensities are of interest.) Often, it is convenient to only consider the relative phase $d = j_y - j_x$ of the two components and normalize the light intensity to unity:

$$\boldsymbol{E} = \frac{1}{\sqrt{E_{0x}^2 + E_{0y}^2}} \begin{pmatrix} E_{0x}\, \hat{\boldsymbol{x}} \\ E_{0y} e^{i\cdot}\, \hat{\boldsymbol{y}} \end{pmatrix}. \tag{2.23}$$

It is obvious from (2.7) that the Jones vector can be expressed in terms of the complex polarization variable χ (the unit vectors are omitted):

$$\boldsymbol{E} = \begin{pmatrix} 1 \\ \chi \end{pmatrix}. \tag{2.24}$$

TABLE 2.1 Normalized Jones Vectors of Selected Polarization States

Polarization State	Jones Vector
Linear parallel to x (horizontal)	$\begin{pmatrix} 1 \\ 0 \end{pmatrix}$
Linear parallel to y (vertical)	$\begin{pmatrix} 0 \\ 1 \end{pmatrix}$
Linear at 45° to x	$\frac{1}{\sqrt{2}}\begin{pmatrix} 1 \\ 1 \end{pmatrix}$
Linear at −45° to x	$\frac{1}{\sqrt{2}}\begin{pmatrix} 1 \\ -1 \end{pmatrix}$
Linear at angle θ to x	$\begin{pmatrix} \cos\theta \\ \sin\theta \end{pmatrix}$
Left-handed circular (E_y lags E_x by 90°)	$\frac{1}{\sqrt{2}}\begin{pmatrix} 1 \\ -i \end{pmatrix}$
Right-handed circular (E_y leads E_x by 90°)	$\frac{1}{\sqrt{2}}\begin{pmatrix} 1 \\ i \end{pmatrix}$

Two Jones vectors $\boldsymbol{E}_1$ and $\boldsymbol{E}_2$ are said to be normal if they satisfy the condition $\boldsymbol{E}_1\ \boldsymbol{E}_2 = 0$, $\boldsymbol{E}_2\ \boldsymbol{E}_1 = 0$. Polarization states corresponding to orthogonal Jones vectors correspond to a pair of antipodes on the Poincaré sphere. If, in addition, the vectors satisfy the conditions $\boldsymbol{E}_1\ \boldsymbol{E}_1 = 1$ and $\boldsymbol{E}_2\ \boldsymbol{E}_2 = 1$, they are called an orthonormal pair of Jones vectors. Table 2.1 depicts the normalized Jones vectors of selected polarization states.

2.7.2 Jones Matrices

The Jones matrix $\boldsymbol{J}$ of a polarization element is a 2×2-matrix of complex elements that relates the Jones vector of the transmitted light field $\boldsymbol{E}_t$ to the Jones vector of the incident field $\boldsymbol{E}_i$:

$$\boldsymbol{E}_t = \boldsymbol{J}\,\boldsymbol{E}_i \tag{2.25}$$

or in expanded notation

$$\begin{pmatrix} \boldsymbol{E}_{tx} \\ \boldsymbol{E}_{ty} \end{pmatrix} = \begin{pmatrix} a & b \\ c & d \end{pmatrix} \begin{pmatrix} \boldsymbol{E}_{ix} \\ \boldsymbol{E}_{iy} \end{pmatrix}. \tag{2.26}$$

The diagonal matrix elements a, d represent transmission ratios without polarization change, whereas the off-diagonal elements b, c describe cross-coupling of the orthogonal fields. If the light passes a series of optical elements with matrices $\boldsymbol{J}_1, \boldsymbol{J}_2, \ldots \boldsymbol{J}_n$, the transmitted field is given by

$$\boldsymbol{E}_t = \boldsymbol{J}_n \ldots \boldsymbol{J}_2\,\boldsymbol{J}_1\,\boldsymbol{E}_i. \tag{2.27}$$

It is important to note that the matrices J_i do not commute and must be applied in proper order. Inserting the input and output Jones vectors from (2.24) into (2.25) gives a relationship between the input and transmitted polarization states, χ_i and χ_t [28]:

$$\chi_t = \frac{c + d\,\chi_i}{a + b\,\chi_i}. \tag{2.28}$$

Table 2.2 lists the Jones matrices of important polarization elements for orientations typically utilized in optical sensors, both in a Cartesian (linear) basis and a circular basis (the transformation between bases is described further below) [28–31].

In particular, the Jones matrix of a linear retarder with retardation $d = d_y - d_x$ (2.20) and a principal axis parallel to x reads in a linear basis [30]:

$$\boldsymbol{J} = \begin{pmatrix} e^{i\delta/2} & 0 \\ 0 & e^{-i\delta/2} \end{pmatrix}. \tag{2.29}$$

(A positive argument indicates a relative phase advance). The Jones matrix of a quarter-wave retarder (d = p/2) with the fast axis parallel to x is

$$\boldsymbol{J} = \begin{pmatrix} e^{i\pi/4} & 0 \\ 0 & e^{-i\pi/4} \end{pmatrix} e^{i\pi/4} \begin{pmatrix} 1 & 0 \\ 0 & -i \end{pmatrix}. \tag{2.30}$$

If $\boldsymbol{J}$ is the Jones matrix of a polarization element in a Cartesian coordinate system (x, y, z), the Jones matrix $\boldsymbol{J}_{\text{rot}}$ of the same element rotated about z by an angle θ with respect to its initial orientation (counterclockwise looking into the beam), the Jones matrix $\boldsymbol{J}_{\text{rot}}$ of the rotated element reads [28]

$$\boldsymbol{J}_{\text{rot}} = \boldsymbol{R}(-\theta)\,\boldsymbol{J}\,\boldsymbol{R}(\theta), \tag{2.31}$$

where $\boldsymbol{R}(\theta)$ is the rotation matrix:

$$\boldsymbol{R}(\theta) = \begin{pmatrix} \cos\theta & \sin\theta \\ -\sin\theta & \cos\theta \end{pmatrix}. \tag{2.32}$$

For example, the Jones matrix of a quarter-wave retarder with the fast axis at an angle θ to x reads

$$\begin{aligned} \boldsymbol{J}_{\text{QWR}}(\theta) &= \begin{pmatrix} \cos\theta & -\sin\theta \\ \sin\theta & \cos\theta \end{pmatrix} e^{\frac{i\pi}{4}} \begin{pmatrix} 1 & 0 \\ 0 & -i \end{pmatrix} \begin{pmatrix} \cos\theta & \sin\theta \\ -\sin\theta & \cos\theta \end{pmatrix} \\ &= e^{i\pi/4} \begin{pmatrix} \cos^2, -i\sin^2\theta & (1+i)\sin,\ \cos\theta \\ (1+i)\sin\theta\cos\theta & \sin^2\theta - i\cos^2\theta \end{pmatrix}. \end{aligned} \tag{2.33}$$

TABLE 2.2 Jones Matrices in Linear (Cartesian) and Circular Bases of Important Polarization Elements

	Jones Matrix	
Optical Element	**Linear Basis**	**Circular Basis**
Free space	$\begin{pmatrix} 1 & 0 \\ 0 & 1 \end{pmatrix}$	$\begin{pmatrix} 1 & 0 \\ 0 & 1 \end{pmatrix}$
Linear polarizer at 0° (parallel to x)	$\begin{pmatrix} 1 & 0 \\ 0 & 0 \end{pmatrix}$	$\frac{1}{2}\begin{pmatrix} 1 & 1 \\ 1 & 1 \end{pmatrix}$
Linear polarizer at 90°	$\begin{pmatrix} 0 & 0 \\ 0 & 1 \end{pmatrix}$	$\frac{1}{2}\begin{pmatrix} 1 & -1 \\ -1 & 1 \end{pmatrix}$
Linear polarizer at 45°	$\frac{1}{2}\begin{pmatrix} 1 & 1 \\ 1 & 1 \end{pmatrix}$	$\frac{1}{2}\begin{pmatrix} 1 & i \\ -i & 1 \end{pmatrix}$
Linear polarizer at −45°	$\frac{1}{2}\begin{pmatrix} 1 & -1 \\ -1 & 1 \end{pmatrix}$	$\frac{1}{2}\begin{pmatrix} 1 & -i \\ i & 1 \end{pmatrix}$
Quarter-wave retarder with fast axis at 0°	$e^{i\pi/4}\begin{pmatrix} 1 & 0 \\ 0 & -i \end{pmatrix}$	$\frac{1}{\sqrt{2}}\begin{pmatrix} 1 & i \\ i & 1 \end{pmatrix}$
Quarter-wave retarder with fast axis at 90°	$e^{i\pi/4}\begin{pmatrix} 1 & 0 \\ 0 & i \end{pmatrix}$	$\frac{1}{\sqrt{2}}\begin{pmatrix} 1 & -i \\ -i & 1 \end{pmatrix}$
Quarter-wave retarder with fast axis at 45°	$\frac{1}{\sqrt{2}}\begin{pmatrix} 1 & i \\ i & 1 \end{pmatrix}$	$\frac{1}{\sqrt{2}}\begin{pmatrix} 1 & -1 \\ 1 & 1 \end{pmatrix}$
Half-wave retarder with fast axis at 0°	$\begin{pmatrix} 1 & 0 \\ 0 & -1 \end{pmatrix}$	$\begin{pmatrix} 0 & 1 \\ 1 & 0 \end{pmatrix}$
Half-wave retarder with fast axis at 45°	$\begin{pmatrix} 0 & i \\ \mathrm{i} & 0 \end{pmatrix}$	$\begin{pmatrix} 0 & -1 \\ 1 & 0 \end{pmatrix}$
Ideal mirror (normal incidence)	$\begin{pmatrix} -1 & 0 \\ 0 & 1 \end{pmatrix}$	$\begin{pmatrix} 0 & -1 \\ -1 & 0 \end{pmatrix}$
Optical activity: polarization rotation by angle α_r (cw for $\alpha_r > 0$, $n_L > n_R$; ccw for $\alpha_r < 0$, $n_L < n_R$)	$\begin{pmatrix} \cos a_r & \sin a_r \\ -\sin a_r & \cos a_r \end{pmatrix}$	$\begin{pmatrix} e^{-ia_r} & 0 \\ 0 & e^{ia_r} \end{pmatrix}$

Stated in words, the orthogonal constituents of the input field parallel to the fast and slow retarder axes f and s (coinciding with x and y before rotation) are projected onto the fast and slow axes of the rotated retarder f' and s'; the resulting fields parallel to f' and s' accumulate a differential phase shift of p/4 in the retarder and finally are projected back onto the axes x and y. At $\theta = 45°$, the matrix reduces to

$$J_{\mathrm{QWR}}\left(\theta = 45^{\circ}\right) = \frac{1}{\sqrt{2}}\begin{pmatrix} 1 & i \\ i & 1 \end{pmatrix}. \tag{2.34}$$

Similarly, if

$$J_P = \begin{pmatrix} 1 & 0 \\ 0 & 0 \end{pmatrix} \tag{2.35}$$

is Jones matrix of a linear polarizer oriented parallel to x ($\theta = 0$), the Jones matrix of the same polarizer oriented at an angle θ to x reads

$$J_P(\theta) = \begin{pmatrix} \cos^2\theta & \sin\theta\cos\theta \\ \sin\theta\cos\theta & \sin^2\theta \end{pmatrix}. \tag{2.36}$$

2.7.3 Linear and Circular Base

The Jones matrices in the left column of Table 2.2 are valid in a Cartesian or linear basis, i.e., a given SOP is represented as a superposition of orthogonal linear SOP:

$$\boldsymbol{E} = \begin{pmatrix} \boldsymbol{E}_x \\ \boldsymbol{E}_y \end{pmatrix}, \tag{2.37}$$

where $\boldsymbol{E}_x$, $\boldsymbol{E}_y$ are the complex fields from (2.23). In case of circular or elliptical birefringent optical elements, it is often convenient to work with a circular basis and express the input and output optical fields as a linear combination of left circular and right circular polarization states $\boldsymbol{E}_L$ and $\boldsymbol{E}_R$:

$$\boldsymbol{E}^{(\text{circ})} = \begin{pmatrix} \boldsymbol{E}_L \\ \boldsymbol{E}_R \end{pmatrix}. \tag{2.38}$$

The light intensity in terms of $\boldsymbol{E}_L$ and $\boldsymbol{E}_R$ is

$$I = \boldsymbol{E}_L^* \boldsymbol{E}_L + \boldsymbol{E}_R^* \boldsymbol{E}_R. \tag{2.39}$$

A base change matrix $\boldsymbol{F}$ and its inverse $\boldsymbol{F}^{-1}$ relate the field components in the circular base to the components in the linear base, and vice versa [28]:

$$\begin{pmatrix} \boldsymbol{E}_x \\ \boldsymbol{E}_y \end{pmatrix} = \boldsymbol{F} \begin{pmatrix} \boldsymbol{E}_L \\ \boldsymbol{E}_R \end{pmatrix}, \tag{2.40}$$

$$\begin{pmatrix} \boldsymbol{E}_L \\ \boldsymbol{E}_R \end{pmatrix} = \boldsymbol{F}^{-1} \begin{pmatrix} \boldsymbol{E}_x \\ \boldsymbol{E}_y \end{pmatrix}, \tag{2.41}$$

where $\boldsymbol{F}$ and $\boldsymbol{F}^{-1}$ are given by

$$\boldsymbol{F} = \frac{1}{\sqrt{2}}\begin{pmatrix} 1 & 1 \\ -i & i \end{pmatrix}, \tag{2.42}$$

$$\boldsymbol{F}^{-1} = \frac{1}{\sqrt{2}}\begin{pmatrix} 1 & i \\ 1 & -i \end{pmatrix}. \tag{2.43}$$

$\boldsymbol{F}$ is a unitary matrix satisfying the condition $\boldsymbol{F}^{-1} = \boldsymbol{F}^{\dagger}$. The matrix columns of $\boldsymbol{F}$ are so-called basis vectors and correspond to the Jones vectors of left and right circular polarization states (see Table 2.1).

The Jones matrix $\boldsymbol{J}_{\text{circ}}$ in the circular base is obtained from the Jones matrix $\boldsymbol{J}_{\text{cart}}$ in the Cartesian base and vice versa by the following transformations [28]:

$$\boldsymbol{J}_{\text{circ}} = \boldsymbol{F}^{-1}\boldsymbol{J}_{\text{cart}}\boldsymbol{F}, \tag{2.44}$$

$$\boldsymbol{J}_{\text{cart}} = \boldsymbol{F}\boldsymbol{J}_{\text{circ}}\boldsymbol{F}^{-1}. \tag{2.45}$$

If we express $\boldsymbol{J}_{\text{cart}}$ and $\boldsymbol{J}_{\text{circ}}$ as

$$\boldsymbol{J}_{\text{cart}} = \begin{pmatrix} a & b \\ c & d \end{pmatrix} \tag{2.46}$$

and

$$\boldsymbol{J}_{\text{circ}} = \begin{pmatrix} A & B \\ C & D \end{pmatrix}, \tag{2.47}$$

the elements of the circular Jones matrix in terms of the elements of the corresponding Cartesian Jones matrix are [28]

$$J_{\text{circ}} = \begin{pmatrix} A & B \\ C & D \end{pmatrix} = \frac{1}{2}\begin{pmatrix} (a+d)-i(b-c) & (a-d)+i(b+c) \\ (a-d)-i(b+c) & (a+d)+i(b-c) \end{pmatrix}, \tag{2.48}$$

and, conversely, the elements of the Cartesian Jones matrix follow from the elements of the circular Jones matrix as

$$J_{\text{cart}} = \begin{pmatrix} a & b \\ c & d \end{pmatrix} = \frac{1}{2}\begin{pmatrix} (A+B+C+D) & i(A-B+C-D) \\ -i(A+B-C-D) & (A-B-C+D) \end{pmatrix}. \tag{2.49}$$

For example, the circular Jones matrix of a circularly birefringent polarization element (optical rotator) reads

$$J_{\text{circ}}(\alpha) = \begin{pmatrix} e^{-ia_r} & 0 \\ 0 & e^{ia_r} \end{pmatrix}, \tag{2.50}$$

where a_r is the polarization rotation angle (corresponding to half the differential phase shift between the LHC and RHC polarization states); also see Table 2.2. Note that other than in the linear basis (Table 2.2), there are no off-diagonal matrix elements. Their absence indicates that the left and right circular basis states propagate through the rotator unchanged, apart from a relative phase shift of $2a_r$. By contrast, a linear retarder couples the two circular states in a circular basis as indicated by the non-vanishing off-diagonal matrix elements [28]:

$$J_{\text{circ}}^{(\text{lin.retarder})} = \begin{pmatrix} \cos(\delta/2) & ie^{i2\theta}\sin(\delta/2) \\ ie^{-i2\theta}\sin(\delta/2) & \cos(\delta/2) \end{pmatrix} \tag{2.51}$$

Here, δ is again the retardation and θ the angle of the fast axis to x.

Note that the Jones matrices of lossless polarization elements such as retarders are unitary and have the form [28, 38]

$$J_{\text{cart}} = \begin{pmatrix} a & -b^* \\ b & a^* \end{pmatrix}, \tag{2.52}$$

$$J_{\text{circ}} = \begin{pmatrix} A & -B^* \\ B & A^* \end{pmatrix}, \tag{2.53}$$

with $|a|^2 + |b|^2 = 1$ and $|A|^2 + |B|^2 = 1$. The asterisks again indicate the complex conjugate.

2.7.4 Eigenpolarizations and Eigenvalues

The eigenpolarizations of a so-called homogeneous polarization element are orthogonal states of polarization, χ_{e1} and χ_{e2}, that, apart from a phase shift, propagate unaltered through the element and are obtained by setting $\chi_{\text{out}} = \chi_{\text{in}}$ in (2.28). Note, however, that the SOP of the overall light field may change during the propagation. The eigenpolarizations of a linear retarder, e.g., a quarter-wave retarder, are orthogonal linear polarization states parallel to the fast and slow retarder axes, and the eigenpolarizations of a circularly birefringent element (circular retarder) are LHC and RHC states. An element with both linear and circular birefringence (elliptic retarder) has orthogonal elliptical SOP as eigenpolarizations. In contrast to homogeneous polarization elements, non-homogeneous elements do not have orthogonal eigenpolarizations [37]. For example, skew rays through optical systems or some meta-materials can be inhomogeneous [36]. Also, certain combinations of homogeneous elements are inhomogeneous such as a linear polarizer combined with quarter-wave retarder to a circular polarizer, which has only one eigenpolarization.

The orthogonal eigenvectors of the (Cartesian) Jones matrix, $\boldsymbol{E}_{e1} = (\boldsymbol{E}_{e1.x}, \boldsymbol{E}_{e1.y})$ and $\boldsymbol{E}_{e2} = (\boldsymbol{E}_{e2.x}, \boldsymbol{E}_{e2.y})$, represent the field constituents of the eigenpolarizations, χ_{e1} and χ_{e2}, respectively, and satisfy the following relationships [28, 35, 36] (The analog holds for the eigenvectors and eigenvalues in the circular basis):

$$\boldsymbol{J}\boldsymbol{E}_{e1} = \lambda_1 \boldsymbol{E}_{e1}, \tag{2.54}$$

$$\boldsymbol{J}\boldsymbol{E}_{e2} = \lambda_2 \boldsymbol{E}_{e2}, \tag{2.55}$$

and

$$\boldsymbol{E}_{e1}^{\dagger}\, \boldsymbol{E}_{e2} = 0, \boldsymbol{E}_{e2}^{\dagger}\, \boldsymbol{E}_{e1} = 0 \tag{2.56}$$

Here, the dagger symbol † again denotes the conjugate transpose (Hermitian conjugate); λ_1, λ_2 are the commonly complex eigenvalues of the Jones matrix ($\lambda_1 = \lambda_2^*$ and $\lambda_1\lambda_2 = 1$). Analog to (2.24), the Jones eigenvectors can be expressed in terms of the orthogonal eigenpolarizations $\chi_{e1,2}$ as

$$\boldsymbol{E}_{e1,2} = \begin{pmatrix} 1 \\ \chi_{e1,2} \end{pmatrix}. \tag{2.57}$$

$\chi_{e1,2}$ follow from (2.28):

$$\chi_{e1,2} = (1/2b)\left[(d-a) \pm \sqrt{(d-a)^2 + 4bc}\right], \tag{2.58}$$

with $\chi_{e1}\chi_{e2}^* = \chi_{e1}^*\chi_{e2} = -1$. The commonly complex eigenvalues λ_1, λ_2 are given by

$$\lambda_{1,2} = (1/2)\left[(d+a) \pm \sqrt{(d-a)^2 + 4bc}\right]. \tag{2.59}$$

The arguments of the eigenvalues give the relative phase shift that the two eigenpolarizations accumulate in the polarization element:

$$\delta = -i \arg(\lambda_1/\lambda_2). \tag{2.60}$$

By reversing the above procedure, one can construct the Jones matrix of a polarization element from the Jones eigenvectors and eigenvalues of the element [28, 30]:

$$\boldsymbol{J} = \mathbf{Q}\Lambda\,\mathbf{Q}^{-1}. \tag{2.61}$$

The orthogonal Jones eigenvectors from (2.57) represent the columns of $\boldsymbol{Q}$:

$$\boldsymbol{Q} = \begin{pmatrix} 1 & 1 \\ \chi_{e1} & \chi_{e2} \end{pmatrix}, \tag{2.62}$$

$$\boldsymbol{Q}^{-1} = \frac{1}{\chi_{e2} - \chi_{e1}} \begin{pmatrix} \chi_{e2} & -1 \\ -\chi_{e1} & 1 \end{pmatrix}. \tag{2.63}$$

Λ is the diagonal eigenvalue matrix

$$\Lambda = \begin{pmatrix} \lambda_1 & 0 \\ 0 & \lambda_2 \end{pmatrix}. \tag{2.64}$$

The Jones matrix in terms of χ_{ei} and λ_i then reads

$$\boldsymbol{J} = \left[1/\left(\chi_{e1} - \chi_{e2}\right)\right] \begin{pmatrix} \lambda_2 \chi_1 - \lambda_1 \chi_2 & \lambda_1 - \lambda_2 \\ \chi_1\, \chi_2 \left(\lambda_2 - \lambda_1\right) & \lambda_1 \chi_1 - \lambda_2 \chi_2 \end{pmatrix}. \tag{2.65}$$

2.7.5 Jones Matrix in Terms of Its Eigenvalues

Naturally, the Jones matrix in the basis of the eigenpolarizations of a given polarization element (eigenbasis) is a diagonal matrix with the eigenvalues representing the matrix elements:

$$\boldsymbol{J}_{\text{eigen}} = \begin{pmatrix} \lambda_1 & 0 \\ 0 & \lambda_2 \end{pmatrix}. \tag{2.66}$$

For example, the eigenvalues of a linear retarder in the Cartesian basis are $\lambda_1 = \exp\left(\frac{i\delta}{2}\right)$ and $\lambda_2 = \exp\left(\frac{-i\delta}{2}\right)$; (the fast axis is parallel to x). The Jones matrix in another basis, e.g., the circular basis, is obtained from $\boldsymbol{J}_{\text{eigen}}$, analog to (2.44) and (2.45), by the transformation

$$\boldsymbol{J} = \boldsymbol{F}^{-1}\, \boldsymbol{J}_{\text{eigen}}\, \boldsymbol{F}, \tag{2.67}$$

where the columns of $\boldsymbol{F}$ are formed by the basis vectors of the new basis.

2.7.6 *N*-Matrix

So far, we have assumed that the properties of the considered optical element are constant along the propagation direction z of the light. For example, the orientation of the birefringent axes of a linear retarder and its birefringent phase shift per unit length can be taken as independent of z. In optically anisotropic media, this assumption is no longer valid. An example is an optical fiber that is both linearly and circularly birefringent. The problem is also well-known from liquid crystals. The two-component Jones vector $\boldsymbol{E}$ of the propagating light wave then becomes a function of z. The variation of the field in an infinitesimally thin slab of thickness dz of the medium can be written as [28]

$$\mathbf{E}(z+dz)=\boldsymbol{J}(z)\mathbf{E}(z), \tag{2.68}$$

where $\boldsymbol{J}(z)$ is the (local) Jones matrix of the slab. The variation of $\boldsymbol{E}$ as a function of z can be expressed by a differential equation

$$\frac{d\mathbf{E}(z)}{dz}=\mathbf{N}(z)\mathbf{E}(z). \tag{2.69}$$

In expanded form and a Cartesian basis, this set of coupled differential equations reads

$$\frac{d\mathbf{E}_x}{dz}=n_{11}\mathbf{E}_x+n_{12}\mathbf{E}_y, \tag{2.70}$$

$$\frac{d\mathbf{E}_y}{dz}=n_{21}\mathbf{E}_x+n_{22}\mathbf{E}_y. \tag{2.71}$$

The 2×2 **N**-matrix with elements n_{ij}, also introduced by Jones, is called differential propagation Jones matrix [28], and relates to the Jones matrix as follows:

$$\frac{d\boldsymbol{J}(z)}{dz}=\mathbf{N}(z)\boldsymbol{J}(z). \tag{2.72}$$

If the **N**-matrix elements are independent of z, the overall Jones matrix $\boldsymbol{J}$(z) of the optical device is easily obtained by integration of (2.72); see [28] for details. Otherwise, the coupled equations are commonly solved numerically. In Chapter 7, we will encounter **N**-matrices and corresponding Jones matrices in the context of optical fiber current sensors.

2.7.7 Jones Matrices for Backward Propagation

Some optical sensors of the following chapters are operated in a reflective mode, i.e., after having passed the sensing element, the light reflects off a mirror and traverses the sensing element a second time, now in backward direction ($-z$-direction). The Jones matrix of an elliptical retarder (with linear and circular retarders being special cases) has the general form as given by (2.52) and (2.53), i.e., the Cartesian Jones matrix for light propagation in $+z$-direction reads

$$J_{\text{cart}}^{(\text{forward})}=\begin{pmatrix} a & -b^* \\ b & a^* \end{pmatrix}. \tag{2.73}$$

The Cartesian Jones matrix of a reciprocal elliptical retarder (linear birefringence and optical activity) for propagation in $-z$-direction is obtained by replacing the off-diagonal matrix elements by their complex conjugate [36]:

$$J_{\text{cart}}^{(\text{backward})} = \begin{pmatrix} a & -b \\ b^* & a^* \end{pmatrix}. \tag{2.74}$$

Furthermore, if

$$J_{\text{circ}}^{(\text{forward})} = \begin{pmatrix} A & -B^* \\ B & A^* \end{pmatrix} \tag{2.75}$$

is the circular basis Jones matrix for forward propagation, the circular Jones matrix of a reciprocal retarder for backward propagation reads [38]

$$J_{\text{circ}}^{(\text{backward})} = \begin{pmatrix} A & B \\ -B^* & A^* \end{pmatrix} \tag{2.76}$$

and is the transpose of $J_{\text{circ}}^{(\text{forward})}$. Other than optical activity (e.g., of a sugar solution), magnetic field-induced circular birefringence (Faraday effect) is non-reciprocal, and the signs of the corresponding phase angles in the Jones matrices for backward propagation must be adapted accordingly; their signs switch opposite to those of optical activity.

Finally, it should be mentioned that Stokes vectors and Mueller matrices are alternatives to Jones vectors and Jones matrices for describing propagation of polarized light through an optical system [28–31]. The 4-element Stokes vector characterizes a polarization state — including the state of partially polarized light — in terms of light intensities, and the 4×4 Mueller matrix of a system relates the input and output Jones vectors to each other. One of the many applications of the Mueller calculus include, for example, the light transmission through scattering and depolarizing media such as biological tissue. For a description of the relationship between Mueller and Jones matrices, the reader is referred to [39]. However, we will not make use of the Mueller calculus in this book.

CHAPTER 3

Sensing Methods

3.1 POLARIMETRIC SENSORS

In polarimetric sensors, the external field to be measured induces linear or circular birefringence in a sensing medium or modifies the birefringence of naturally birefringent materials. Measurands that can generate linear birefringence include mechanical stress, hydrostatic or acoustic pressure, temperature, or electric fields [40–42]. By contrast, mechanical torsion or a magnetic field induces circular birefringence [42, 43]. The induced linear birefringence causes a phase shift between two orthogonal linear polarization states, whereas circular birefringence can be observed as a rotation of a linear polarization state. The external field may act directly on the sensing medium, for instance, an electro-optic crystal for electric field and voltage sensing, or the field may induce birefringence via a transducer. An example of the latter is a quartz disk that transmits a voltage-induced piezoelectric deformation onto a polarization-maintaining optical fiber (Chapter 16). Figure 3.1a depicts a basic polarimetric sensor set-up for measuring induced linear birefringence. Chapter 5 will depict a slightly modified set-up for circular birefringence. A combination of a polarizer (P1) and a quarter-wave retarder (QWR) circularly polarizes the probe light before the light enters the sensing medium. P1 is oriented, for example, at 0° to the horizontal. The principal axes of the retarder and induced birefringence are at ±45° to the polarizer. The sensing medium converts the circular input SOP into an elliptical SOP depending on the magnitude of the measurand. After a beam splitter (BS), one fraction of the elliptical state passes polarizer P2, and the other fraction passes polarizer P3. P2 and P3, oriented at 0° and 90°, respectively, bring the two orthogonal constituents (oriented at ±45°) of the elliptical SOP to interference.

The interference intensities I_1 and I_2 at the photodetectors are [31]

$$I_1 = \left(I_0/2\right)\left(1 + K\cos\varphi\right), \tag{3.1}$$

$$I_2 = \left(I_0/2\right)\left(1 - K\cos\varphi\right). \tag{3.2}$$

DOI: 10.1201/9781003100324-5

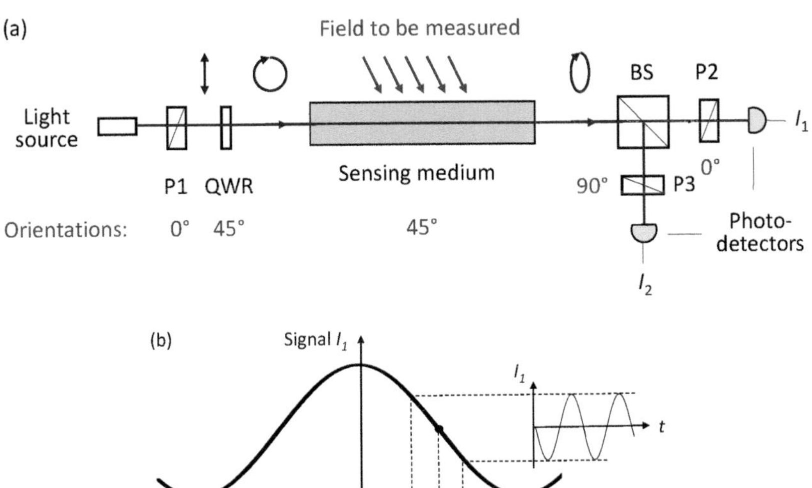

FIGURE 3.1 (a) Basic set-up of a polarimetric sensor; P: polarizer, QWR: quarter-wave retarder, BS: beam splitter; (b) Signal I_1 vs. birefringent phase shift φ and modulation $\Delta\phi$ by an external ac field with the operating point at quadrature.

Here, I_0 is proportional to the light source power; φ is the total phase difference between the two orthogonal SOP:

$$\varphi = \theta + \frac{\pi}{2} + \Delta\phi, \tag{3.3}$$

where θ is the phase shift due to intrinsic birefringence of the sensing element, if any, the $(p/2)$-term is the QWR retardation, and $\Delta\phi$ is the birefringent phase shift induced by the measurand. (It is assumed that principal axes of the intrinsic and induced birefringence coincide.) The fringe contrast K, also called fringe visibility, is defined as

$$K = \frac{I_{max} - I_{min}}{I_{max} + I_{min}}, \tag{3.4}$$

where I_{max} and I_{min} correspond to the interference intensities at constructive and destructive interference, respectively. K equals unity at ideal conditions, i.e., if the interfering fields are fully coherent and have equal amplitudes. Partial coherence reduces K. Often, the field to be measured is an ac field that produces a sinusoidal birefringent phase modulation $\Delta\phi = \phi_0 \sin(\varpi t)$ with amplitude ϕ_0 and frequency ϖ: The difference of the two detector signals divided by their sum gives a signal independent of the light intensity I_0:

$$S = \frac{I_1 - I_2}{I_1 + I_2} = K \cos\varphi. \tag{3.5}$$

With the assumptions $\theta = 0$ and $\phi_0 \ll 1$, S can be approximated as

$$S = K\phi_0 \sin(\varpi t), \tag{3.6}$$

i.e., the signal is proportional to the measurand. Note that to this end, the p/2-phase shift by the retarder moves the operating point of the sensor into the linear range of the cosine transfer function (Figure 3.1b).

Parts II and III will present polarimetric current and voltage sensors with both bulk and fiber-optic sensing media.

3.2 TWO BEAM INTERFEROMETERS

Figure 3.2 depicts three examples of interferometric optical fiber sensors [44, 45]. In the fiber Mach-Zehnder interferometer, a 1×2 directional fiber coupler splits the input light into the two fiber arms of the interferometer. The external field to be measured modulates the optical phase in one fiber arm (sensing arm), here via a transducer. The other arm serves as a reference. The light waves in the two arms interfere in a second fiber coupler. For maximum interference contrast, the interfering waves must have the same polarization states. This requires the use of polarization-maintaining fibers or active polarization control. The signals I_1 and I_2 are the same as in a polarimetric sensor; Eqs. (3.1)–(3.6), but the phase terms now refer to phase differences between the two interferometer arms. The quasi-static phase difference θ depends on the optical length difference of the fiber arms and typically varies with temperature and other disturbances such as fiber bending. Various detection techniques are known to avoid signal fading at changing θ. (Signal fading occurs when the operating point of the interferometer approaches a maximum or minimum of the cosine transfer function.) With the homodyne phase tracking method, the signal processor applies a phase tracking signal to a piezoelectric transducer (PZT) in one arm of the interferometer. The phase tracker keeps the operating point at quadrature, that is, at θ = p/2 modulo 2p, by keeping the difference of the quasi-static components of

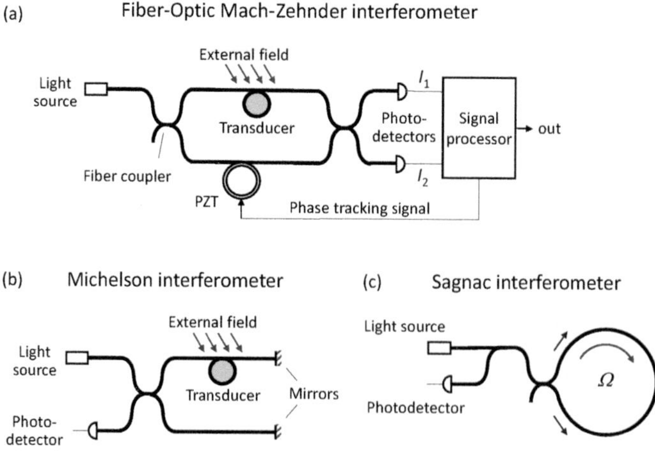

FIGURE 3.2 Interferometric optical fiber sensors.

I_1 and I_2 at zero [46]. Another method is phase-generated carrier demodulation [47]. Here, the phase difference of the two interferometer arms is modulated at a frequency ϖ_{mod} well above the frequency ϖ of the measurand. The modulation is achieved, for example, by means of a piezoelectric transducer or by modulating the wavelength of the light source. The interferometer output (I_1 or I_2) then contains two sets of sideband signals centered about even and odd multiples of ϖ_{mod}, respectively. The amplitudes of the two signal sets vary in proportion to $\cos\theta\ (t)$ and $\sin\theta\ (t)$, i.e., if one set fades away, the other reaches a maximum. Relatively simple signal processing then yields the measurand induced phase shift $\Delta\phi = \phi_0 \sin(\varpi t)$ independent of variations in θ. Further techniques to overcome signal fading have been demonstrated in [48].

The fiber Michelson interferometer (Figure 3.2b) differs from the Mach-Zehnder interferometer in that the two fiber branches are terminated by mirrors. Hence, the light waves return to the input fiber coupler, where they interfere. As a result, the sensitivity doubles compared to an equivalent Mach-Zehnder interferometer. The fiber Sagnac interferometer (Figure 3.2c) comprises a fiber coil with two counter-propagating light waves. A rotation of the coil produces a differential phase shift of the counter-propagating waves (Sagnac effect). Sagnac interferometers form the basis of fiber-optic gyroscopes for rotation sensing [45]. In Sagnac interferometer current sensors, the phase shift is produced by the magnetic field of a current passing through the coil aperture (Faraday effect). We will encounter corresponding sensors and signal processing methods in Chapter 8.

Still other fiber interferometers make use of the interference of the fundamental fiber mode LP_{01} with at least one higher order mode [49, 50]. Voltage sensors employing an interferometric two-mode fiber will be presented in Chapter 16.

3.3 FABRY-PÉROT INTERFEROMETER

A Fabry-Pérot interferometer comprises an optical cavity between two parallel partially reflecting mirrors and relies on multiple beam interference. If the cavity consists of a solid medium — here the sensing medium — with partially reflecting surfaces, it is called Fabry-Pérot etalon (Figure 3.3, left). A light field with amplitude E_0 impinging on the cavity reflects multiply back and forth within the cavity. Assuming a lossless medium, the total intensity I_t of the transmitted fields as a function of the round-trip phase shift φ and normalized by the input intensity I_i is known as the Airy function (Figure 3.3, right) and given by [31]

$$\frac{I_t}{I_i} = \frac{1}{1 + \hat{F}\sin^2\left(\frac{\varphi}{2}\right)}. \tag{3.7}$$

The coefficient of finesse $\hat{F}$ is given by

$$\hat{F} = \left(\frac{2r}{1-r^2}\right)^2. \tag{3.8}$$

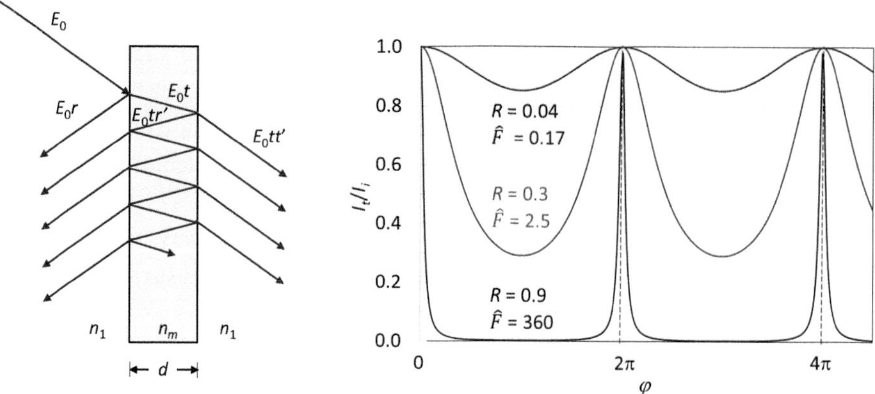

FIGURE 3.3 Multiple reflections in a Fabry-Pérot etalon, r, r' and t, t' are amplitude reflection and transmission coefficients, respectively (left); Airy function vs. phase shift φ for various mirror reflectances R and corresponding coefficients of finesse $\hat{\boldsymbol{F}}$ (right).

Here, r is the amplitude reflection coefficient (r^2 gives the reflectance R). At normal incidence, the round-trip phase shift φ (phase shift between two successively transmitted waves) reads

$$\varphi = \frac{4\pi}{\lambda} nd, \tag{3.9}$$

where n and d are the refractive index and thickness of the etalon, respectively. The relative intensity I_r/I_i reflected from the cavity is the complementary of the Airy function: $I_r/I_i = 1 - I_t/I_i$. The finesse F of the cavity is the ratio of the separation of adjacent maxima of the Airy function to the full width at half maximum of the transmission peaks and given by

$$F = \left(\pi/2\right)\sqrt{\hat{F}}. \tag{3.10}$$

It is obvious that, depending on the cavity finesse, Fabry-Pérot interferometers can reach much higher sensitivity (that is, intensity change per change in the optical length nd) than two beam interferometers but at the expense of a smaller unambiguous measurement range. Illumination of the interferometer with a broad-band spectrum results in an Airy-function-like transmission spectrum. The transmission peaks at normal incidence are separated in wavelength by the free spectral range of the cavity:

$$\text{FSR} \approx \frac{\lambda_0^2}{2nd}, \tag{3.11}$$

where l_0 is the center wavelength of the spectrum. A change in the optical cavity length can then be observed as a shift of the spectral fringes. Another frequently used parameter to characterize a Fabry-Pérot cavity is its Q-factor, $Q = n_o/Dn$ for high Q-cavities, where n_o and Dn are the frequency and line width of the resonance (full width at half maximum,

FWHM), respectively [31]. Parts II and III will show how certain types of optical current and voltage sensors make use of Fabry-Pérot etalons.

3.4 COHERENCE

Not being monochromatic, all practical light sources have a finite coherence length l_c. For high fringe visibility, the optical path imbalance ΔL of the interfering waves in polarimetric and interferometric sensors should be well within l_c, and the waves should have equal amplitudes E_{01}, E_{02}. The fringe visibility is given by [31, 51]

$$K = \frac{2\sqrt{\tilde{I}_1 \tilde{I}_2}}{\tilde{I}_1 + \tilde{I}_2} \left|\gamma_{12}\left(\Delta L\right)\right|. \tag{3.12}$$

Here, $\tilde{I}_1, \tilde{I}_2$ are the intensities of the interfering waves: $\tilde{I}_1 = E_{01}^2/2$ and $\tilde{I}_2 = E_{02}^2/2$. $\gamma_{12}(\Delta L)$ is the complex degree of temporal coherence, that is, the normalized self-coherence function: $\gamma_{12}(\tau) = \Gamma(\tau)/\Gamma(0)$, where $\tau = \Delta L/c$ is the relative time delay of the waves. The self-coherence function $\Gamma(\tau)$ is the inverse Fourier transform of the power spectral density of the light source. The absolute value $|\gamma_{12}(\Delta L)|$ equals unity at $DL = 0$, when the interfering waves are fully coherent and becomes zero at $DL \to \infty$, when the waves are incoherent. In between the two limits, i.e., when $0 < (|\gamma_{12}(\Delta L)| < 1)$, the waves are said to be partially coherent. With $E_{01} = E_{02}$, K reduces to

$$K = \left|\gamma_{12}\left(\Delta L\right)\right|. \tag{3.13}$$

Frequently, the coherence length l_c of a light source is defined as the path length difference at which $|\gamma_{12}|$ has dropped to one half. This length depends to some extent on the shape of the source spectrum. For a Gaussian spectrum — $|\gamma_{12}(\Delta L)|$ is then also a Gaussian —, one obtains [52]

$$l_c = \frac{4\ln 2}{\pi} \frac{\lambda_0^2}{\Delta\lambda}, \tag{3.14}$$

where l_0 is the center wavelength and Dl the full width at half maximum (FWHM). A Lorentzian line shape gives half this value, and $|\gamma_{12}(\Delta L)|$ decays exponentially [52]. (The reader should be aware, though, that the definitions of l_c vary in the literature.) The coherence length of stabilized narrow line width lasers can exceed 100 km. By comparison, the coherence length of simpler single longitudinal mode laser diodes is on the order of some 100 m or shorter. Many of the polarimetric sensors in the following chapters work with broad band sources, particularly superluminescent diodes (SLED) with coherence lengths of a few tens of mm. As an example, the coherence length of a 1310-nm SLED with a Gaussian spectrum and a FWHM of 40 nm is 38 mm; Eq. (3.14). Typically, the group delay of the interfering waves, e.g., in a polarimetric sensing fiber, well exceeds this length. (The relative delay of the orthogonal SOP in conventional PM fibers is a few hundred mm per meter of fiber.) The delay is then compensated in an appropriately oriented second fiber section. A short coherence length helps to suppress disturbing interference signals from parasitic waves that are generated, for instance, by undesired polarization cross-coupling, back-reflections at interfaces, or Rayleigh scattering.

3.5 SENSOR COMPONENTS

This subsection briefly describes the functions and gives some specifications of optical components of the optical current and voltage sensors (OCS and OVS) discussed in more detail in the subsequent chapters. More details can be found in those chapters and the literature cited below.

3.5.1 Light Sources

Typical light sources of OCS and OVS are light-emitting diodes (LED), laser diodes, and most importantly, superluminescent diodes (SLED) [53, 54]. Occasionally, fluorescent fiber sources have also been employed [55]. LED are economic solutions for sensors with bulk optic sensing elements that are pigtailed with multimode fiber leads, but high insertion loss essentially excludes them from single-mode fiber sensors. By contrast, laser diodes and SLED are much better adapted to the small SMF cores. Most early sensors operated at wavelengths in the 780–850 nm window. With increasing availability of inexpensive optical fiber products from telecom suppliers and corresponding light sources, operating wavelengths have shifted to 1310 nm and 1550 nm. SLED at those wavelengths are based on InP and provide fiber-coupled output powers up to several 10 mW. Typical output powers in OCS and OVS are between 5 mW and 10 mW, and typical spectral widths (FWHM) are between 30 nm and 40 nm. SLED emitting flattop spectra with widths up to 150 nm, having lower coherence lengths, are available for optical coherence tomography (OCT). The output power of uncooled SLED (and other semiconductor sources, for that matter) degrades towards elevated temperatures, and the emission wavelength exhibits a significant redshift (by several 100 pm/K in case of 1310-nm SLED). Frequently, a thermoelectric cooler stabilizes the SLED chip temperature; the whole assembly is then mounted in a fiber pigtailed so-called butterfly housing. Uncooled SLEDs mostly are packaged inside TO style cans.

3.5.2 Lyot Fiber Depolarizers

Commonly, the emission from a SLED chip is linearly polarized with a polarization extinction ratio of roughly 20 dB. At the end of the SLED's fiber pigtail, however, the polarization is often an arbitrary elliptical state, subject to fiber bending and other stresses. A fiber Lyot depolarizer inserted into the optical path after the SLED serves to prevent random power fluctuations after a polarizer further down the line. The depolarizer consists of two sections of PM fiber with a length ratio of 1:2 and the slow axis of the second section at 45° to the slow axis of the first section [19, 56, 57]. The differential group delay of the orthogonal polarization states in both PMF sections is much longer than the source coherence length. As a result, the orthogonal constituents of the input elliptical SOP become incoherent; the light is then said to be pseudo depolarized (not to be confused with unpolarized light). Typical lengths of the depolarizer fiber sections range from tens of centimeters to several meters.

3.5.3 Fiber Polarizers

The prism polarizers discussed in Section 2.5 are often used in laboratory experiments but are too bulky and too expensive for commercial OCT (Optical Current Transducers) and OVT (Optical Voltage Tranducers). Sensors commonly employ thin polarizing platelets with fiber pigtails or polarizing fiber. Platelet polarizers can be as thin as 30 mm and contain oriented elongated metal particles (e.g., from silver) in a glass matrix. Typical polarization extinction ratios (PER) are around 25 dB. Polarizing fibers are mostly coiled PMF operated in a wavelength window where only one of the orthogonal SOP of the fundamental mode is guided, while the other mode leaks out [19, 58]. Polarizing fibers achieve PER of at least 30–35 dB. PER as high as 62 dB have been reported in [58].

3.5.4 Directional Fiber Couplers

Directional fiber couplers are fiber-optic beam splitters. In a 2×2-coupler, the claddings of two single-mode fibers are fused together so that over some distance the fiber cores are in close proximity and the fiber modes couple via their evanescent fields [59]. The couplers can be designed for virtually any splitting ratio. Most common are 3 dB couplers with a splitting ratio of 1:1. 3x3-couplers are employed in passively interrogated fiber interferometers (Section 8.4) [60].

3.5.5 Sensing Fibers, Fiber Retarders, and Sensing Materials

Specific sensing fibers, fiber retarders as well as magneto-optic, electro-optic, and piezoelectric sensing materials will be discussed in context of the corresponding sensors.

3.5.6 Fiber Mirrors

Many of the sensors in the following chapters are operated in reflective mode. The end of the sensing fiber is then coated with a metallic coating (e.g., Cr/Au) or a multiple layer dielectric coating. Typical reflectance values are well over 90%.

3.5.7 Phase Modulators

Simple fiber-optic phase modulators consist of a piezoelectric disk or hollow cylinder with several loops of pre-strained fiber wrapped around the circumferential surface [46]. A piezoelectric deformation of the transducer shifts the optical phase of the guided light via a change in the physical length of the fiber and associated refractive index change due to the elasto-optic effect. Piezoelectric modulators are employed to control the operating point of polarimetric and interferometric sensors or for sinusoidal phase modulation, for example, in interferometric sensors with phase generated carrier demodulation. Piezoelectric modulators have the benefit of simplicity, but the usable frequency and amplitude ranges are limited. Higher-performing integrated-optic lithium niobate phase modulators will be discussed in Chapter 8.

3.5.8 Faraday Rotators, Rotator Mirrors, and Optical Isolators

A Faraday rotator mirror rotates linear polarization states upon reflection by 90°. Rotator mirrors are often used in fiber sensors operated in reflective mode such as fiber Michelson interferometers [61]. Birefringent optical phase shifts that the light accumulated on the forward path cancel on the return path. As a result, the input SOP is again restored after the round-trip. Rotator mirrors consist of a magneto-optic material, typically a high Verdet constant garnet, in a permanent magnet. One-way polarization rotation is 45°. Optical isolators serve, for instance, to prevent back reflections out of a sensor from disturbing the light source, particularly in case of laser diode sources. Simple isolators consist of 45°-rotators between polarizers oriented at 45°.

3.5.9 Fiber Bragg Gratings

Fiber Bragg gratings (FBG) are periodic modulations in the refractive of the fiber core, typically over distances from a few mm to a few cm [62]. Most often, the gratings are laser-written from the side by means of a phase mask. When illuminated with broad band light, a FBG reflects a narrow spectral band centered at the Bragg wavelength l_B. The Bragg wavelength equals twice the optical period of the grating: $l_B = 2nL$, where n is the effective refractive index of the fiber and L the physical grating period. The peak reflectivity can reach close to 100%. Typical spectral widths range from below 100 pm to several 100 pm (FWHM). FBG serve, for example, as spectral filters or narrow-band mirrors in fiber lasers. Chirped or aperiodic gratings can be used for dispersion compensation. FBG applications also include strain and temperature sensing [63]. Strain and temperature change the optical grating period and hence cause a shift in l_B. FBG in fused silica fibers have a sensitivity to strain of 1.2 pm/mstrain at 1550 nm; the sensitivity to temperature is 10 pm/K. Particularly attractive features of FBG-based sensors are the wavelength-encoded output and the fact that multiple gratings with the same or different Bragg wavelengths can be multiplexed along a common fiber [63]. The following chapters will show how certain types of OCTs and OVTs employ FBG for measuring magnetic or electric field-induced strains of piezoelectric or magneto-strictive transducer materials.

3.5.10 Fiber Fusion Splices and Mechanical Fiber Connectors

Different fiber sections of a sensor are commonly joined by fusion splices. Fusion splicers for PM fibers automatically orient the principal fiber axes with accuracy to within <±2°. Polarization cross-coupling at well-done splices is below −30 dB (Chapter 11). By contrast, the fiber leads of a sensor are often connected to the opto-electronics and signal processing unit by mechanical connectors for convenient installation. This is relatively uncritical as long as the fibers are standard single-mode or multimode fibers. Learned from longtime experience, though, there is always a risk of improper connector handling by service personnel and contamination of the connector ferrules and fiber end faces by dirt. Therefore, fusion splices are preferred whenever possible. Connectors for PMF, even though commercially available, are especially critical. Typically, the polarization extinction ratio of a PM connector joint is hardly better than 25 dB and varies with temperature (Chapter 11).

References

1. F. P. Kapron, D. B. Keck, and R. D. Maurer, "Radiation losses in glass optical waveguides," *Appl. Phys. Lett.* 17(10), 423–425 (1970).
2. J. Hecht, *City of Light, the Story of Fiber Optics* (Oxford University Press, New York, 1999).
3. N. S. Kapany, *Fiber Optics: Principles and Applications* (Academic Press, San Diego, CA, 1967).
4. T. Miya, Y. Terunuma, T. Hosaka, and T. Miyoshita, "Ultimate low-loss single-mode fibre at 1.55 mm," *Electron. Lett.* 15(4), 106–108 (1979).
5. L. B. Jeunhomme, *Single-Mode Fiber Optics* (Marcel Dekker, 1983).
6. A. W. Snyder and J. D. Love, *Optical Waveguide Theory* (Chapman and Hall, London, 1983).
7. D. Marcuse, *Theory of Dielectric Optical Waveguides*, 2nd ed. (Academic Press, San Diego, CA, 1991).
8. G. P. Agrawal, *Fiber-Optic Communication Systems* (John Wiley & Sons, 2012).
9. P. J. Winzer, D. T. Neilson, and A. R. Chraplyvy, "Fiber-optic transmission and networking: the previous 20 and the next 20 years [Invited]," *Opt. Express* 26, 24190–24239 (2018).
10. D. Gloge, "Weakly guiding fibers," *Appl. Opt.* 10(10), 2252–2258 (1971).
11. R. B. Dyott, *Elliptical Fiber Wavesguides* (Artech House, 1995).
12. C. D. Poole, J. M. Wiesenfeld, D. J. DiGiovanni, and A. M. Vengsarkar, "Optical fiber-based dispersion compensation using higher order modes near cutoff," *J. Lightwave Technol.* 12(10), 1746–1758 (1994), doi: 10.1109/50.337486
13. M. J. Digonnet, *Rare-Earth-Doped Fiber Lasers and Amplifiers, Revised and Expanded* (CRC Press, 2001).
14. G. Tao, H. Ebendorff-Heidepriem, A. M. Stolyarov, et al., "Infrared fibers," *Adv. Opt. Photonics* 7(2), 379–458 (2015).
15. P. St. J. Russell, "Photonic-crystal fibers," *J. Lightwave Technol.* 24(12), 4729–4749 (2006).
16. A. M. Pinto and M. Lopez-Amo, "Photonic crystal fibers for sensing applications," *J. Sens.* (2012) Article ID 598178, 2012.
17. A. A. Stolov, D. A. Simoff, and J. Li, "Thermal stability of specialty optical fibers," *J. Lightwave Technol.* 26(20), 3443–3451 (2008).
18. C. Emslie, "Harsh environment optical fiber coatings: beauty is only skin deep," *Laser Focus World* 51(4), 41–47 (2015).
19. J. Noda, K. Okamoto, and Y. Sasaki, "Polarization-maintaining fibers and their applications," *J. Lightwave Technol.* 4(8), 1071–1089 (1986).
20. A. Ortigosa-Blanch, J. C. Knight, W. J. Wadsworth, et al., "Highly birefringent photonic crystal fibers," *Opt. Lett.* 25(18), 1325–1327 (2000).
21. M. P. Varnham, D. N. Payne, and J. D. Love, "Fundamental limit to the transmission of linearly polarized light by birefringent optical fibers," *Electron. Lett.* 20, 55–56 (1984).

22. A. Kumar and R. K. Varshney, "Propagation characteristics of highly elliptical core waveguides: a perturbation approach," *Opt. Quant. Electron.* 16, 349–354 (1984).
23. M. Legre, M. Wegmuller, and N. Gisin, "Investigation of the ratio between phase and group birefringence in optical single-mode fibers," *J. Lightwave Technol.* 21(12), 3374–3378 (2003).
24. T. Geisler and S. Herstrøm, "Measured phase and group birefringence in elliptical core fibers with systematically varied ellipticities," *Opt. Express* 19(26), B283–B288 (2011).
25. R. Ulrich, S. C. Rashleigh, and W. Eickhoff, "Bending-induced birefringence in single-mode fibers," *Opt. Lett.* 5(6), 273–275 (1980).
26. D. Clarke, "Nomenclature of polarized light: linear polarization," *Appl. Opt.* 13(1), 3–5 (1974).
27. D. Clarke, "Nomenclature of polarized light: elliptical polarization," *Appl. Opt.* 13(2), 222–224 (1974).
28. R. M. A. Azzam and N. M. Bashara, *Ellipsometry and Polarized Light* (North-Holland Pub. Co., sole distributors for the U.S.A. and Canada, Elsevier North-Holland, North-Holland, Amsterdam, 1979).
29. E. Collett, *Polarized Light: Fundamentals and Applications* (Marcel Dekker, New York, 1993).
30. D. Goldstein, *Polarized light*, 2nd ed. (Marcel Dekker, New York, 2003).
31. E. Hecht and A. Zajac, *Optics* (Addison-Wesley Publishing Company, 1974).
32. J. N. Nye, *Physical Properties of Crystals* (Oxford University Press, London, 1967).
33. R. C. Jones, "A new calculus for the treatment of optical systems. I. Description and discussion of the calculus," *J. Opt. Soc. Am.* 31(7), 488–493 (1941).
34. A. Gerald and J. M. Burch, *Introduction to Matrix Methods in Optics* (John Wiley & Sons, 1975), ISBN 978-0471296850.
35. S. N. Savenkov, O. I. Sydoruk, and R. S. Muttiah, "Conditions for polarization elements to be dichroic and birefringent," *J. Opt. Soc. Am. A* 22(7), 1447–1452 (2005).
36. C. Menzel, C. Rockstuhl, and F. Lederer, "Advanced Jones calculus for the classification of periodic metamaterials," *Phys. Rev. A* 82, 053811 (2010).
37. S.-Y. Lu and R. A. Chipman, "Homogeneous and inhomogeneous Jones matrices," *J. Opt. Soc. Am. A* 11(2), 766–773 (1994).
38. R. Dändliker, "Rotational effects of polarization in optical fibers," in *Optical Wave Sciences and Technology, Anisotropic and Nonlinear Optical Waveguides*, C. G. Someda and G. Stegeman, Eds. (Elsevier, 1992), pp. 39–76, ISSN 09275479, ISBN 9780444884893.
39. D. G. Anderson and R. Barakat, "Necessary and sufficient conditions for a Mueller matrix to be derivable from a Jones matrix," *J. Opt. Soc. Am. A* 11(8), 2305–2319 (1994).
40. S. C. Rashleigh, "Acoustic sensing with a single coiled monomode fiber," *Opt. Lett.* 5(9), 392–394 (1980).
41. N. Fürstenau, M. Schmidt, W. J. Bock, and W. Urbanczyk, "Dynamic pressure sensing with a fiber-optic polarimetric pressure transducer with two-wavelength passive quadrature readout," *Appl. Opt.* 37(4), 663–671 (1998).
42. G. A. Massey, D. C. Erickson, and R. A. Kadlec, "Electromagnetic field components: their measurement using linear electrooptic and magnetooptic effects," *Appl. Opt.* 14(11), 2712–2719 (1975).
43. A. Papp and H. Harms, "Magnetooptical current transformer. 1: principles," *Appl. Opt.* 19(22), 3729–3734 (1980).
44. T. G. Giallorenzi, J. A. Bucaro, A. Dandridge, et al., "Optical fiber sensor technology," *IEEE Trans. Microw. Theory Tech.* 30(4), 472–511 (1982).
45. D. A. Jackson, "Monomode optical fibre interferometers for precision measurement," *J. Phys. E: Scient. Instrum.* 18(12), 981 (1985).
46. D. A. Jackson, R. Priest, A. Dandridge, and A. B. Tveten, "Elimination of drift in a single-mode optical fiber interferometer using a piezoelectrically stretched coiled fiber," *Appl. Opt.* 19(17), 2926–2929 (1980).

47. A. Dandridge, A. B. Tveten, and T. G. Giallorenzi, "Homodyne demodulation scheme for fiber optic sensors using phase generated carrier," *IEEE Trans. Microw. Theory Tech.* 30(10), 1635–1641 (1982).
48. S. K. Sheem, T. G. Giallorenzi, and K. Koo, "Optical techniques to solve the signal fading problem in fiber interferometers," *Appl. Opt.* 21(4), 689–693 (1982).
49. J. N. Blake, S. Y. Huang, B. Y. Kim, and H. J. Shaw, "Strain effects on highly elliptical core two-mode fibers," *Opt. Lett.* 12, 732–734 (1987).
50. H. Wang, S. Pu, N. Wang, S. Dong, and J. Huang, "Magnetic field sensing based on singlemode-multimode-singlemode fiber structures using magnetic fluids as cladding," *Opt. Lett.* 38, 3765–3768 (2013).
51. J. W. Goodman, *Statistical Optics* (Wiley, New York, 1985), Chaps 5 and 7.
52. C. Akcay, P. Parrein, and J. P. Rolland, "Estimation of longitudinal resolution in optical coherence imaging," *Appl. Opt.* 41(25), 5256–5262 (2002).
53. T.-P. Lee, C. Burrus, and B. Miller, "A stripe-geometry double-heterostructure amplified-spontaneous-emission (superluminescent) diode," *IEEE J. Quantum Electron.* 9(8), 820–828 (1973), doi: 10.1109/JQE.1973.1077738
54. C. Holtmann, P.-A. Besse, and H. Melchior, "High power superluminescent diodes for 1.3 μm wavelengths," *Electron. Lett.* 32, 1705–1706 (1996).
55. P. F. Wysocki, M. J. F. Digonnet, B. Y. Kim, and H. J. Shaw, "Characteristics of erbium-doped superfluorescent fiber sources for interferometric sensor applications," *J. Lightwave Technol.* 12(3), 550–567 (1994).
56. K. Bohm, K. Petermann, and E. Weidel, "Performance of Lyot depolarizers with birefringent single-mode fibers," *J. Lightwave Technol.* 1(1), 71–74 (1983), doi: 10.1109/JLT.1983.1072097
57. W. Burns, "Degree of polarization in the Lyot depolarizer," *J. Lightwave Technol.* 1(3), 475–479 (1983), doi: 10.1109/JLT.1983.1072136
58. M. P. Varnham, D. N. Payne, A. J. Barlow, and E. J. Tarbox, "Coiled-birefringent-fiber polarizers," *Opt. Lett.* 9(7), 306–308 (1984).
59. P. D. McIntyre and A. W. Snyder, "Power transfer between optical fibers," *J. Opt. Soc. Am.* 63, 1518–1527 (1973).
60. R. G. Priest, "Analysis of fiber interferometer utilizing 3 x 3 fiber coupler," *IEEE Trans. Microw. Theory Tech.* 30(10), 1589–1591 (1982), doi: 10.1109/TMTT.1982.1131294
61. A. D. Kersey, M. J. Marrone, and M. A. Davis, "Polarisation-insensitive fibre optic Michelson interferometer," *Electro. Lett.* 27, 518–520 (1991).
62. K. O. Hill and G. Meltz, "Fiber Bragg grating technology – fundamentals and overview," *J. Lightwave Technol.* 15(8), 1263–1276 (1997).
63. A. D. Kersey, M. A. Davis, H. J. Patrick, et al., "Fiber grating sensors," *J. Lightwave Technol.* 15(8), 1442–1463 (1997).

PART II

Magnetic Field and Current Sensors

CHAPTER 4

Magneto-Optic Effects

4.1 FARADAY EFFECT

Michael Faraday, in 1845, observed that in a piece of glass the plane of polarization of linearly polarized light experienced a rotation in the presence of a magnetic field parallel to the propagation direction of the light (Figure 4.1) [1]. The Faraday effect, also called magneto-optic effect, occurs in solids, liquids, and gases. The angle of rotation φ_F is given by

$$\varphi_F = V' \int_0^l \boldsymbol{B} \cdot \boldsymbol{dl}, \tag{4.1}$$

where V' is the Verdet constant in units of rad/(Tm) [2], B the magnetic flux density, and l the length of the traversed medium. Instead of V', literature on optical current sensors often states the term $V = m_0 V'$ (with units of rad/A); $m_0 = 1.256 \times 10^{-6}$ Vs/(Am) being the free space permeability, for V indicates the polarization rotation per unit current. The polarization rotation is the result of field-induced circular birefringence, $Dn_c = n_R - n_L$ (see Section 2.4):

$$\varphi_F = \frac{\pi (n_R - n_L)}{\lambda} l, \tag{4.2}$$

where n_R and n_L are the refractive indices of the right and left circular constituents, respectively. Note that (4.1) and (4.2) assume that the medium is free of linear birefringence and optical activity. In wavelength regions of normal dispersion ($\partial n/\partial l < 0$), the Verdet constant of diamagnetic materials is positive by convention. The polarization rotation is then counterclockwise (left-rotary, $n_L < n_R$) for an observer looking against the source and a magnetic field that points in the light propagation direction. φ_F is counted as positive in this case; see introduction to Chapter 2 on conventions. The polarization rotation is clockwise (right-rotary, $n_L > n_R$) and j_F is negative if the field points against the propagation direction.

The Faraday effect varies with wavelength l and temperature T and is strongest near absorption lines or electronic bandgaps, where the refractive index dispersion $n(l)$ is

DOI: 10.1201/9781003100324-7

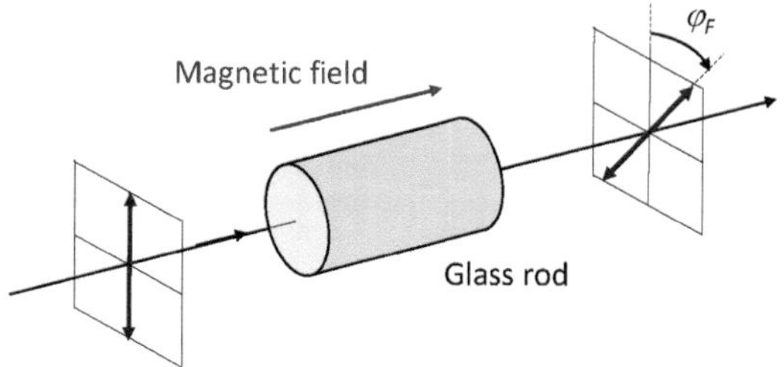

FIGURE 4.1 Faraday effect.

strongest. An (approximate) relationship for the Verdet constant of diamagnetic materials versus wavelength was first given by Becquerel and reads in the cgs system [3]

$$V' = -\frac{e}{2mc}\lambda\frac{\partial n(\lambda)}{\partial \lambda}. \tag{4.3}$$

Here, e and m are the electron charge and mass, respectively, and c is the speed of light in vacuum. An explicit expression for the Verdet constant dispersion of fused silica can be found, for example, in [4]. Note that the presence of dopants like in fused silica fiber somewhat modifies the effective Verdet constant [5]. At wavelengths well on the red side of the resonances, V' varies in good approximation in proportion to inverse square of the wavelength, that is,

$$V'(\lambda) = V'(\lambda_0)\frac{\lambda_0^2}{\lambda^2}, \tag{4.4}$$

where λ_0 is a reference wavelength [6–8]. More sophisticated phenomenological dispersion models for both diamagnetic and paramagnetic materials can be found in [9] and references therein.

In a simple classical explanation of the Faraday effect, the magnetic field induces a precession of the oscillating charges of the medium at the Larmor frequency $w_L = (e/m)\,B$. The LHC and RHC components then experience dispersion curves shifted by $-w_L$ and w_R, respectively, compared to the case without applied magnetic field. Consequently, the refractive indices, n_L and n_R, and phase velocities of the two circular components differ, that is, the medium becomes circularly birefringent: $n_L - n_R = n(w + w_L) - n(w - w_L) = 2(\partial n/\partial w)w_L$. The different phase velocities of the circular constituents show up as a polarization rotation of the resultant linear light. A more rigorous theoretical treatment of the Faraday effect involves a quantum mechanical description of the magnetic splitting of the electronic energy states of the medium (Zeeman effect) [10–12]. Whereas the sign of the Verdet constant in diamagnetic materials is positive (in regions of normal dispersion), it is negative in paramagnetic and ferrimagnetic materials. The difference is due to the opposite alignment of the magnetic moments in diamagnetic and paramagnetic materials with respect to the field direction. A theoretical expression for the Verdet constant of paramagnetic materials is given in [13] and references therein. As already mentioned in Part I, the Faraday effect is a non-reciprocal effect, i.e., the polarization rotation doubles, if the light, after reflection at a mirror, traverses the medium a second time in the opposite direction. This is in contrast to optical activity in chiral materials, e.g., sugar molecules dissolved in water, which undo the polarization rotation on the return trip.

4.2 OTHER MAGNETO-OPTIC EFFECTS

It should be noted that besides the Faraday effect (and underlying Zeeman effect), there are further magneto-optic effects, in particular the Cotton-Mouton and Voigt effects, the magneto-optic Kerr effect and magneto-optic dichroism [8, 14, 15]. The Cotton-Mouton effect is the magnetic counterpart of the Kerr electro-optic effect and gives rise to linear birefringence at light propagation perpendicular to the field direction. The birefringence is proportional to the square of the magnetic field strength but very small under most circumstances. The still smaller Voigt effect describes the same macroscopic phenomenon but is of different origin (transverse Zeeman effect). The magneto-optic Kerr effect is a reflection anisotropy, and magneto-optic dichroism refers to anisotropic absorption of left and right circular light in some ferrimagnetic materials.

4.3 MAGNETO-OPTIC MATERIALS

Table 4.1 lists selected diamagnetic, paramagnetic, and ferrimagnetic materials and their Verdet constants that have been used in magnetic field and current sensors. Diamagnetic glasses such as fused silica have relatively small Verdet constants that vary relatively little

TABLE 4.1 Verdet Constants and Temperature Dependences of Selected Magneto-Optic Materials

Material	Type of Magnetism	Verdet Constant (rad T^{-1} m^{-1})	Wavelength (nm)	$(1/V_0)$ (dV/dT) (10^{-4} K^{-1})	References
Fused silica glass	diamagnetic	3.6	633	0.69	[28]
Borosilicate glass, BK-7	diamagnetic	4.30	633	0.63	[28, 29]
Flint glass, SF-57	diamagnetic	20.1	633	1.26	[28]
Fused silica fiber	diamagnetic	6.11	458	0.69	[28, 30]
($Ge:SiO_2$ core)		3.25	633		
		1.83	825		
		0.54	1523		
Flint glass fiber	diamagnetic	3.2	1550	1.2	[31, 32]
$B_{12}SiO_{20}$ (BSO)	diamagnetic	29.1	870	1.4	[33]
$B_{12}GeO_{20}$ (BGO)	diamagnetic	101	633	-	[34]
		55	850		[35]
$Bi_4Ge_3O_{12}$ (BGO)	diamagnetic	30.8	633	-	[36]
Tb-doped flint glass, FR-5	paramagnetic	−71.3	633	−38	[37]
		−32	870		[35]
Tb-doped silica fiber	paramagnetic	−15.5	1300	-	[38]
Tellurite glass fiber	paramagnetic	−28.1	633	-	[39]
Terbium gallium garnet (TGG, $Tb_3Ga_5O_{12}$)	paramagnetic	−128	633	−34.7	[39–41]
Ga:Yttrium iron garnet (Ga:YIG, Y_3Fe_4 Ga O_{12})	ferrimagnetic	−86900[1]	1300	-	[42]

[1] Samples with a demagnetization factor $N_d = 0.08$.

(but by no means negligibly) as a function of temperature, and the materials can easily be drawn into fiber. Appropriate glass dopants or special glass compositions can significantly enhance the Verdet constant but commonly lead to higher optical attenuation. Flint glass with a high lead oxide content is of interest due to its small stress optic coefficients, as Section 7.10 will explain in more detail. Ferrimagnetic crystals such as yttrium iron garnets (YIG) and related materials are typically two to three orders more sensitive than diamagnetic glasses. However, ferrimagnetic domain movement at varying field strength often leads to a nonlinear response, hysteresis, and magnetic saturation at high fields. Furthermore, the geometric demagnetization factor N_d of the samples must be accounted for ($0 < N_d < 1$), that is, the sample dimensions influence the effective internal magnetic field and polarization rotation [16, 17]. The internal field is $H_{in} = H - N_d M$, where H is the external field and M is the sample magnetization. A long cylinder has a demagnetization factor $N_d \ll 1$ along its axis, whereas a thin plate has a demagnetization factor close to unity for fields normal to the plate. Apart from magnetic field sensors, ferrimagnetic materials are employed, for example, in optical isolators and rotators [9]. Further information on Verdet constants, Verdet constant measurement, dispersion, and temperature dependence can be found, for example, in [18–42].

CHAPTER 5

Current Sensing with Local Magneto-Optic Field Sensors

5.1 POINT MAGNETIC FIELD SENSORS

Point magneto-optic field sensors infer the current from the magnetic field at a single point in the vicinity of the electric conductor and represent a simple way of optical current measurement. Exploratory work on such sensors for use in electric power systems began at several institutions in the 1960s and early 1970s [43–59]. In the basic set-up, a magneto-optic material, e.g., a glass rod, is arranged between a polarizer and an analyzer and placed near the conductor such that the light path is parallel to the field, that is, the light propagation is perpendicular to the conductor direction (Figure 5.1). The polarization directions of the polarizer and analyzer are at 45° to each other. Often, a Wollaston prism instead of a single analyzer — or equivalently a beam splitter and two orthogonal analyzers — translate the field-induced polarization rotation into two anti-phase signals

$$I_1 = \left(I_0/2\right)\left[1 + K\sin\left(2\varphi_F\right)\right], \tag{5.1}$$

$$I_2 = \left(I_0/2\right)\left[1 - K\sin\left(2\varphi_F\right)\right], \tag{5.2}$$

where I_0 is proportional to the light source power, and K is the fringe contrast defined in Section 3.1. The signal $S = (I_1 - I_2) / (I_1 + I_2)$ is independent of I_0 and provides the rotation angle φ_F and hence the current [47, 56]:

$$S = K\sin\left(2\varphi_F\right). \tag{5.3}$$

Typically, φ_F remains sufficiently small, so that S varies in proportion to φ_F. Sensors for alternating current often work with only a single detector and normalize the signal by dividing the alternating component by the signal's dc offset $I_0/2$ or by controlling the light source power by means of a feedback circuit such that the dc offset remains constant.

DOI: 10.1201/9781003100324-8

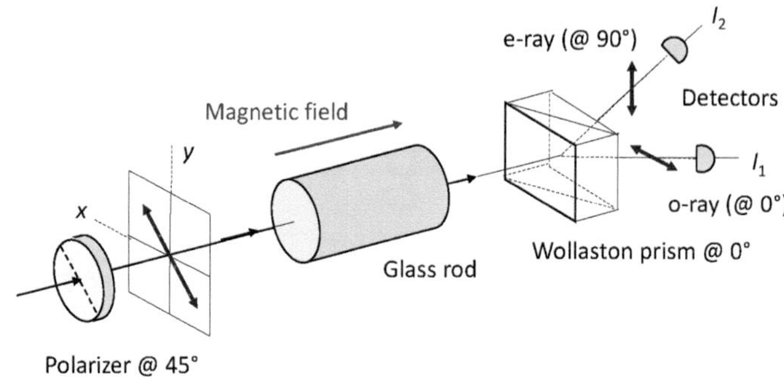

FIGURE 5.1 Faraday effect magnetic field sensor.

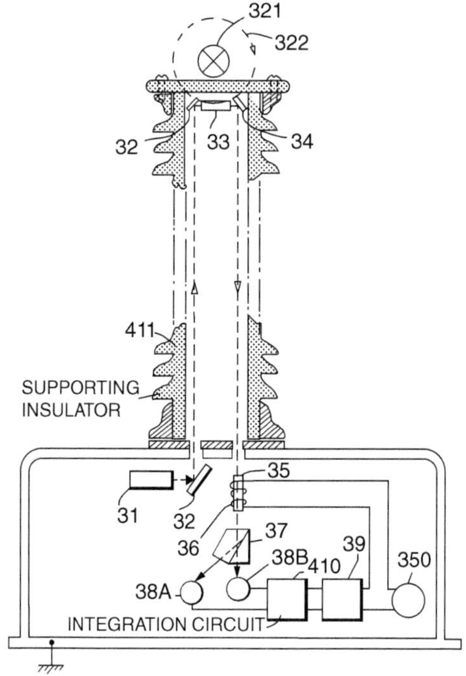

FIGURE 5.2 Early magneto-optic current sensor by Saito et al. (from US patent 3,597,683 [46].)

Already in the 1960s, Saito et al. at the University of Tokyo were among the first who demonstrated such a sensor, designed for current measurement at voltages up to 500 kV and beyond (Figure 5.2) [44–46]. The Faraday element was a 5-cm-long flint glass rod (33) in proximity to the current carrying conductor (321) at the top of a hollow-core support insulator. The Helium-neon laser source (31) and a Glan-Thompson analyzer (37) resided at ground potential. The laser beam propagated through free space, as optical fibers were still in an embryonic state at the time. A second flint glass rod (35) in a closed-loop feedback circuit compensated the Faraday rotation of the sensing rod. The circuit generated a secondary current in a wire coil (36) around the compensating rod and kept the difference of the two anti-phase detector signals at zero. Hence, drifts in the optical power, electronic gains, etc., and the temperature dependence of the Faraday effect were automatically compensated. Moreover, the signal varied linearly with the applied current, also at rotation angles approaching ±45° and beyond. At the time, sensor tests were performed at the Shiobara Laboratory of the Central Research Institute of Japan's Electric Power Industry. It is worth noting that the researchers also explored interrogation of the sensing medium by microwaves [45, 46].

In parallel to Saito et al.'s work, similar systems were developed by Pelenc et al. at Merlin & Gerin SA, Grenoble, France [47, 48] and G. Lesueur at Alstom, France [51, 52]. Somewhat later, in the early 1970s, A. J. Rogers, then at the Central Electricity Research Laboratories in Leatherhead, UK, explored such sensors, again with interrogation via free space, but with direct rather than closed-loop detection and with the sensing flint glass rod operated in reflective mode [56–58]. K. Mollenbeck at Siemens, Germany, invented a system where conventional means including an analog-to-digital converter first converted the high voltage current into a series of electric pulses with amplitude proportional to the instantaneous current amplitude. The pulses controlled the transmission of a YIG magneto-optic modulator interrogated by a free-space laser beam [50].

Sensor interrogation via free space became obsolete, after low-loss fibers became available during the 1970s. Moreover, LED and semiconductor laser light sources replaced incandescent sources and gas lasers. In 1982, K. Kyuma demonstrated a multimode fiber coupled flint-glass-based sensor using AlGaAs-based LED and laser diode sources, achieving accuracy to within ±0.5% between −25°C and 80°C. The magnetic fields ranged from 1.6×10^3 to 40×10^3 A/m and corresponded to currents up to 5 kA [35, 60]. A drawback of the above sensors is that the Faraday rotation is often relatively small, resulting in a modest signal-to-noise ratio at low currents. Arrangements for multiple light passes through the sensing medium helped to enhance the sensitivity [53, 61–64]. N. Inoue et al. demonstrated a current sensor utilizing a highly sensitive Bi-doped garnet optimized for low temperature dependence (-2.6×10^{-4}°C^{-1} between 0°C and 70°C) [65]. Besides glasses and garnets, diamagnetic crystals such as $Bi_{12}SiO_{20}$ and $Bi_4Ge_3O_{12}$ were also explored [33, 66–71]. The materials have relatively high Verdet constants (see Table 4.1) and can be employed for simultaneous current and voltage measurement [36]. Mihailovic et al. exploited the intrinsic optical activity of $Bi_{12}SiO_{20}$ for temperature compensation of the current signal [71].

Obvious downsides of point magneto-optic current sensors are a signal that varies with geometry, in particular the sensor's distance to the current conductor, and cross-sensitivity to neighbor currents. The effects can be mitigated with appropriate sensor arrangements and calibration strategies [56, 72–77]. A method to lessen cross-sensitivity is to place two sensors at opposite sides of the conductor, where the field directions are also opposite. By contrast, the signals from adjacent currents are to a large extent common mode and thus can be eliminated by taking the difference of the two signals [56, 73]. Perciante et al. utilized a modified conductor geometry with the magnetic field at the sensor location essentially orthogonal to the fields from neighbor phases [74]. Di Rienzo et al. and Ma et al. investigated crosstalk reduction by means of circular (albeit non-magneto-optic) sensor arrays around the conductor [75, 76]. In a three-phase sensor system, cross-sensitivity is commonly accounted for by appropriate calibration. The instantaneous signal of each sensor is determined by the instantaneous currents of all three phases, i.e., the signals S_i are related to the currents I_i by a matrix of sensitivity coefficients $s_{i,j}$, $i, j = 1...3$, obtained by calibration [77]:

$$S_i = \sigma_{ij} I_j. \tag{5.4}$$

Naturally, the geometry must not change during sensor operation, and crosstalk from potential further currents nearby is not excluded. The influence of an *inhomogeneous* field at the sensor location and the resulting Cotton-Mouton effect were studied by Li et al. and Perciante [78, 79].

Even though there have been demonstrations of point magneto-optic current sensors in high voltage power transmission (besides the references above, see, e.g., [80–83]), the residual errors due to crosstalk and sensitivity to position (e.g., at temperature variations) and other errors limit the achievable accuracy and have prevented wider use in demanding protection and metering applications at transmission voltages, that is, at voltages from about 70 kV to 800 kV and beyond. In this context, G. W. Day et al. presented a detailed analysis of the limitations of early electro-optic and magneto-optic sensors in a 1987 National Bureau of Standards (now NIST) report [84]. Point sensors, however, have been successfully introduced for monitoring, diagnostics, and fault localization in medium voltage distribution grids and at power cables (medium voltages are typically from about 1 kV to 35 kV) [85–87]. The passive nature of the sensors allows utilities to operate sensors for fault localization kilometers away from the nearest substation without needing any local electric infrastructure. Another benefit is the option of designing the sensors as clamp-on devices for easy retrofit installation [88, 89]. In particular, the Danish company PowerSense (later acquired by Landis+Gyr) has developed sensors for distribution grids, including clamp-on versions [88]. A challenge for medium voltage systems is to reach sufficiently low cost. Zubia et al. reported an example of a low-cost sensor [90] that consists of a flint glass rod (length of 20 mm and diameter of 5 mm), a 650 nm LED light source, and otherwise low-cost plastic optical components (Figure 5.3). Tests up to 800 A at constant temperature

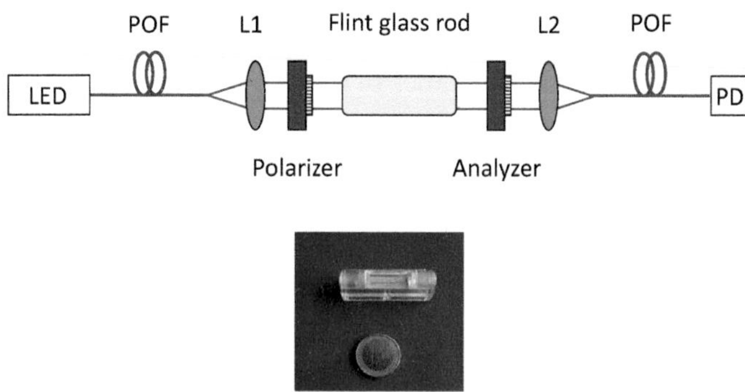

FIGURE 5.3 Low-cost magneto-optic current sensor by Zubia et al.: (a) Set-up with 650 nm LED light source, plastic optical fiber leads (POF), collimating lenses (L1, L2), polarizer, flint glass rod, analyzer, and photodetector (PD); (b) Close-up view of the magneto-optic rod (adapted from Zubia et al. [90].)

showed accuracy to within ±1% above 175 A and ±5% between 30 A and 175 A. Methods for compensation of the temperature dependence of the Verdet constant of such sensors were reported in [34, 91–93].

Besides current measurement, magnetic field sensors have numerous uses in other areas such as automation, automotive, aerospace, data storage, or mineral prospecting, to name a few. However, compared to magneto-resistive, Hall effect, fluxgate sensors, and others, magneto-optic sensors play only a minor role. For reviews of magnetic field sensors and their applications, see, e.g., [94–96]. Magneto-optic sensors can be beneficial in environments with strong electro-magnetic interference. They have also been used for magnetic field mapping on integrated electronic circuits [97], sub-surface material defect evaluation [98], plasma science [99], and power semiconductors [100]. Deeter et al. studied yttrium-iron-garnet-based sensors for sensing of high-frequency fields and, using a YIG rod with a length of 3 mm and a diameter of 5 mm, achieved a hysteresis-free response up to 60 mT, a minimum detectable field of 10 nT/$\sqrt{\text{Hz}}$, and a usable bandwidth from dc to 700 MHz [16, 101–103]. Supplementing the sensor with magnetic concentrators improved the noise equivalent field to 6 pT/$\sqrt{\text{Hz}}$, albeit at a reduced 3dB-bandwidth of 10 MHz [103]. Similar work was reported in [104, 105].

5.2 FIELD SENSORS IN MAGNETIC CORE

E. J. Casey and C. H. Titus at General Electric Corp., US, in 1963, invented a current transducer that combined traditional technology and a magneto-optic sensing element (Figure 5.4) [106]. In one version (Figure 5.4, left), a circular magnetic core (30) around the current conductor (10) concentrated the magnetic field in a gap of the core. The Faraday

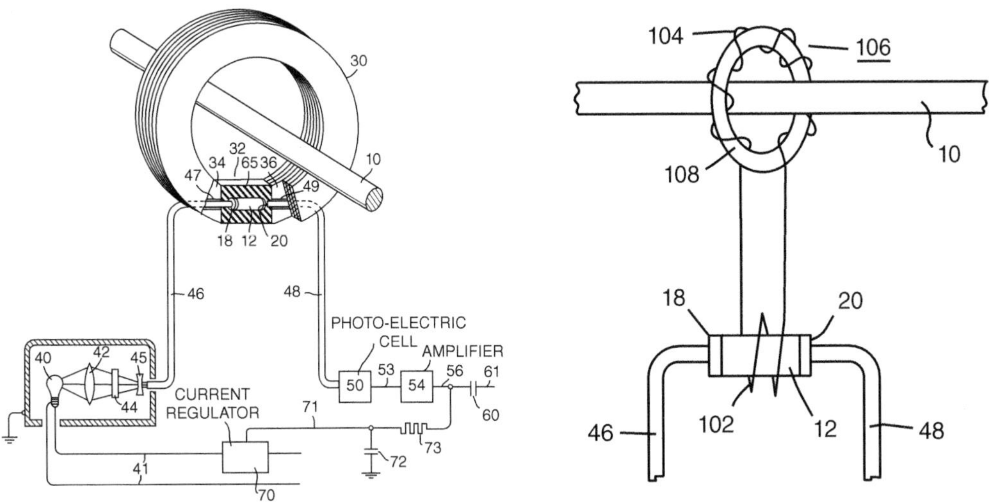

FIGURE 5.4 Early magneto-optic current sensors with magnetic core. Left: A magneto-optic flint glass sensor element (12) measures the magnetic field in the gap of a magnetic core (30) around the current conductor (10). Right: The magnetic core (108) around the conductor (10) has a secondary winding (104). The induced current produces a magnetic field at magneto-optic sensor element (12) (from US Patent 3,324,393 [106].)

element of flint glass (12) with thin glass laminated polarizers (18, 20) resided in that gap. Fiber-optic "light conveyors" (fibers as known today were still to come) guided the monochromatically filtered light from an incandescent lamp (40) on ground to the Faraday element and back to a photoelectric cell (30). A feedback circuit controlled the light source such that the dc optical power received at the photocell remained constant. In another version (Figure 5.4, right), a closed magnetic core featured a secondary electric winding (104); the current conductor being the primary. The induced secondary current flowed through a serial wire coil (102) which contained the Faraday element (12). In later years, many other researchers also reported magnetic-core-assisted sensors, e.g., [61, 67, 85, 104, 107, 108–119].

The magnetic concentrator (Figure 5.4, left) leads to a significant enhancement of the sensor signal. Furthermore, the method strongly reduces sensitivity to the conductor position and crosstalk. However, the core material and shape should be chosen for small hysteresis and small eddy current loss. Laminated cores limit the latter. Moreover, no saturation should occur within the specified current range. Core materials have included alloys of FeNi, FeSi, and others. Cores consisting of a compressed powder of multiple alloys (e.g., from Fe, Ni, Mo) in a dielectric matrix are another option. For a given core material, larger core diameter and cross section enhance the saturation field. Narrower core gaps result in higher gap fields and reduce residual crosstalk but leave less space for the sensing element. The magnetic flux density B_g on the central axis of the (air) gap of a ring-shaped core is approximately given by [61]

$$B_g = \frac{\mu_0 \mu_r I}{2\pi r + (\mu_r - 1)d}. \tag{5.5}$$

Here, m_r, r, and d are the relative core permeability, core radius, and gap width, respectively. For large permeabilities and sufficiently small core radii, B_g can be further approximated as

$$B_g = \mu_0 \frac{I}{d}. \tag{5.6}$$

The flux density is then enhanced by a factor of $2p\ r/d$ compared to a sensor at a distance r from the conductor without a field concentrator.

Itoh et al. reported an example of a particularly compact sensor set-up [110]. The magnetic field probe was configured in a separate casing for insertion into the air gap of the core (Figure 5.5). An 880-nm LED served as a light source. The magneto-optic element was a thin film of a high Verdet constant rare-earth iron garnet with a thickness of 60 mm and cross-sectional area of 2 mm^2 sandwiched between 0.5-mm-thick glass polarizers and 90°-reflecting prisms. The total probe width was only 5 mm. The magnetic core from a high permeability material had inner and outer diameters of 15 mm and 35 mm, respectively, and an air gap of 6 mm. The sensor provided a linear output within ±1% up to 300 A with a noise equivalent current of 1.8 mA/$\sqrt{\text{Hz}}$. The isolation ratios against external currents in the immediate vicinity of the core ranged from −53 dB to −75 dB, as indicated in Figure 5.5. The probe

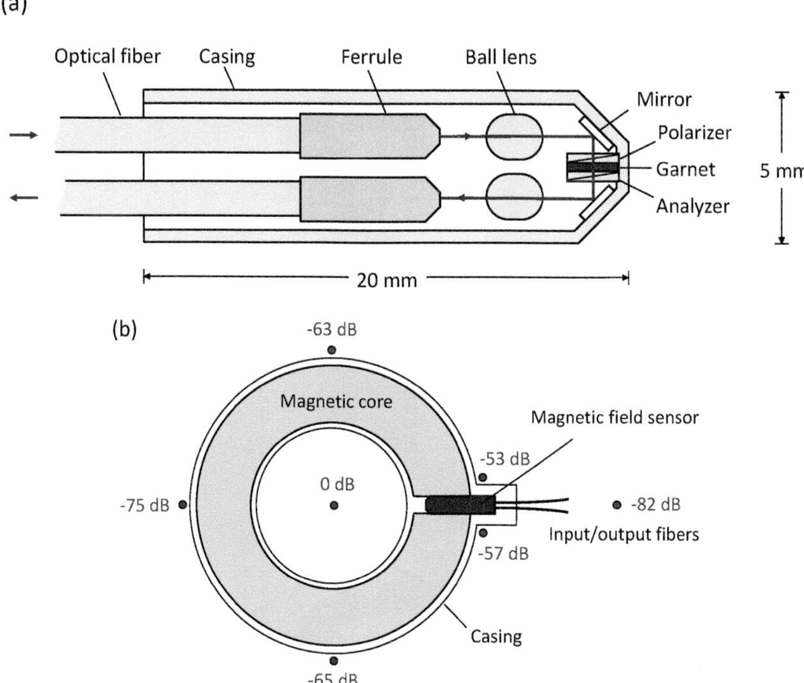

FIGURE 5.5 Faraday current sensor with magnetic core: Field sensor probe (a); magnetic core with sensor probe and isolation ratios in dB for nearby external currents (b) (adapted with permission from [110] © The Optical Society.)

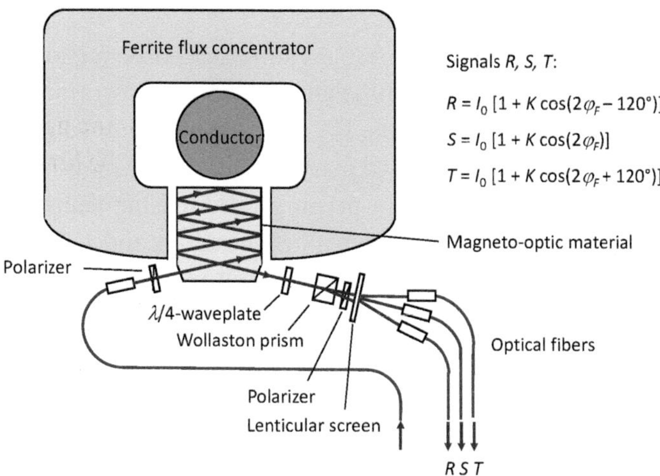

FIGURE 5.6 Faraday current sensor with magnetic core: Multiple round-trips of the light through the sensing element enhance the sensitivity. The Faraday rotation angle j_F follows from three spatially separated signals R, S, T with mutual phase differences of 120° (adapted with permission from [108] © The Optical Society.)

signal (without core) was temperature-stable within ±1% between −20°C and 80°C. Others have worked with lower Verdet constant materials and compensated the smaller Faraday effect with multiple roundtrips of the light through the sample (Figure 5.6) [61, 107, 108, 116].

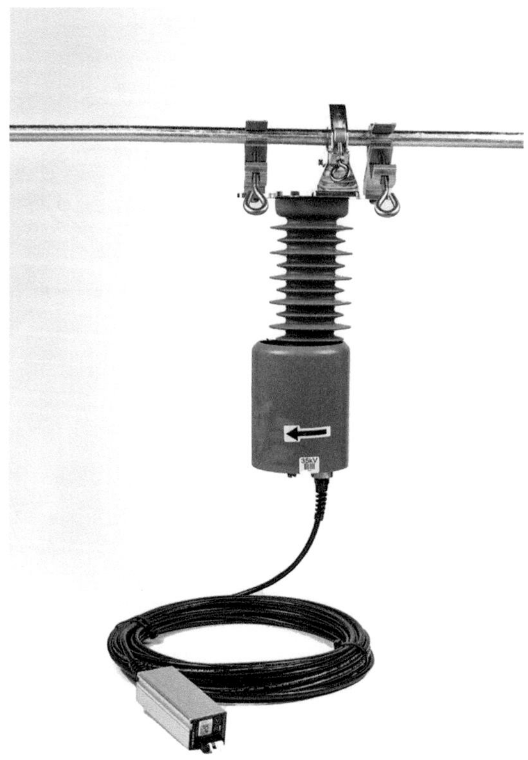

FIGURE 5.7 Clamp-on current sensor for medium voltage distribution grids (courtesy of Micatu, Horseheads, NY, USA).

Of particular practical interest are portable, clamp-on sensor versions that can be easily attached to a power line [109, 114, 115, 119]. Figure 5.7 shows a corresponding line-hanging sensor developed by Micatu (Horseheads, NY, USA) for use in medium voltage distribution grids [119]. The sensor has a D-shaped magnetic core in the movable part of the clamp. Closing the clamp moves the core over the prism-shaped sensing element in the base under the clamp. The sensor is specified for currents between 10 A and 1000 A at voltages up to 35 kV and achieves accuracy to within 0.5% at temperatures between −30°C and 70°C. The assembly also contains an optical voltage sensor.

CHAPTER 6

Glass Block Optical Current Sensors

The glass block current sensors of this chapter, as well as the all-fiber sensors of Chapter 8 further below, are "true" current sensors in the sense that they measure a closed-loop integral of the magnetic field around the current conductor. To this end, the light travels in the magneto-optic medium along one of several closed-loop paths around the conductor. The polarization rotation angle φ_F is then given as

$$\varphi_F = NV'\oint Bdl = NV'\mu_0\oint Hdl, \tag{6.1}$$

and can be rewritten as

$$\varphi_F = NVI \tag{6.2}$$

where N is the number of loops and $V = \mu_0 V'$. The integral $\oint Hdl$ represents, according to Ampère's law, the electric current. Other than in the case of point-like magnetic field sensors, φ_F is now independent of any geometrical parameters, in particular, of the loop diameter and conductor position inside the loop. Moreover, currents outside the loop have no influence. Closed-loop integration significantly enhances the application flexibility of the sensors and also avoids a magnetic core and its limitations.

Research and development of bulk glass current sensors, often called MOCT — Magneto-Optic Current Transformers — primarily took place in the 1980s and 1990s [120–169]. Several electric power equipment manufacturers designed and made such devices commercially available, as an optical equivalent to the so-called current transformers commonly used up to that point to measure high voltage electric currents . At the time, all-fiber current sensors still suffered from limitations of the fibers, and bulk sensor types allowed researchers to circumvent those limitations. Typically, the packaged glass

DOI: 10.1201/9781003100324-9

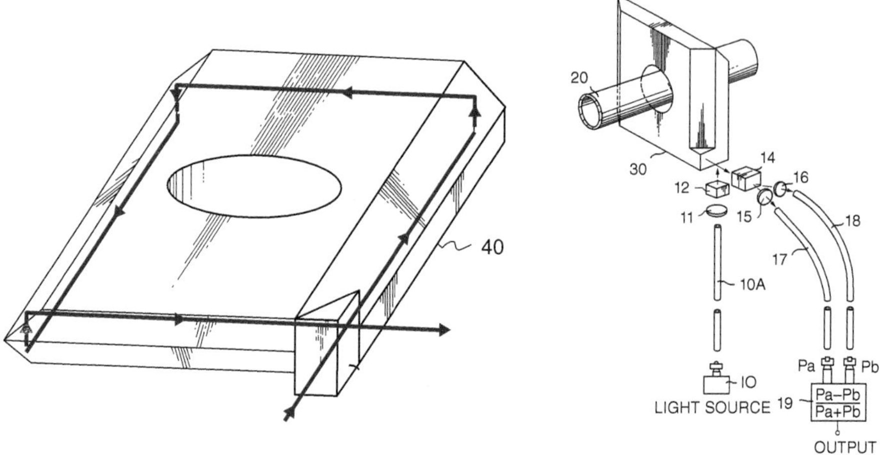

FIGURE 6.1 Magneto-optic current transformer (MOCT) with ortho-conjugate total internal light beam reflections: MOCT body (40, 30), current conductor (20), connecting optical fibers (10A, 17, 18), collimators (11, 15, 16), polarizer (12), input/output prism (41), Wollaston prism (14) (adapted from US Patent 4,564,754 [121].)

block with attached input/output polarizers and collimators is mounted at the high voltage current conductor. Multimode fibers run through a high voltage insulator to ground and connect the sensor head to the light source and one or two photo-detectors. T. Sato et al. at Hitachi invented the classical MOCT design in 1982 (Figure 6.1) [121, 122].

The sensing element was a square block (40), e.g., of fused silica, crown glass, or flint glass, with a central aperture for the current conductor (20). The light (red path) entered the block at a corner through a collimator (11) and polarizer (12), and experienced two ortho-conjugate total internal reflections at each of the other three corners, before again leaving the block at the entrance corner. A Wollaston prism generated two output signals according to (5.1) and (5.2), from which the rotation angle φ_F and current I followed. Total internal reflection occurs at angles of incidence larger than the critical angle θ_c, given by $\theta_c = a\sin(n_1/n_2)$, where n_2 is the glass refractive index and n_1 the index of the outer medium (air) [170]. The critical angle of a fused silica/air interface, for instance, corresponds to 43.5° (n_2 = 1.454) at 780 nm. The two consecutive total internal reflections at each corner served to maintain the linear light polarization. Upon the first reflection, the polarization components parallel and perpendicular to the plane of incidence (p- and s-polarizations) experience a relative phase shift; hence, the light is elliptically polarized along the corner path segments parallel to the conductor. The p- and s-polarized components of the first reflection become s- and p-polarizations at the second reflection, respectively, so that the phase shift is reversed and linear polarization restored.

Since the path segments of elliptical polarization contribute only with reduced sensitivity, the closed-loop integration is, strictly speaking, imperfect. In most practical cases, resulting errors, for example, due to higher cross-sensitivity, are negligible, though, because the current magnetic fields are largely orthogonal to those segments. The effects of non-polarization

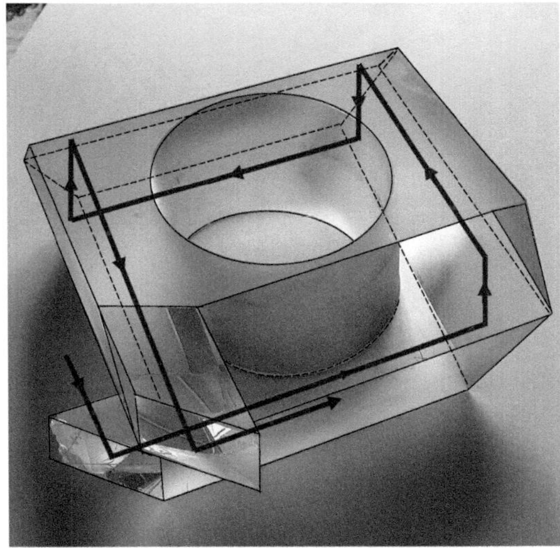

FIGURE 6.2 MOCT glass block (polarizers, collimators, and fiber leads are omitted). The light path is schematically indicated [128].

preserving reflections in MOCT were numerically investigated by Bush et al., in particular, the resulting non-linearity in the response, sensitivity to the conductor position, and crosstalk [129, 131]. Related work was reported by Wang et al. [141]. It should be noted that a square-shaped bulk glass current sensor had already been proposed by R. Lenz (Transformation Union Ltd, Germany) in 1970, albeit with non-polarization-preserving reflections at the corners instead of ortho-conjugate reflections [120].

Square-shaped bulk glass sensors with ortho-conjugate beam steering reflections were developed for commercial use by further companies including the Westinghouse T&D division (acquired by ABB in 1989) (Figure 6.2) [125–128], Siemens [144–146], Areva T&D / GEC Alsthom [158, 159], Square D Company [123, 124], and Japanese companies [150]. (Square D's sensor was not a monolithic block, however, but was composed of several individual glass segments.) A MOCT field installation will be illustrated further below.

As an alternative to the design of Figure 6.1, Chu et al. at the University of Kent, UK, demonstrated a triangular-shaped sensing element of flint glass with total internal reflections at the critical angle (Figure 6.3) [132, 133]. Flint glass has a refractive index n of 1.799 at 633 nm, which results in a critical angle θ_c of 33.774°. At this angle, the light is reflected without losing its linear polarization state. The method eliminates the need for ortho-conjugate reflections and thus results in a significantly simpler sensing element. On the other hand, the sensor signal is strongly sensitive to deviations from the correct angle. At smaller angles, the reflection coefficients for s- and p-polarizations differ; at larger angles the coefficients are equal but the phases of the two components differ [170]. Already small deviations from the critical angle significantly reduce the sensor's response to current, and the sensor becomes sensitive to the conductor position within the triangle and crosstalk. According to [132], the sensitivity reduces by about 20% if the angle of incidence is 0.5° lower than the critical angle, and by 6% if the angle is 0.5° higher. The variation of the glass

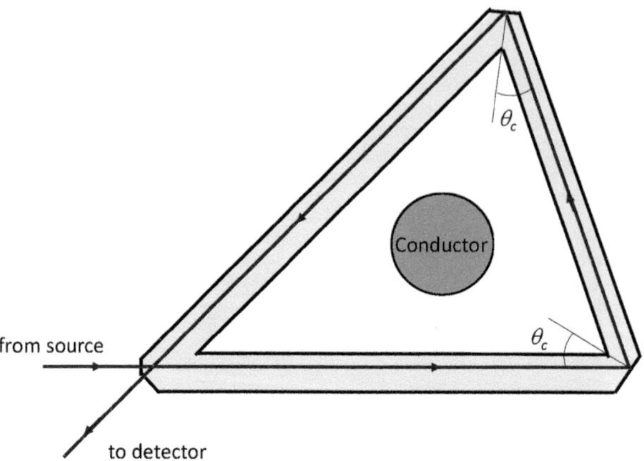

FIGURE 6.3 Triangular sensing element with polarization-maintaining reflections at critical angle θ_c (adapted with permission from [132] © The Optical Society.)

refractive index and critical angle with temperature introduced a small extra temperature dependence of 0.16%/100°C (in addition to the contribution of Verdet constant of 1.26%/100°C [28]). The sensor had a sensitivity of 2.35×10^{-5} rad/A at 633 nm and a noise equivalent current of 20 mA/$\sqrt{\text{Hz}}$. A split triangular sensing element for clamp-on mounting was demonstrated in [133].

Figure 6.4 shows a circular sensing element as a further alternative, demonstrated by Y. N. Ning et al., also at Kent University [130]. The light entered the flint glass ring with inner and outer diameters of 35 mm and 80 mm, respectively, through a roof-top prism and experienced 15 internal reflections at the critical angle before again leaving the ring through the

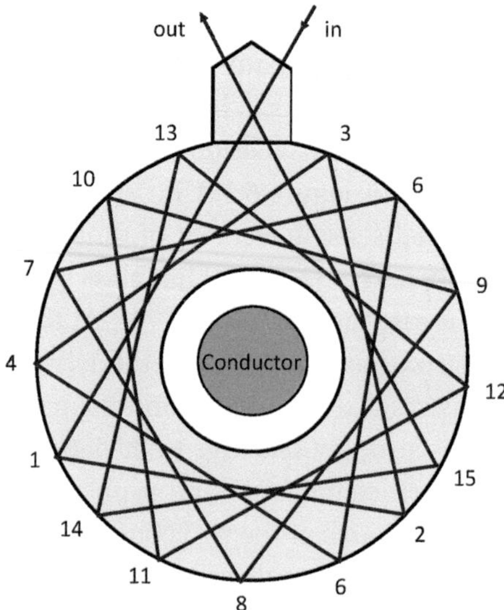

FIGURE 6.4 Glass ring sensing element with 15 internal, polarization-maintaining reflections at critical angle. The light encircles the current conductor five times (adapted with permission from [130] © The Optical Society).

prism. The light encircled the conductor five times with correspondingly enhanced polarization rotation. Another version was with 44 internal reflections and 14 light loops. The latter had a sensitivity of 3.9×10^{-4} rad/A (0.022 deg/A) at 780 nm.

Besides internal reflections that imperfectly maintain the linear polarization state of the light, another potential error source of MOCT is that the optical path between input and output polarizers is inherently not fully closed. In the glass ring of Figure 6.4, for example, input and output beams intersect inside the roof top prism, whereas the polarizers (not shown) are located not at that point but outside the prism. Several researchers investigated dedicated reflecting layers and MOCT designs to overcome those deficits [151, 153, 155, 168, 169]. Yoshino et al. explored a 60-mm-square flint glass MOCT with dielectrically coated, polarization preserving total reflection surfaces on three sides (Figure 6.5) [153]. The coating consisted of Ta_2O_5 and SiO_2 films and permitted some tolerances to deviations from the 840-nm-design-wavelength and the nominal 45°-angles of incidence. An input/output prism of small size and less field-sensitive BK7 glass kept signal contributions of unclosed optical path segments negligible. Crosstalk from currents at 60 mm distance from the MOCT center varied between 0.1% and 0.66%, with the highest contribution when external conductor faced the input/output prism. Wang et al. conceived a bulk glass current sensor immune to reflection phase shifts consisting of a magneto-optic glass ring with inner and outer graded index boundary layers [166].

A noteworthy miniature MOCT for the measurement of small currents in wide frequency ranges was reported by Rockford and Rose et al. [42], [168]. The sensor consisted of four YIG rods with a diameter of 2 mm and a length of 2.5 arranged as a square and joined by corner prisms with low birefringent reflective coatings. The sensor, operated at 1300 nm,

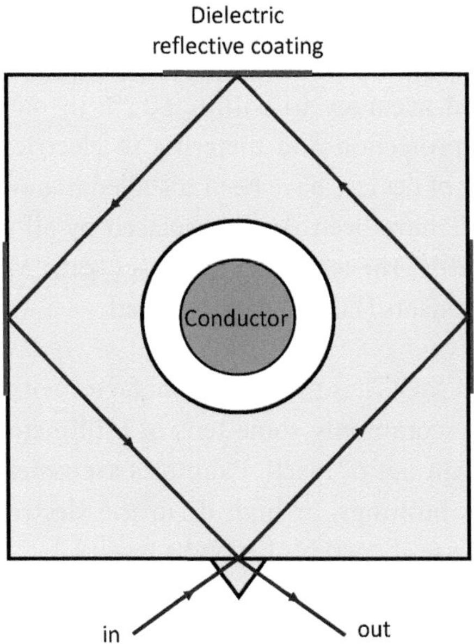

FIGURE 6.5 MOCT with dielectrically coated, polarization-preserving total reflection layers (adapted with permission from [153] © The Optical Society).

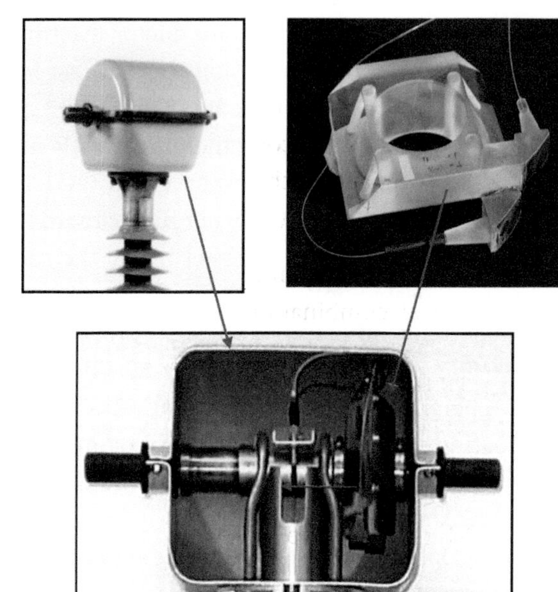

FIGURE 6.6 Commercial MOCT: sensor installation in substation (left); sensor housing on high voltage insulator (middle top), MOCT glass body with input/output prims, collimators, and fiber pigtails (right top), interior of sensor housing (right bottom) [128].

achieved a −3-dB bandwidth of 500 MHz with a noise-equivalent current of 840 nA/Hz/ 2 at 1.8 kHz and linear range of 10 A.

Finally, Figure 6.6 illustrates the installation of commercial MOCT from ABB Ltd in a high voltage substation. Details can be found in [125–128].

6.1 SOME CONCLUSIONS ON GLASS BLOCK CURRENT SENSORS

MOCT have demonstrated accuracy to within ±0.2% in outdoor temperature ranges, appropriate for high-end protection and metering in electric power transmission [125, 126]. A significant number of devices have been installed in power grid substations worldwide. Nevertheless, MOCT have been largely replaced by all-fiber current sensors from about the mid-2000s onwards. The following discusses some MOCT limitations, many of which fiber-optic current sensors (FOCS) have resolved:

- A severe limitation of MOCT is the fixed form factor with a limited aperture for the current conductor — commonly some tens of millimeters. Applications requiring larger apertures remain out of reach. Examples are generator circuit breakers, gas-insulated switchgear, bushings, or high dc in the electro-chemical industry. Also, thermal issues may arise at currents beyond a few kA.
- Most MOCT are with only a single optical loop. Hence, compared to FOCS, the sensitivity is smaller and cannot be adapted to particular rated currents. High Verdet constant glasses such as flint glass, which is about six times more sensitive than fused

silica (see Table 4.1), mitigate the problem. However, while on one hand a high Verdet constant enhances the signal-to-noise ratio at low currents, on the other hand, it reduces the maximum measurable current. For MOCT of fused silica, polarization rotations of ±45° correspond to currents of about ±300 kA at 825 nm. (The usable linear range is much smaller of course.) The minimum detectable current is typically around 1 A/$\sqrt{\text{Hz}}$. (The corresponding polarization rotation is 2.7 mrad/$\sqrt{\text{Hz}}$).

- Temperature-dependent birefringence in the glass block, e.g., due to intrinsic stress, material combinations of unmatched thermal expansion, or inadequate packaging, has been a frequent source of instability, which may result in poor production yield [145, 149, 162–165]. Here again, flint glass, with its low stress-optic coefficients, is beneficial. Other mitigation measures have been thermal annealing of the glass block, special signal processing [145, 163], compensation of birefringent phase shifts by operating the sensor in reflective mode utilizing a Faraday rotator mirror (FRM) [164] or a quadruple-reflection prism [149], and others [124, 168].
- The typically multimode fiber links of MOCT are sensitive to mechanical shock and vibration. Corresponding signal perturbations are mostly uncritical in electricity metering, as signal filtering can be applied, but may not be acceptable in relaying, for a shock signal is difficult to distinguish from a sudden current surge due to a fault. Researchers have developed MOCT with two independent counterpropagating beams, making use of the non-reciprocity of the Faraday effect. Taking the difference of the two corresponding signals doubles the current signal and largely cancels common-mode mechanical effects [136, 143, 146]. However, the method adds significant complexity and cost. In particular, optical crosstalk between the two channels must be accounted for, for example, via modulating the intensity of the beams at two different frequencies or working with two wavelengths. A mechanical noise canceling scheme utilizing a pulsed source and time multiplexing has been reported in [135].
- Cross-sensitivity to neighbor currents as a result of imperfect closed-loop integration of the magnetic field is commonly not a serious error source but should be kept in mind when choosing the sensor location.
- MOCT for accurate current measurement (to within <±1%) at outdoor temperatures commonly need a temperature sensor or other means to account for the influence of temperature. This of course also applies to the all-fiber current sensors.
- Last but not least, the production and assembly cost of a bulk glass sensor head is significantly higher than the cost of FOCS fiber coils.

CHAPTER 7

Optical Fibers for Current Sensing

7.1 INTRODUCTION

In fiber-optic current sensors (FOCS), also called fiber-optic current transducers (FOCT), fibers not only serve to guide the source light to and from the sensor head, but the sensing element itself is also an optical fiber. This does not exclude discrete non-fiber components such as polarizer platelets, beam splitters, modulators, polarization rotators, and others. All-fiber current sensors have fascinated researchers in the electric power world since the advent of low-loss optical fibers around 1970. The sensors promised to replace massive inductive current transformers by simple optical fibers.

The operating principle is straightforward. One or several loops of sensing fiber must enclose the current conductor. The degree of polarization rotation of light traveling through the fiber, caused by the Faraday effect, serves a proportional measure of the electronic current flowing in the encircled conductor (Figure 7.1). Like the bulk glass sensors of the previous section, the sensors measure a closed loop integral to the magnetic field around the current conductor according to (6.2). In the ideal case, the signal is only given by the current and independent of any geometrical parameters such as the loop diameter or the conductor position inside the fiber coil. Currents outside the coil have no influence.

The early investigations soon identified linear fiber birefringence as a serious obstacle on the path to viable sensors. Notable causes of birefringence were mechanical stress from fiber bending, inadequate fiber coatings, and fiber imperfections such as deviations from circular core shape [171–180]. Other hurdles were sensitivity to temperature, mechanical shock and vibration [179, 181], and, to a lesser extent, cross-sensitivity to neighbor currents [182, 183]. Over time, dedicated fibers and sensor topologies were developed to overcome those limitations. The following illustrate the various developments towards viable sensors that took place over several decades.

DOI: 10.1201/9781003100324-10

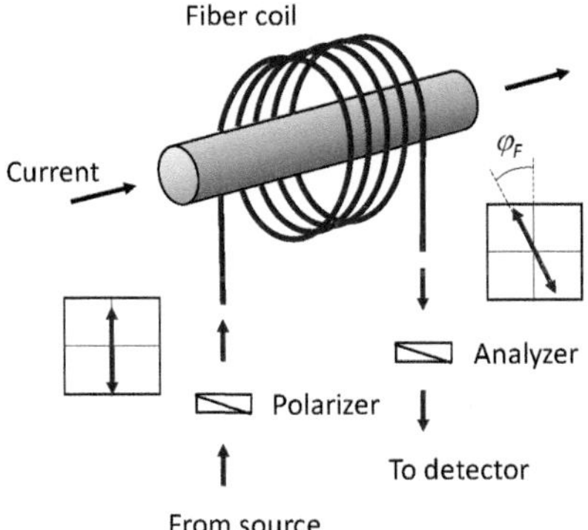

FIGURE 7.1 Basic concept of a fiber-optic current sensor. The Faraday polarization rotation angle is $\varphi_F = VNI$, where N, V, and I are the number of fiber loops, the Verdet constant, and the current, respectively.

7.2 LIGHT WAVE PROPAGATION IN PERTURBED SINGLE-MODE FIBERS

The propagation of light through dielectric waveguides such as optical fibers is governed by the wave equation which follows from Maxwell's equations [184]:

$$\nabla^2 \boldsymbol{E} = \epsilon \mu \frac{\partial^2 \boldsymbol{E}}{\partial t^2}. \tag{7.1}$$

Here, ∇^2 is the Laplacian operator, $\boldsymbol{E}$ is the electric field vector, $e = e_0 e_r$ is dielectric permittivity with e_0 being the vacuum permittivity, and e_r the relative permittivity (or dielectric constant); $m = m_0\, m_r$ is the magnetic permeability, where m_0 is the vacuum permeability and m_r the relative permeability, which is close to unity in dielectric materials such as optical fibers.

The equation can be rewritten as

$$\nabla^2 \boldsymbol{E} = \varepsilon_0 \mu_0 \frac{\partial^2 \boldsymbol{E}}{\partial t^2} + \mu_0 \frac{\partial^2 \boldsymbol{P}}{\partial t^2}. \tag{7.2}$$

Here, $\boldsymbol{P}$ is the induced electric polarization: $\boldsymbol{P} = e_0 \chi_{el} \boldsymbol{E}$, χ_{el} is the electric susceptibility.

In an ideal fiber, $e(r)$ is a scalar function of the radial distance r from the fiber axis and defines the refractive index profile of the fiber; $n(r) = \sqrt{e_r(r)}$. With appropriate boundary conditions, i.e., the refractive index difference between the fiber core and cladding ($n_{co} - n_{cl}$), the core diameter, and the wavelength, the solution of the wave equation is a single guided mode (HE_{11} mode), as we have already seen in Chapter 1 [184]. In the limit of a weakly guiding fiber ($n_{co} \approx n_{cl}$), the modal field approaches a plane wave solution. The field can be represented as the vector sum of two degenerate orthogonal polarization eigenmodes. We assume that the fiber axis corresponds to the z-axis of a Cartesian coordinate system, and the modal fields vibrate in x- and y-directions. The superposition of the two modes results in a transverse electric field, which, at a given location z along the fiber, can be expressed as [185–190]

$$\boldsymbol{E}(x,y,z) = \boldsymbol{E}_x(x,y,z) + \boldsymbol{E}_y(x,y,z), \tag{7.3}$$

where $\boldsymbol{E}_x(x, y, z)$ and $\boldsymbol{E}_y(x, y, z)$ read in complex notation

$$\boldsymbol{E}_x(x,y,z) = E_{0x}A_x(x,y)\exp\left[i\left(\varpi t - \beta_x z\right)\right]\hat{\boldsymbol{x}}, \tag{7.4}$$

$$\boldsymbol{E}_y(x,y,z) = E_{0y}A_y(x,y)\exp\left[i\left(\varpi t - \beta_y z\right)\right]\hat{\boldsymbol{y}}. \tag{7.5}$$

Here, the E_{0i} are the mode amplitudes, the terms $A_i(x,y)$, normalized to unity, describe the variation of the two mode fields in transverse directions and are commonly approximate Gaussians; the b_i denote the modal propagation constants ($b_x = b_y$ for degenerate modes), w is the angular frequency, and $\hat{\boldsymbol{x}}$ and $\hat{\boldsymbol{y}}$ are unit vectors. Modal phase offsets δ_m such as in (2.2) and (2.3) are assumed as zero. The propagation constants are defined as

$$\beta_i = n_i k = n_i\left(2\pi/\lambda\right), \tag{7.6}$$

where the n_i are the effective refractive indices of the modes, and $k = 2p/l$ is the wave number with l being the wavelength in vacuum.

In an ideal unperturbed fiber, the mode amplitudes E_{0i} and propagation constants b_i can be assumed as independent of z. A linear or circular polarization state is then maintained while the light propagates through the fiber. Mechanical perturbations of the fiber, e.g., due to fiber bending, fiber twist, lateral forces, etc., as well as deviations from a perfectly circular fiber core, break the circular symmetry of an ideal fiber and lift the degeneracy of the orthogonal modal fields. The modified dielectric permeability tensor e' can be expressed as

$$\varepsilon' = \varepsilon + \Delta\varepsilon. \tag{7.7}$$

In case of mechanical perturbations, the elements of the perturbation tensor $De\ (x,y,z)$ are given by [191, 192]

$$\Delta\left(\frac{1}{\varepsilon}\right)_{ij} = p_{ijkl}S_{kl}, \tag{7.8}$$

where p_{ijkl} is the fourth-rank photoelastic tensor, and the S_{kl} represent the strain tensor. The permeability changes induced by a magnetic field H_z are $De_{12} = -De_{12} = -2in_eV/k$ [186].

The perturbation gives rise to an extra dielectric polarization D$\boldsymbol{P}$ = e_0De $\boldsymbol{E}$, so that the total polarization is given as

$$\boldsymbol{P} = \varepsilon_0 \left(\chi_{el} \boldsymbol{E} + \Delta\varepsilon\, \boldsymbol{E} \right). \tag{7.9}$$

Typically, the perturbation can be assumed as weak (De ≪ *e*). The extra polarization *D***P** and corresponding changes in the effective refractive indices lift the degeneracy of the orthogonal modes and/or couples the modes. The complex amplitudes of the modes $\boldsymbol{E}_i = E_{0i}$ exp($-ib_iz$) $\hat{\boldsymbol{u}}_i = \boldsymbol{E}_{0i}$ exp($-ib_i z$) (with $\hat{\boldsymbol{u}}_i$ standing for the unit vectors) then become functions of *z* and are described by a set of coupled mode equations [186–189]:

$$\partial \boldsymbol{E}_x(z)/\partial z = i\left[\kappa_{11}\boldsymbol{E}_x(z) + \kappa_{12}\boldsymbol{E}_y(z)\right], \tag{7.10}$$

$$\partial \boldsymbol{E}_y(z)/\partial z = i\left[\kappa_{21}\boldsymbol{E}_x(z) + \kappa_{22}\boldsymbol{E}_y(z)\right]. \tag{7.11}$$

For simplicity, we have dropped the oscillatory terms exp(iωt). In short form, the equations read

$$\frac{d}{dz}\begin{pmatrix} \boldsymbol{E}_x \\ \boldsymbol{E}_y \end{pmatrix} = i\boldsymbol{K}\begin{pmatrix} \boldsymbol{E}_x \\ \boldsymbol{E}_y \end{pmatrix}, \tag{7.12}$$

with

$$\boldsymbol{K} = \begin{pmatrix} \kappa_{11} & \kappa_{12} \\ \kappa_{21} & \kappa_{22} \end{pmatrix}. \tag{7.13}$$

The coupling coefficients κ_{mn} depend on the nature of the perturbation De(*x,y*) and the overlap of the extra dielectric polarization DP(*x,y*) with the mode fields $A_i(x,y)$ and can be written as [189]:

$$\kappa_{mn} = \left[-k/\left(2nW_n\right)\right] \int dxdyA_n^*(x,y)\Delta\varepsilon(x,y)A_m(x,y), \tag{7.14}$$

where *n* is the effective refractive index of the unperturbed fiber. The terms W_n are defined via the orthogonality relation

$$\int dxdyA_n^*(x,y)A_m(x,y) = W_n\delta_{nm}. \tag{7.15}$$

d_{nm} is the Kronecker symbol. (The asterisk indicates complex conjugation.) Here, De is assumed as independent of *z*. The diagonal coefficients κ_{11} and κ_{22} describe changes in the propagation constants b_x, b_y of the orthogonal modes. Unequal coefficients κ_{11}, κ_{22} lift the degeneracy of the modes and result in spatial mode beating with a beat length of $L_B = l/[2(\kappa_{11} - \kappa_{22})]$. The off-diagonal coefficients κ_{12} and κ_{21} ($\kappa_{12} = \kappa_{21}^*$) describe the mutual coupling of the modes that results in an interchange of power. For example, an applied magnetic field H_z results in $\kappa_{12} = -\kappa_{21} = -i2VH_z$. In a lossless fiber, the total power of the modes is preserved and the matrix $\boldsymbol{K}$ is Hermetian, i.e., κ_{11} and κ_{22} are real numbers and $\kappa_{12} = \kappa_{21}^*$. It should be remembered that the mode fields have, besides the transverse field components along x and y, a small longitudinal component in z-direction. Fiber twist, for example, mutually couples the transverse components of a given mode to the longitudinal component of the respective orthogonal mode resulting in circular birefringence [186].

The coupled-mode equations thus describe the evolution of the local polarization along a perturbed fiber. In order to obtain, for a given input polarization state at $z = 0$, the corresponding output polarization state after a fiber length z, the coupled mode equations must be integrated. The integration results in the Jones matrix $\boldsymbol{J}$ of the fiber piece (2.7.2). The Jones matrix relates to the matrix $\boldsymbol{K}$ of coupling coefficients as follows:

$$\frac{d}{dz}\boldsymbol{J} = i\boldsymbol{KJ} = \boldsymbol{NJ}. \tag{7.16}$$

The product $i\boldsymbol{K}$ is the $\boldsymbol{N}$-matrix introduced in 2.7.2. One should mention that in case of monochromatic light, only the state of polarization varies in a perturbed fiber but not the degree of polarization. By contrast, in case of non-monochromatic light with frequency-dependent propagation constants, the two orthogonal modes gradually lose their coherence and the light becomes depolarized, as we will discuss in more detail further below.

In the following, we will first consider how linear birefringence, e.g., from fiber bending, deteriorates magneto-optic current measurement. The subsequent sections will then present special fiber types and fiber arrangements for current sensing. Some of those fibers suppress linear birefringence by intentional perturbations De.

7.3 FARADAY EFFECT IN THE PRESENCE LINEAR BIREFRINGENCE

Bending a single-mode fiber to a coil increases the effective refractive index for light polarized parallel to the coil normal (slow axis, index n_s) and decreases the index for light polarized in the plane of the coil (fast axis, index n_f), i.e., the fiber becomes birefringent. The linear birefringence quenches the Faraday polarization rotation, particularly if the fiber length l approaches and exceeds a quarter of the linear beat length L_b. The index difference of the orthogonal polarization modes is [173, 193]

$$\Delta n_b = n_s - n_f = (1/2)EC\left(r^2 / R^2\right), \tag{7.17}$$

where, r is the fiber radius, and R is the fiber loop radius. E and C are Young's modulus and the stress-optic coefficient of the fiber, respectively; $E = 7.5 \times 10^{10}$ Pa and $C = 3.28 \times 10^{-12}$ Pa^{-1}

for fused silica fiber [194]. The corresponding difference in propagation constants or birefringent phase retardation per unit length is

$$\beta_b = \beta_y - \beta_x = (2\pi / \lambda)(n_s - n_f). \tag{7.18}$$

Here, it has been assumed that the x-coordinate axis is aligned parallel to the fast fiber axis. A linear input state of polarization (SOP) with arbitrary orientation with respect to a principal axis of the birefringence is not maintained but evolves through a series of elliptical polarizations. We will discuss the SOP trajectory on the Poincaré sphere further below. In a Cartesian basis, the matrix of coupling coefficients in the presence of a magnetic field H_z is [186, 189]

$$\boldsymbol{K} = \begin{pmatrix} \beta_b / 2 & -iVH_z \\ iVH_z & -\beta_b / 2 \end{pmatrix}, \tag{7.19}$$

where, in a fiber free of linear birefringence, $2VH_z$ is the phase retardation per unit length between right-handed circular SOP (RHC) and left-handed circular SOP (LHC) as a result of the Faraday effect:

$$2VH_z = (2\pi / \lambda)(n_R - n_L). \tag{7.20}$$

The signs in (7.19) indicate that the fast and slow fiber axes are parallel to x and y, respectively (the x-polarized field component leads the y-polarized component), and the magnetic field is in the light propagation direction ($n_L < n_R$, LHC leads RHC). Integration of the coupled mode equation gives the Jones matrix of the fiber of length l [171, 189, 195]:

$$\boldsymbol{J} = \begin{pmatrix} \cos\left(\frac{\gamma}{2} l\right) + i \frac{\beta_b}{\gamma} \sin\left(\frac{\gamma}{2} l\right) & \frac{2VH_z}{\gamma} \sin\left(\frac{\gamma}{2} l\right) \\ -\frac{2VH_z}{\gamma} \sin\left(\frac{\gamma}{2} l\right) & \cos\left(\frac{\gamma}{2} l\right) - i \frac{\beta_b}{\gamma} \sin\left(\frac{\gamma}{2} l\right) \end{pmatrix}, \tag{7.21}$$

where γ is the elliptical birefringence that results from the superposition of the linear birefringence b_b and the magnetic field-induced circular birefringence $2VH_z$:

$$\gamma = \left[\beta_b^2 + (2VH_z)^2 \right]^{1/2}. \tag{7.22}$$

The reader can easily verify that by setting b_b or $2VH_z$ to zero in (7.21), one obtains the Jones matrix of a linear retarder or circular retarder (polarization rotator), respectively; see Section 2.7. Utilizing the conversion equations of that section, one obtains the Jones matrix in a circular basis as [189]

$$\boldsymbol{J}_{\text{circ}} = \begin{pmatrix} \cos\left(\frac{\gamma}{2}l\right) + i\frac{2VH_z}{\gamma}\sin\left(\frac{\gamma}{2}l\right) & i\frac{\beta_b}{\gamma}\sin\left(\frac{\gamma}{2}l\right) \\ i\frac{\beta_b}{\gamma}\sin\left(\frac{\gamma}{2}l\right) & \cos\left(\frac{\gamma}{2}l\right) - i\frac{2VH_z}{\gamma}\sin\left(\frac{\gamma}{2}l\right) \end{pmatrix}. \tag{7.23}$$

(Note: The reader should be aware that definition of the linear fiber birefringence in [189] differs by a factor of 2 from the definition in the present text and others. The present equations differ from [189] accordingly.) If the fiber is operated in a polarimetric sensor setup according to Figure 5.1, the electric fields $\boldsymbol{E}_1$, $\boldsymbol{E}_2$ arriving at two detectors are given by

$$\boldsymbol{E}_1 = \boldsymbol{J}_{P45°}\,\boldsymbol{J}\,\boldsymbol{J}_{P0°}\boldsymbol{E}_{\text{in}}, \tag{7.24}$$

$$\boldsymbol{E}_2 = \boldsymbol{J}_{P-45°}\,\boldsymbol{J}\,\boldsymbol{J}_{P0°}\boldsymbol{E}_{\text{in}}. \tag{7.24}$$

Here, $\boldsymbol{E}_{\text{in}}$ is the field arriving at the input polarizer; $\boldsymbol{J}_{P0°}$ is the Jones matrix of the input polarizer, oriented at 0° to a principal fiber axis, and $\boldsymbol{J}_{P45°}$ and $\boldsymbol{J}_{P-45°}$ are the Jones matrices of the output polarizers at ±45° to the input polarizer. (In Figure 5.1, the output polarizers are represented by a Wollaston prism.) The detected intensities I_1 and I_2 are proportional to square of the field amplitudes:

$$I_i = \boldsymbol{E}_i^* \boldsymbol{E}_i \; i = 1, 2. \tag{7.26}$$

The resulting sensor signal $S = (I_1 - I_2) / (I_1 + I_2)$ is given by [171, 196, 197]

$$S = 2VH_z l \frac{\sin(\gamma l)}{\gamma l}. \tag{7.27}$$

Often, the Faraday rotation is much smaller than the linear birefringence ($2VH_z \ll b_b$). S can then be approximated by

$$S = 2VH_z l \frac{\sin(\beta_b l)}{\beta_b l}. \tag{7.28}$$

In the opposite case ($2VH_z \gg Db_b$), the signal becomes $S = \sin(2VH_z l)$, i.e., approaches the signal of an ideal fiber. As an example, the retardation $d_b = b_b l$ of a coil with 10 windings, a coil diameter of 150 mm, and a fiber diameter of 125 μm corresponds to about 110 deg at 1310 nm. The corresponding beat length $L_b = 2p/b_b$ is 15.4 m. By comparison, a current of 1000 A produces a Faraday rotation of 0.65 deg. It should be noted that additional birefringence arises if the fiber is coiled under tension onto a drum [174–176].

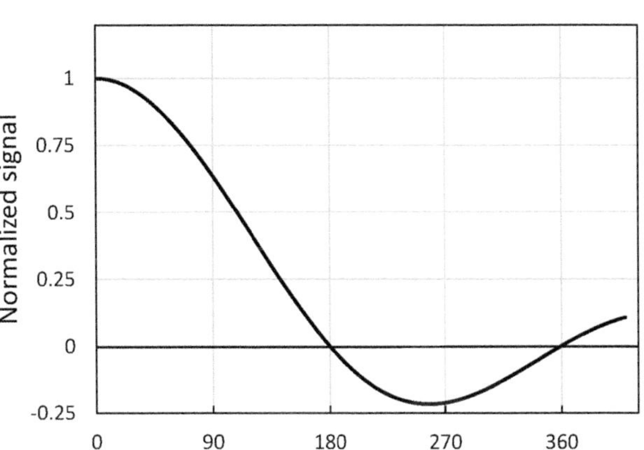

FIGURE 7.2 Normalized sensor signal, $S_{norm} = S/(2VH_z l)$, as a function fiber birefringence ($2VH_z \ll b_b$).

It is obvious from (7.27) and (7.28) that the signal normalized to the signal of an ideal fiber coil, $S_{norm} = S/(2VH_z l)$, oscillates as a function of the fiber length and vanishes, where $b_b l$ corresponds to multiples of p (Figure 7.2). Hence, the maximum usable fiber length for current sensing corresponds to roughly a quarter of the beat length L_b. Moreover, the signal is no longer strictly proportional to $\sin(2VH_z)$, and the sensor becomes sensitive to the diameter of the fiber coil and conductor position within the coil cross-section.

7.3.1 Attempts of Linear Birefringence Compensation

From early on, researchers have tried various methods to compensate bend-induced birefringence or lessen its impact:

- Feldtkeller et al. at Siemens compensated birefringence by winding consecutive fiber loops in two planes orthogonal to each other. The respective second loops reversed the birefringent phase retardation of the preceding loops and prevented build-up of excessive birefringent phase shifts [198–201]. Thirty-five years later, non-planar fiber winding was again reported by Perciante et al. [202].
- Others compensated linear birefringence by operating the fiber in reflection-mode employing a 90°-Faraday-polarization-rotator mirror at the fiber end [203–205].
- Ben-Kish and Tur reported a sensor where the fiber formed a square with 270°-loops at three of the four corners. Whereas the straight sections were free of birefringence, the loops of 10-mm radius acted as 2p-waveplates and preserved the polarization state [206].
- J. W. Dawson et al. disclosed an arrangement with three straight fiber sections that formed a triangle around the current conductor and were connected by pieces of polarizing fiber [207].

- Ulmer showed that the disturbing effects of birefringence could be minimized by choosing an appropriate angle of the input polarization relative to a principal axis of the birefringence. That angle depended, however, on the amount of birefringence and current [124]. Similar work was reported by Forman et al. [208]. Menke et al. combined the method with ac-dc signal evaluation [209, 210], and Liu et al. separated the Faraday effect from linear birefringence by a full analysis of the transmitted polarization state [211]. Similar work was done by Flores-Nunez et al. [212] and Fisher et al. [213].

- Day et al. maintained efficient Faraday rotation by choosing the circumference of the fiber coil equal to the beat length of the birefringence caused by bending [214]. Similarly, Stolen et al. showed that efficient Faraday rotation in birefringent media could be obtained by switching the magnetic field direction at spatial intervals corresponding to the linear beat length or by spacing sections of the medium with and without field at such intervals [215].

- Perciante et al. compensated bend-induced birefringence by applying lateral pressure onto the fiber loops. To this end, the fiber was wound on an aluminum cylinder, and a compressed spring co-wound with the fiber exerted pressure in a direction orthogonal to the loop plane [216].

- Ren et al. separated the Faraday rotation and linear birefringence by alternatingly switching between linearly and circularly polarized input light [197].

- In more recent work, Huang et al. proposed a closed loop system with a sensing fiber coil and an additional compensation coil [217].

Most of the above methods had no lasting impact, as they did not truly solve the underlying problem and were not compatible with efficient sensor fabrication and flexible usage. The following sections present alternative solutions, some of which led to commercial sensors with robust performance.

7.4 TWISTED FIBER

Fiber twisting was an early solution for suppressing bend-induced linear birefringence in fiber-optic current sensors [189, 218–233]. The twist-induced shear strains ($S_{yz,}$ S_{zx}) mutually couple the small longitudinal field component of one fiber mode with the transverse field of the respective orthogonal mode. Since the two fields are p/2 out of phase, twisting produces circular birefringence (optical activity) given by [186]

$$\alpha = g\tau. \tag{7.29}$$

Here, τ is the twist rate (twist angle per unit length), and the unitless parameter g is given as $g = -n^2(p_{11} - p_{12})/2$, p_{11}, p_{12} being the relevant photoelastic coefficients ($g \sim 0.16$ for fused silica fiber). Positive t stands for right-handed twist and produces left-rotary optical activity ($n_R > n_L$), assuming the observer looks towards the source [186]. An observer looking in the light propagation direction will notice that the polarization rotation follows the twist, albeit at a rate reduced by a factor of $g/2$. Note that the sense of twist expresses a "snapshot picture" of the twist helix and is therefore, other than the convention for the polarization rotation sense, independent of the viewing direction. The eigenpolarizations of a twisted fiber free of linear birefringence are right and left circular light waves. They accumulate a relative phase shift per unit length of

$$\alpha = (\alpha_R - \alpha_L) = (2\pi / \lambda)(n_R - n_L) = g\tau, \tag{7.30}$$

where, a_L and a_R are the propagation constants of the eigenpolarizations. The corresponding polarization rotation angle is $a_r = a/2$. a is usually taken as positive for positive τ (left-rotary optical activity, $n_R > n_L$) opposite to the sign convention for optical activity in optics textbooks, for example, [170]; see introduction to Chapter 2.

Figure 7.3 illustrates the trajectories ***C(z)*** on the Poincaré sphere of selected input polarization states propagating through fibers with both linear and circular birefringence [186, 234]. According to (2.7), the polarization states are defined in a linear basis as

$$\chi = \frac{E_{0y}}{E_{0x}} e^{-i(\beta_y - \beta_x)z} \tag{7.31}$$

and in a circular basis as (2.12)

$$\chi_{\text{circ}} = \frac{E_R}{E_L} e^{-i(\alpha_R - \alpha_L)z}. \tag{7.32}$$

The latitude and azimuth angles of the polarization states, 2θ and 2ψ, on the Poincaré sphere are then given by (2.10) and (2.11), respectively. For a given input polarization state, the evolution of the field amplitudes as function of z follows from the coupled mode equations and corresponding Jones matrix. First, Figure 7.3a shows the case of a fiber coil according to the previous section with only bend-induced linear birefringence. It is assumed that the slow axis is parallel to x and corresponds to horizontal polarization on the Poncaré sphere. The fiber's slow and fast eigenpolarizations then correspond to horizontal and vertical polarization, respectively, and are indicated by red dots.

In order to find the evolution of an arbitrary elliptical input polarization state, the birefringence is represented by the birefringence vector $\boldsymbol{b}_b$ (7.18). The vector points from the origin of the sphere to the slow eigenpolarization. Polarization states evolve along circles (red) on the sphere's surface centered about the direction of $\boldsymbol{b}_b$. (The normal of the circle planes is parallel to $\boldsymbol{b}_b$.) As an example, the red great circle illustrates the evolution of a linear input polarization state oriented at $\theta = 45°$ to x at $z = 0$ (fiber entrance). The trajectory starts at the corresponding +45°-point on the equator (azimuth $2\theta = 90°$), evolves through RHC polarization (North pole, E_y leads E_x by 90°), again reaches linear polarization but now at −45°, passes through LHC (south pole), and finally again assumes its initial linear

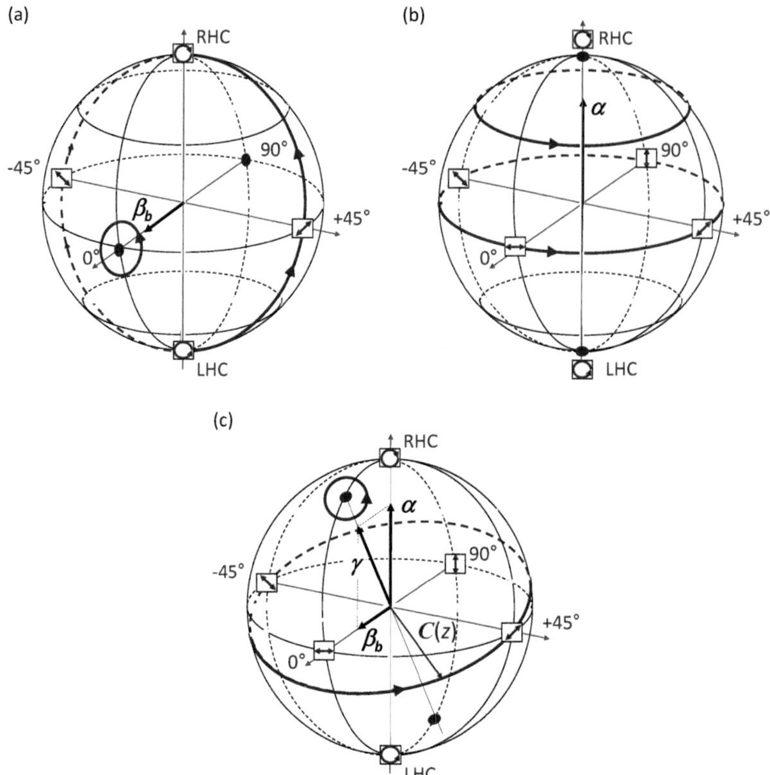

FIGURE 7.3 Polarization state evolution on Poincaré sphere in fibers with various types of birefringence: (a) fiber with only linear (bend-induced) birefringence; (b) straight twisted fiber with only circular birefringence; (c) coil of twisted fiber with linear and circular birefringence. The twist is right-handed. The red dots indicate the slow and fast eigenpolarizations. The polarization states evolve along circles (red) centered about the slow eigenvector of the corresponding birefringence. A full 2p-revolution corresponds to a beat length of the corresponding eigenpolarizations.

polarization after a full revolution (when E_y and E_x have accumulated a differential phase shift of 2p). The smaller red circle indicates the evolution of a linear input state oriented at $z = 0$ at an angle in the range $0 < \theta < 45°$.

Figure 7.3b shows the case of a straight twisted fiber with only circular birefringence. It is assumed that the optical activity is left-rotary ($n_R > n_L$, α is positive, the twist is right-handed), i.e., the RHC state is the slow eigenpolarization and the LHC-state is the fast eigenpolarization. The circular birefringence vector $\boldsymbol{a}$ points along the positive polar axis from the origin to the slow circular eigenpolarization (North pole). The polarization states now evolve along circles centered about $\boldsymbol{a}$ (i.e., the circle planes are parallel to the equatorial plane). In particular, linear input states evolve along the equator, i.e., they remain linearly polarized. The smaller red circle shows the evolution of a right-handed elliptical SOP in the northern hemisphere.

Finally, Figure 7.3c shows the polarization evolution in a fiber with both linear and circular birefringence (slow axis of linear birefringence again horizontal). The eigenpolarizations are now right and left elliptical SOP. The elliptical birefringence vector $\boldsymbol{\gamma}$ is the sum of

$\boldsymbol{b}_b$ and $\boldsymbol{a}$, $\boldsymbol{\gamma} = \boldsymbol{b}_b + \boldsymbol{a}$, and points to the slow elliptical eigenpolarization (right-handed eigenpolarization). The circular trajectories are now centered about $\boldsymbol{\gamma}$. For example, a linear input state oriented at 45° to the fast axis starts at the 45°-point on the equator, evolves through right-handed elliptical states in the northern hemisphere, then assumes −45°-linear-polarization, and further evolves through left-handed elliptical states in the southern hemisphere, before the state again returns to its initial 45°-linear-polarization; the two underlying orthogonal elliptical eigenpolarizations have then accumulated a 2p-phase relative shift. Note that the graphs are somewhat unphysical in that the radius of the Poincaré sphere is, other than the birefringence vectors, unitless. However, only the relative lengths of the vectors are relevant for the illustration.

The direction and velocity of the SOP evolution are determined by the vector product of $\boldsymbol{\gamma}$ and $\boldsymbol{C}(z)$:

$$\frac{d\boldsymbol{C}(z)}{dz} = \boldsymbol{\gamma} \times \boldsymbol{C}(z), \tag{7.33}$$

where the unit vector $\boldsymbol{C}(z)$ points from the origin to the given SOP on the corresponding trajectory. The magnitude of the elliptical birefringence, which is the relative phase shift per unit length of the elliptical eigenpolarizations, is

$$\gamma = \left(\beta_b^2 + \alpha^2\right)^{1/2}. \tag{7.34}$$

The corresponding elliptical beat length is $L_E = 2\text{p}/\gamma$. A full 2p-revolution around any circle corresponds to a length of fiber equal to the beat length of the corresponding birefringence; for instance, the fiber length corresponding to a revolution around the $\boldsymbol{\gamma}$-vector equals L_E. It is obvious from Figure 7.3c, that if $|a|$ is made much larger than $|\beta_b|$, a twisted fiber essentially becomes circularly birefringent and largely preserves a linear input polarization state (or any other SOP), apart from an azimuth rotation. Roughly speaking, the linear birefringence of a given fiber section having a length of a quarter of the twist period is cancelled in the following section of the same length as a consequence that, on average, such section is rotated by p/2 as compared to the first section [222].

In the presence of an applied magnetic field, the additional circular birefringence $2VH_z$ simply adds to the twist-induced optical activity, so that γ reads

$$\gamma = \left[\beta_b^2 + \left(2VH_z + \alpha\right)^2\right]^{1/2}. \tag{7.35}$$

The optical activity and magnetic field-induced circular birefringence are both left-rotatory (ccw, $n_R > n_L$), if $2VH_z$ is positive (field points in the light propagation direction) and a is also positive (right-handed twist). Commonly, $|2VHz|$ is much smaller than $|a|$, so that the field only moderately modifies the positions of the eigenpolarizations on the Poincaré sphere and the ecliptic of the circular contours. Note, however, that a full revolution around a contour corresponds to somewhat different fiber lengths with and without field. The difference between the azimuth rotations at a fixed fiber length z with and without field represents the current to be measured.

In a linear basis, the mode coupling coefficients of a twisted fiber with bend-induced birefringence and an applied magnetic field are

$$\boldsymbol{K} = \begin{pmatrix} \beta_b / 2 & -i\left(VH_z + \alpha / 2\right) \\ i\left(VH_z + \alpha / 2\right) & -\beta_b / 2 \end{pmatrix}. \tag{7.36}$$

Since the eigenmodes of a twisted fiber are elliptically polarized, it is often more convenient to work in a basis of circular polarizations. The coupling coefficients are then [189]:

$$\boldsymbol{K}_{\text{circ}} = \begin{pmatrix} VH_z + \alpha / 2 & \beta_b / 2 \\ \beta_b / 2 & -\left(VH_z + \alpha / 2\right) \end{pmatrix}. \tag{7.37}$$

The corresponding Jones matrix reads [189]:

$$\boldsymbol{J}_{\text{circ}} = \begin{pmatrix} \cos\left(\frac{\gamma}{2} l\right) + i \frac{\left(2VH_z + \alpha\right)}{\gamma} \sin\left(\frac{\gamma}{2} l\right) & i \frac{\beta_b}{\gamma} \sin\left(\frac{\gamma}{2} l\right) \\ i \frac{\beta_b}{\gamma} \sin\left(\frac{\gamma}{2} l\right) & \cos\left(\frac{\gamma}{2} l\right) - i \frac{\left(2VH_z + \alpha\right)}{\gamma} \sin\left(\frac{\gamma}{2} l\right) \end{pmatrix}. \tag{7.38}$$

With $|\alpha|$»$|\beta_b|$, the phase shift *Df* between the left and right near-circular polarizations at the output of a fiber of length l can be approximated as [189]:

$$\Delta\phi = \gamma l \approx 2\left(VH_z + \alpha\right) l. \tag{7.39}$$

Apart from the twist induced phase offset *al*, the phase shift corresponds to the magneto-optic phase shift $Df_F = 2VH_z l$ of an ideal fiber. A general expression for the signal of a polarimetric current sensor with linear and twist-induced circular birefringence can be found, for example, in [222, 233].

It should be noted that high twist rates are necessary to sufficiently suppress the linear birefringence. Twist rates up to several tens of 2p-turns per meter have been applied [172, 176, 189]. Still higher rates risk fiber breakage. According to (7.29), 30 turns per meter result in circular birefringence of about 30 rad/m. The corresponding beat length is 21 cm. For comparison, the phase retardation and linear beat length due to bend-induced birefringence are 0.41 rad/m and 15.4 m, respectively, at 1310 nm for the fiber and coil radii in the previous section (r = 62.5 mm, R = 75 mm). With decreasing coil diameters, fiber twisting becomes more and more ineffective, as the linear birefringence increases in proportion to R^{-2}.

It is obvious from Figure 7.3c, that at finite twist rates a linear input polarization state evolves through a series of elliptical SOP with linear polarization being restored at intervals of $L_E/2$ (or L_E if the input polarization is at 0° or 90°). As a result, the magneto-optic sensitivity varies along the fiber with the same periods. Since the elasto-optic coefficients change with temperature, the SOP ellipticity at the fiber end (SOP latitude on Poincaré sphere) varies as a function of temperature, which adds to the temperature dependence of the Verdet constant. The phenomenon is discussed in more detail in context of spun sensing fibers below.

Furthermore, the twist-induced polarization rotation (SOP azimuth) also varies with temperature – $(1/g)\partial g/\partial T = 4.95 \times 10^{-4}\ \mathrm{K}^{-1}$ at 633 nm for fused silica fiber — causing shifts in the signal's dc offset [180, 235]. Compensation measures include operating the fiber in reflective mode (the reciprocal twist-induced polarization rotation is undone on the return path, while the non-reciprocal rotation by the Faraday effect doubles) or utilizing two serial fiber sections of equal length but opposite sense of twist [236, 237]. Other downsides of twisted fiber include complicated coil fabrication, mechanical means for holding the torsion which introduce new stress and birefringence, and poor application flexibility. Fiber twisting was largely abandoned after spun highly birefringent fibers became available in the late 1980s and 1990s. It is worth mentioning that in a 1979 patent application, Barkmann and Winterhoff at AEG (a former German company) proposed to hold the twist by the fiber coating. To this end, the fiber was to be spun during the coating application at the draw tower, a method that came close to the fabrication procedure of spun fibers [238].

7.5 SPUN HIGHLY BIREFRINGENT FIBERS

In view of the inadequate performance of fiber-optic current sensors at the time, David Payne and co-workers at the University of Southampton, U.K., began in the early 1980s to develop spun highly birefringent fiber for current sensing [239–241]. Since then, numerous theoretical and experimental studies on spun fibers have been published [242–272]. Meanwhile, spun fiber has become the fiber of choice for most commercial FOCS. Spun fiber is fabricated by rapidly spinning the preform of a highly birefringent fiber during the drawing process. As a result, the slow and fast principal fiber axes rotate along the fiber at a rate $\xi = 2p/p$, where p is the spin pitch (Figure 7.4). By convention, the spin rate ξ is positive for a right-handed spin and negative for a left-handed spin. An internal stress field like in bowtie or Panda fibers, an elliptical core, or an appropriate fiber micro-structure generates the embedded linear birefringence. In a local coordinate system $S'(x', y', z)$ that co-rotates with the birefringent fiber axes, the combination of circular and linear birefringence results in two orthogonal elliptically polarized (quasi) eigenmodes with opposite handedness and different phase velocities, that is, the propagating modes preserve their ellipticity, and the azimuths of the polarization ellipses coincide with the birefringent axes. The modes are also referred to as helical polarization modes [234, 263]. In contrast to a linearly polarization-maintaining fiber or twisted fiber, a spun fiber has no eigenmodes in the laboratory coordinate system $S(x, y, z)$, i.e., in a fixed coordinate system, the SOP

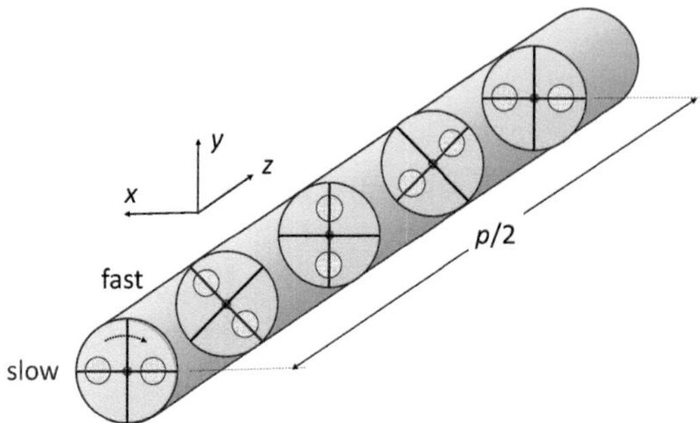

FIGURE 7.4 Spun highly birefringent Panda fiber with right-handed spin: the birefringent fiber axes rotate along the fiber at a rate $\xi = 2p/p$, where p is the spin pitch.

continuously change along the fiber [189]. An illustrative description of the physics behind the formation of the quasi eigenmodes can be found in [234]. It is worth mentioning that, unlike in twisted fiber, shear strain does not significantly contribute to the circular birefringence of spun fibers owing to the state of low viscosity during the drawing and spinning process.

Rather, the circular birefringence is a consequence of the rotating linear birefringence and does not exist independently of the linear birefringence.

The choice of the spin pitch and embedded linear birefringence (birefringence of an equivalent unspun fiber) determines the ellipticity of the quasi eigenmodes, the fiber's apparent magneto-optic sensitivity, and robustness against external perturbations. Commonly, the spin pitch is a few mm; a lower practical limit is about 2.5 mm. With the spin pitch fixed, the beat length L_B of the embedded linear birefringence then determines the crucial fiber properties. The ratio $x = 2L_B/p$ is referred to as spin ratio. A small beat length (comparable to the spin pitch or smaller, i.e., $x < \approx 2$) enhances the robustness of the fiber against bending-induced birefringence and other external perturbations such as low-temperature stress from the fiber coating. On the other hand, a small beat length reduces the ellipticity e of the quasi eigenmodes and hence lowers the fiber's sensitivity to magnetic fields. (Note that $e = 0$ corresponds to linear polarization, and $e = \pm 1$ corresponds to right and left circular polarizations.) A further parameter to consider is the temperature dependence of the linear beat length. Especially at relatively small spin ratios, the contribution of the linear birefringence to the temperature dependence of the sensor signal can be significantly higher than the contribution of the Faraday effect alone (Section 8.6). By comparison, the spin pitch is almost independent of temperature due to small thermal expansion coefficient of fused silica fiber (0.5×10^{-6} K^{-1}). In many cases, a reasonable compromise is a linear beat length of 3–10 times the spin pitch [249]. It is interesting to note that a spun fiber with a spin ratio that gradually increases as function of z from $x = 0$ to $x \gg 1$ acts a polarization converter that converts linearly polarized light into circularly polarized light on the forward path (+z-direction) and vice versa on the backward path [247, 273–276].

7.5.1 Coupled Mode Equations

The coupling coefficients of a spun fiber are z-dependent. In the presence of an applied magnetic field H_z and absence of bend-induced linear birefringence, they are given in a circular basis as [189, 245, 251, 263]

$$\boldsymbol{K}_{\text{circ}} = \begin{pmatrix} VH_z & \left(\frac{\beta_B}{2}\right)e^{i2\xi z} \\ \left(\frac{\beta_B}{2}\right)e^{-i2\xi z} & -VH_z \end{pmatrix}. \tag{7.40}$$

The coefficients in a linear basis are more intricate and can be found, for example, in [249, 250]. $b_B = (2p/l)(n_s - n_f)$ is the difference in propagation constants associated with the embedded linear birefringence (slow axis parallel to x); the linear beat length is $L_B = l/B$, $B = Dn_B = n_s - n_f$; ξ is the spin rate: $\xi = 2\pi/p$. The solutions of the coupled-mode equations are left and right elliptically polarized waves with orthogonal azimuths and represent the quasi eigenpolarizations. As already mentioned above, the axes of the modal polarization ellipses rotate at a rate ξ along the fiber and coincide with the principal axes of the embedded birefringence. The optical waves propagating in +z-direction are given by [263]

$$\boldsymbol{E}_1 = \boldsymbol{e}_1 e^{-(i/2)\gamma z}, \tag{7.41}$$

$$\boldsymbol{E}_2 = \boldsymbol{e}_1 e^{(i/2)\gamma z}. \tag{7.42}$$

(For an alternative representation, see [244].) Here, γ is again the elliptical phase birefringence, i.e., the differential phase shift (rad/m) that the two quasi eigenmodes accumulate per unit length:

$$\gamma = \left[\left(2VH_z - 2\xi\right)^2 + \beta_B^2\right]^{1/2}. \tag{7.43}$$

The vectors $\boldsymbol{e}_1$, $\boldsymbol{e}_2$ are

$$\boldsymbol{e}_1 = \frac{1}{\sqrt{1+A^2}}\begin{pmatrix} e^{i\xi z} \\ Ae^{-i\xi z} \end{pmatrix}, \tag{7.44}$$

$$\boldsymbol{e}_2 = \frac{1}{\sqrt{1+A^2}}\begin{pmatrix} -Ae^{i\xi z} \\ e^{-i\xi z} \end{pmatrix}. \tag{7.45}$$

The relative amplitude A is

$$A = x\left(\sqrt{1+x^{-2}} - 1\right), \tag{7.46}$$

where x is the spin ratio introduced further above:

$$x = \frac{2L_B}{p} = \left|\frac{2\xi}{\beta_B}\right|. \tag{7.47}$$

See [263] for sign changes, if the light travels in $-z$-direction. The ellipticity of the modes (ratio of minor and major ellipse axes) equals $e = \tan\psi$ with

$$\psi = \pm(1/2)a\tan(2L_B / p) = \pm(1/2)a\tan(x), \tag{7.48}$$

where 2ψ is the latitude of the respective SOP on the Poincaré sphere. Positive ψ stands for right-handed rotation and negative ψ for left-handed rotation. The absolute value of e can also be written as [263]:

$$\text{abs}(e) = (1-A)/(1+A) = \sqrt{1+x^{-2}} - x^{-1}. \tag{7.49}$$

At long linear beat lengths ($2L_B \gg p$), i.e., small embedded linear birefringence ($|\beta_B| \ll |2\xi|$), abs(e) approaches unity, that is, the modes approach circular polarization, whereas at large linear birefringence ($L_B \ll p$, ($|\beta_B| \gg |2\xi|$), the modes approach linear polarization. Figure 7.5 illustrates the polarization ellipses ($e = \pm 0.72$) of the fast and slow quasi eigenpolarizations of left-handed and right-handed spun fibers with a spin ratio of $x = 3$ [234].

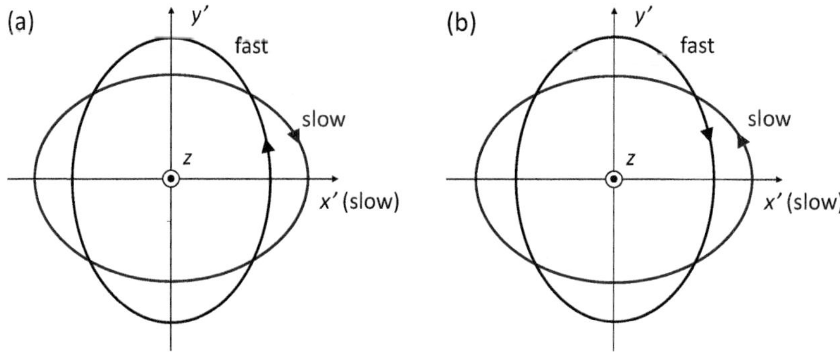

FIGURE 7.5 Elliptical (quasi) eigenpolarizations of a left-handed (a) and right-handed (b) spun fiber with a spin ratio of $x = 3$ in the local coordinate system S' co-rotating with the principal fiber axes. The slow axis is parallel to x'.

The effective refractive indices of the two modes are

$$n_{1,2} = \bar{n} \pm \left(\gamma / 2k\right), \tag{7.50}$$

where $\bar{n}$ is the average refractive index. In the local coordinate system $S'(x', y', z)$, the elliptical beat length L_E', i.e., the length of fiber the light must travel until a given input polarization state repeats itself, is $L_E' = 2p/\gamma$ or [245, 246]:

$$L_E' = \frac{p\,L_B}{\sqrt{p^2 + 4\,L_B^2}}. \tag{7.51}$$

Here, H_z is assumed as zero (typically the Faraday phase shift $2VH_z$ is much smaller than 2ξ). At high spin ratios ($2L_B \gg p$), L_E' can be approximated as $L_E' \approx p/2$. The elliptical phase birefringence in the local system is $Dn_E = (\Delta n_B^2 + \Delta n_c^2)^{1/2}$. The spin rate ξ determines the circular birefringence $\Delta n_c = n_R - n_L$ according to

$$\alpha = -2\xi = \left(\alpha_R - \alpha_L\right) = \left(2\pi / \lambda\right)\left(n_R - n_L\right). \tag{7.52}$$

The circular birefringence is negative ($n_R < n_L$) for spun fibers with right-handed spin (positive ξ) and positive ($n_R > n_L$) for fibers with left-handed spin (negative ξ). Note that this is different from twisted fiber, where right-handed twist results in $n_R > n_L$.

In the laboratory coordinate system, the elliptical phase birefringence is given by $Dn_E = (\Delta n_B^2 + \Delta n_c^2)^{1/2} - Dn_c$ or, in units of rad/m, by $\gamma - 2\xi$. The corresponding elliptical beat length is $L_E = 2p/(\gamma - 2\xi)$ or, in terms of p and L_B [246],

$$L_E = \frac{p\,L_B}{\sqrt{p^2 + 4\,L_B^2} - 2L_B}. \tag{7.53}$$

For $2L_B \gg p$, L_E is in good approximation equal to $L_E = 2xL_B = 4\,L_B^2 / p$. For example, $p = 5$ mm and $L_B = 10$ mm ($x = 4$) result in $L_E = 81$ mm (see Figure 7.8 further below). Other than the elliptical beat length of twisted fiber, L_E is not a length after which a polarization state repeats itself. Nevertheless, L_E is a measure for the robustness of the fiber against external perturbations.

Another important parameter of spun fibers is the elliptical group birefringence $Dn_{g,E}$, in particular when the fiber is operated with broadband light [269]:

$$\Delta n_{g,E} = \Delta n_E - \lambda\left(d\Delta n_E / d\lambda\right). \tag{7.54}$$

Other than the elliptical phase birefringence Dn_E, the elliptical group birefringence $Dn_{g,E}$ is independent of the chosen coordinate system. Koval et al. determined a simple relationship between the linear group birefringence $Dn_{g,B}$ and elliptical group birefringence $Dn_{g,E}$ [269]:

$$\Delta n_{g,E}(\lambda) = \frac{\Delta n_B}{\sqrt{\Delta n_B^2 + \Delta n_c^2}} \Delta n_{g,B}(\lambda), \tag{7.55}$$

where $Dn_{g,B}$ is given by

$$\Delta n_{g,B} = \Delta n_B - \lambda\left(d\Delta n_B / d\lambda\right). \tag{7.56}$$

Elliptical-core fibers and micro-structured fibers often exhibit strong chromatic dispersion, and hence the linear group birefringence may significantly differ from the linear phase birefringence. The same is then true for the elliptical group and phase birefringence [277].

Integration of the coupled mode equations (7.40) gives the Jones matrix of the spun fiber, which reads in a circular basis [189, 251]

$$J_{\text{circ}} = \begin{pmatrix} \left[\cos\left(\frac{\gamma}{2}l\right) - i\frac{\left(2VH_z - 2\xi\right)}{\gamma}\sin\left(\frac{\gamma}{2}l\right)\right]e^{i\xi l} & \left[i\frac{\beta_B}{\gamma}\sin\left(\frac{\gamma}{2}l\right)\right]e^{i\xi l} \\ \left[i\frac{\beta_B}{\gamma}\sin\left(\frac{\gamma}{2}l\right)\right]e^{-i\xi l} & \left[\cos\left(\frac{\gamma}{2}l\right) + i\frac{\left(2VH_z - 2\xi\right)}{\gamma}\sin\left(\frac{\gamma}{2}l\right)\right]e^{-i\xi l} \end{pmatrix}. \tag{7.57}$$

For high spin ratios ($|2\xi| \gg |\beta_B|$), a spun fiber behaves like a circular birefringent fiber. For $2VH_z = 0$, the Jones matrix can then be approximated as [189]

$$J_{\text{circ}} = \begin{pmatrix} e^{i\frac{\hat{\alpha}}{2}l} & 0 \\ 0 & e^{-i\frac{\hat{\alpha}}{2}l} \end{pmatrix}. \tag{7.58}$$

Compared to the spin rate ξ, the apparent optical activity $\hat{\alpha}$ (rad/ m) is small [189]:

$$\hat{\alpha} = \frac{\beta_B^2}{4\xi} = \frac{1}{2}\pi\frac{p}{L_B^2}. \tag{7.59}$$

Reports on the experimental characterization of spun fibers, in particular by spectral polarimetry and Bragg gratings imprinted into the fiber, can be found, for example, in [243, 249, 250, 256, 260, 264, 268, 269].

7.5.2 Polarization State Evolution

An input polarization state with the ellipticity, azimuth, and handedness of one of the two quasi eigenmodes of the fiber is mapped one-to-one onto that mode. For all other input

SOP, the light field $\boldsymbol{E}$ in the fiber is a superposition of the two eigenmodes with amplitudes C_1, C_2 [263]:

$$\boldsymbol{E} = C_1 \boldsymbol{e}_1 e^{-ikn_1 z} + C_2 \boldsymbol{e}_2 e^{-ikn_2 z}. \tag{7.60}$$

The evolution of arbitrary input SOP has been investigated by R. Ulrich [186] and in more detail by R. Dändliker [189], Y. Wang et al. [250], and Y. V. Przhiyalkovsky et al. [234]. Figure 7.6a,b depicts the SOP trajectories in the local system S' co-rotating with the birefringent fiber axes for linear input SOP at 45° and 0° to the slow fiber axis. The sense of spin is assumed as left-handed, i.e., ξ is negative, and the RHC state is the slow circular state

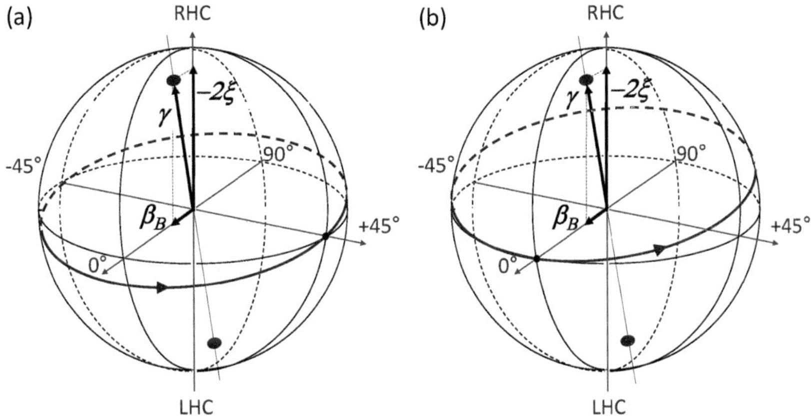

Coordinate system S' co-rotating with principal fiber axes

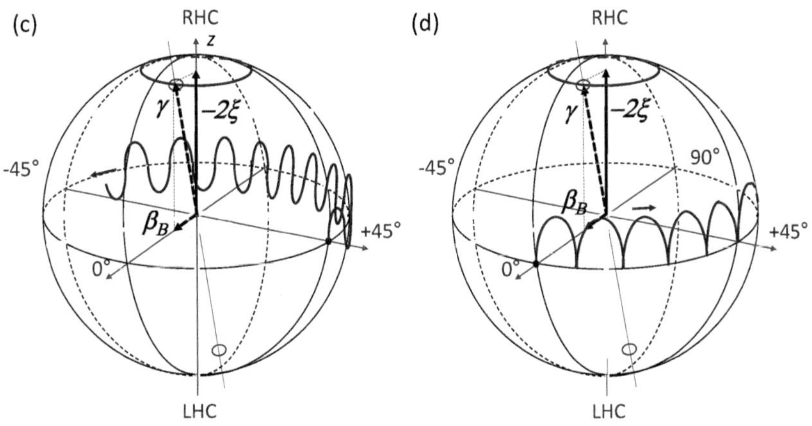

Laboratory system S

FIGURE 7.6 Evolution (schematically) of linear input polarization states oriented at 45° (a), (c) and 0° (b), (d) with respect to the slow fiber axis in a left-handed spun fiber with a spin ratio $x = 3$: (a), (b): SOP in the local coordinate system S' co-rotating with the principal fiber axes (red circles). The red dots indicate the elliptical eigenpolarizations. (c), (d): SOP in the laboratory system S (red cycloids). The blue circles depict the evolution of the right-handed elliptical eigenpolarization. (Note that in S the vectors $\boldsymbol{bB}$ and $\boldsymbol{\gamma}$ rotate at a rate $\boldsymbol{\xi}$, i.e., their azimuths as indicated are only valid at $z = mp$, $m = 0, 1, 2, \ldots$)

(7.52). The Poincaré sphere is again oriented such that the horizontal polarization coincides with the slow fiber axis. The illustration is only schematic but resembles the situation of a spin ratio of $x = |2\xi/\beta_B| = 3$, which is in a range of interest to current sensing. The linear birefringence vector $\boldsymbol{b}_B$ points from the origin to the slow linear polarization state, while the vector $-2\boldsymbol{\xi}$ (= **a**) points to the slow circular polarization state (RHC). The right-handed and left-handed eigenpolarizations (red dots) lie in the directions $\pm\boldsymbol{\gamma}$, respectively; $\boldsymbol{\gamma} = \boldsymbol{b}_B - 2\boldsymbol{\xi}$. Their latitudes 2ψ are ±71.6°, and their ellipticities e are ±0.72. The SOP trajectories in the local system are analog to the ones of a twisted fiber with bend-induced linear birefringence (Figure 7.3c) and are circular contours, with $\boldsymbol{\gamma}$ being the normal to circular planes. The starting points at $z = 0$ are marked by black dots. A full revolution corresponds to a fiber section of length L_E'.

The SOP trajectories on the Poincaré sphere of the laboratory system S (Figure 7.6c,d) are obtained by superimposing to the S'-trajectories a revolution at a rate of $\boldsymbol{\xi}$ about the south-north direction. The eigenpolarizations then evolve along circles parallel to the equator (blue circles), whereas the linear input SOP evolve on cycloids (red lines). The cycloids of the 45°-input polarization are centered about the equator, i.e., the SOP ellipticity alternates between right-handed elliptical, linear, and left-handed elliptical polarization, while the SOP azimuth varies according to the progressing longitude (Figure 7.6c). The linear input SOP at 0° (parallel to the slow axis) periodically evolves through right-handed elliptical polarization states (in the northern hemisphere), becoming linear at points were the cycloids touch the equator (Figure 7.6d). A full 2p-revolution around the sphere corresponds to the beat length L_E, and the period of the cycloids corresponds to L_E'. Accordingly, the (non-integer) number N of periods over a full revolution is $N = L_E / L_E'$; $N = 19.5$ for $x = 3$. N increases with increasing spin ratio and is approximately equal to $2x^2$ for sufficiently large x. For example, N equals 3.4 for $x = 1$ and 52 for $x = 5$.

The amplitude of the cycloids decreases with increasing spin ratio, i.e., the variations in the SOP ellipticity become smaller and disappear in the limit of very large x, when the eigenpolarizations essentially coincide with poles of the sphere (Figure 7.7). An observer in

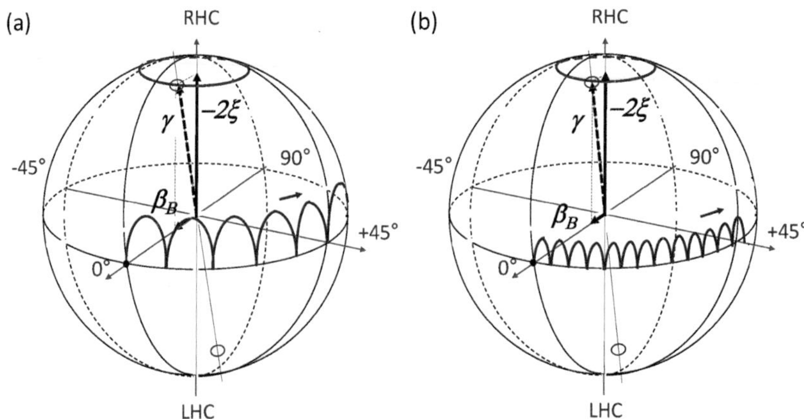

FIGURE 7.7 Evolution (schematically) of a linear input polarization state oriented at 0° to the slow axis in a left-handed spun fiber with spin ratios $x = 3$ (a) and $x = 5$ (b); laboratory system.

the local coordinate system S' advancing in a frozen-in optical field a distance L_E' (L_E' = 0.474 p for $x = 3$) will see an SOP azimuth rotation by $\theta = 180°$, and a rotation of the laboratory system S by 170.8°; S will have rotated by 180° only after a distance $p/2$. Hence, in the laboratory system, the SOP azimuth will have rotated by $\theta = -9.2°$. Obviously, this rotation, quantified by (7.59) for large x, will continue to grow as the observer moves further down the fiber. An observer in the laboratory system looking in the direction of the light propagation will see that, in contrast to twisted fibers, the SOP azimuth as a function of z rotates in a direction opposite to the sense of spin. When x approaches infinity and L_E' reaches $p/2$, the SOP azimuth naturally ceases to rotate in the laboratory system. Further information can be found in [189] and [234].

7.5.3 Polarization State Evolution in Bent Spun Fiber

Bending a spun fiber to a coil of constant radius introduces additional bending-induced birefringence b_b. The elliptical birefringence vector $\boldsymbol{\gamma}$ is then the sum of the vectors $\boldsymbol{b}_\mathrm{B}$, -2ξ, and $\boldsymbol{b}_\mathrm{b}$. Typically, b_b is much smaller than b_B. The direction of $\boldsymbol{b}_\mathrm{b}$ as a function of z remains constant in the laboratory system, while $\boldsymbol{b}_\mathrm{B}$ rotates at a rate of ξ. As a result, the ellipticity of the two orthogonal polarization modes is no longer constant but changes synchronously with the principal axis rotation. Moreover, tight bending radii reduce the mean ellipticity of the modes and hence the magnetic field sensitivity of the fiber. In a circular base, the coupling coefficients of a bent spun fiber read [189, 249, 263]

$$\boldsymbol{K}_\mathrm{circ} = \begin{pmatrix} VH_z & \left(\frac{\beta_B}{2}\right)e^{i2\xi z} + \left(\frac{\beta_b}{2}\right)e^{i2\varphi} \\ \left(\frac{\beta_B}{2}\right)e^{-i2\xi z} + \left(\frac{\beta_b}{2}\right)e^{-i2\varphi} & -VH_z \end{pmatrix} \tag{7.61}$$

where φ is the angle between the xz-plane in Figure 7.4 and the plane of the coiled fiber. The coupled-mode equations have no exact analytical solutions and are commonly solved numerically or by perturbation theory [249, 251, 270, 271]. Approximate analytical solutions were presented by Przhiyalkovskiy et al. [263]. At decreasing R, the ellipticity versus z of the two polarization modes begins to (unharmonically) oscillate about the ellipticity e of a straight fiber; at $x > 1$ with a period of $\approx p/2$. At further decreasing R, the mean ellipticity and hence the sensitivity to current decrease, as Figure 7.9 below will show.

7.5.4 Magneto-Optic Sensitivity of Spun Fibers

7.5.4.1 Intrinsic Magneto-Optic Sensitivity

The non-circularity of the (quasi) eigenpolarizations results in a reduced magneto-optic sensitivity s compared to an ideal fiber with perfectly circular eigenpolarizations and the same Verdet constant:

$$s = \Delta\phi_\mathrm{spun} / \Delta\phi_\mathrm{ideal}. \tag{7.65}$$

Here, Df_{spun} is the differential magneto-optic phase shift per unit length of the elliptical eigenpolarizations, and $Df_{ideal} = 2\varphi_F = 2VH_z$ is the corresponding phase shift of the left and right circular SOP in an ideal fiber; Df_{spun} corresponds to the change of the elliptical birefringence γ in a magnetic field. Assuming small values of the Faraday rotation φ_F and negligible bending-induced birefringence β_b, the relative phase γ (φ_F) per unit length of the elliptical modes can be approximated as $\gamma(\varphi_F) = \gamma\ (\varphi_F = 0) + s\varphi_F$, where $s = d\gamma/d\varphi_F$ is given by [263]

$$s = \frac{1}{\sqrt{1+x^{-2}}} = 2\frac{L_E^{'}}{p} = \frac{2\xi}{\tilde{a}\left(\varphi_F = 0\right)} < 1. \tag{7.66}$$

The sensitivity can also be written as [252]

$$s = \sin\left(\left|2\psi\right|\right). \tag{7.67}$$

According to (7.48), 2ψ is the latitude of the eigenpolarizations on the Poincaré sphere, and $e = \text{atan}(\psi)$ is their ellipticity. At large x, s approaches unity (limit of circular polarization modes), and at small x, s approaches zero (limit of linearly polarized modes), as illustrated in Figure 7.8. The Figure also depicts the elliptical beat length L_E versus x for two different values of the spin pitch.

7.5.4.2 Intrinsic Magneto-Optic Sensitivity of Coiled Fiber

Bending the fiber to a coil reduces the intrinsic magneto-optic sensitivity of a coiled fiber as follows [263]:

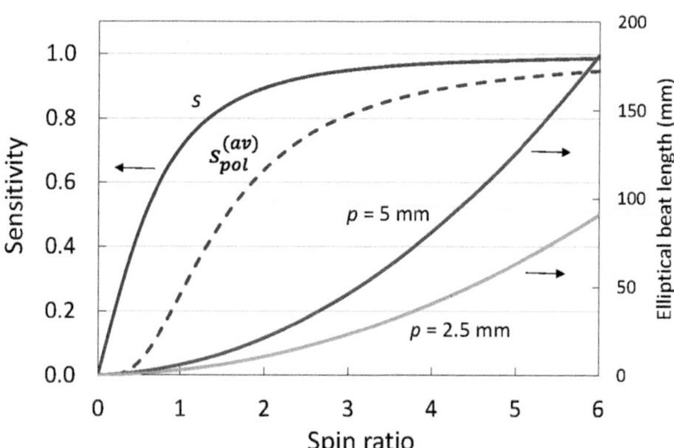

FIGURE 7.8 Intrinsic magneto-optic sensitivity s of a straight spun fiber (solid blue line) as a function of the spin ratio $x = 2L_B/p$ and (average) sensitivity $s_{pol}^{(av)}$ when the fiber is operated in a polarimetric sensor (dashed line). The sensitivity is normalized to the sensitivity of an ideal fiber with the same Verdet constant [263]. The yellow and orange lines indicate the elliptical beat length L_E for spin pitches p of 2.5 mm and 5 mm, respectively.

$$\tilde{s} = \frac{1}{\sqrt{1+x^{-2}}} \frac{1}{\sqrt{1+\tilde{x}^{-2}}}. \tag{7.68}$$

An equivalent expression was also reported in [241]. The first term in (7.68) is the sensitivity of a straight fiber (7.66), and the second term describes the sensitivity reduction as a result of the fiber bending with $\tilde{x}$ equal to

$$\tilde{x} = \left(1+A^2\right)\frac{\left(\gamma-2\xi\right)}{\beta_b}. \tag{7.69}$$

The parameter A has been defined by (7.46). Note that $\tilde{s}$ is the average sensitivity and does not account for the mentioned ellipticity oscillations of the polarization modes of a bent fiber. Figure 7.9 shows $\tilde{s}$ as a function of the bending radius R for a spin pitch of p = 5 mm and two different beat lengths L_B of the embedded linear birefringence; β_b is according to (7.18) at l = 1310 nm. The fiber diameters are 125 mm (solid lines) and 80 mm (dashed lines). It is obvious that smaller linear beat lengths at a given spin pitch allow for smaller bend radii but somewhat reduce the sensitivity at large R, where $\tilde{s}$ flattens out. Still, the fibers tolerate bend radii as small as 30 mm (125-μm fiber) and 20 μm (80-μm fiber), before the sensitivity starts to drop.

7.5.4.3 Magneto-Optic Sensitivity of a Polarimetric Current Sensor with Spun Fiber

The sensitivities s and $\tilde{s}$ refer, as noted, to the differential magneto-optic phase shift of the helical polarization modes of a spun fiber. The actual sensitivity or scale factor of a current sensor with spun fiber varies with the ellipticity and azimuth of the input SOP and, moreover, depends on the detection scheme (polarimetric or interferometric); the detection

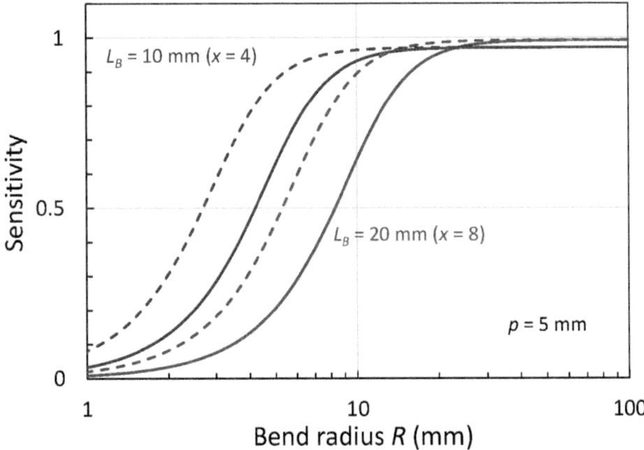

FIGURE 7.9 Intrinsic magneto-optic sensitivity $\tilde{s}$ of a bent spun fiber at 1310 nm as function of the bend radius R for a spin pitch of p = 5 mm and two different values of the beat length L_B. The solid and dashed lines correspond to fiber diameters of 125 μm and 80 μm, respectively.

scheme matters because a spun fiber not only lowers the magneto-optic phase shift but also the interference fringe contrast K. Polarimetric sensors are subject to both effects, whereas interferometric sensors with phase demodulation only react to the phase shift [266].

Assuming a polarimetric detection scheme as in Figure 5.1, monochromatic input light linearly polarized at 45° to a principal fiber axis, and a straight fiber with length equal to a multiple of half the beat length L_E' (i.e., the linear input polarization is restored at the fiber end), the Jones matrix formalism gives a small signal current sensitivity ($j_F \ll 1$) normalized to the sensitivity of an ideal fiber as follows [241]:

$$s_{\text{pol}} = \frac{1}{1+x^{-2}} < 1. \tag{7.70}$$

Note that s_{pol} is the square of the intrinsic sensitivity s and, since $s < 1$, s_{pol} is smaller than s. In case of monochromatic light, the current sensitivity oscillates as a function of the fiber length with a period equal to $L_E'/2$ if the input light is polarized at 45° at $z = 0$; and a period of L_E' if the input light is polarized at 0° or 90° ($L_E' \approx p/2$ for large x). The oscillations reflect the SOP ellipticity variations along the fiber according to the SOP trajectories of Figure 7.6. s_{pol} represents the maxima of this periodic function. The average sensitivity corresponds to $s_{\text{pol}}^{(\text{av})} = s_{\text{pol}}^2$ (dashed line in Figure 7.8):

$$s_{\text{pol}}^{(\text{av})} = \frac{1}{\left(1+x^{-2}\right)^2} < 1. \tag{7.71}$$

For $x \gg 1$, the amplitude of the oscillations relative to the average is approximately x^{-2}, i.e., large spin ratios help to reduce the amplitude. The oscillations also occur at a fixed fiber length but varying temperature, since the beat length L_B and hence L_E' are functions of temperature. For $L_B = 12.5$ mm, $p = 5$ mm ($x = 5$), the peak-to-peak oscillation amplitude corresponds to 8% of the average sensitivity. The oscillation period depends on the fiber parameters and fiber length and often is in the range of a few degrees C. Such behavior would be unacceptable in most sensor applications. However, a broadband light source and fiber operation in reflection mode — the polarization modes swap their handedness upon reflection — helps to average out the oscillations to uncritical levels. The magneto-optic sensitivity is then given by $s_{\text{pol}}^{(\text{av})}$ [241].

These measures will not prevent, however, that the variation of L_B with temperature also changes (monotonously) the average sensitivity $s_{\text{pol}}^{(\text{av})}$. Again, the effect is strongest at small spin ratios and can easily cause signal changes of several percent per 100°C for spun fibers with stress-induced birefringence [266]. By comparison, the contribution of the Verdet constant is only 0.7%/100°C. Section 8.5 will show that interferometric current sensors, in contrast to polarimetric sensors, can be set up in a way that the signal becomes immune against variations in L_B.

The sensitivity of a *bent* spun fiber exhibits, in addition to the oscillations of period L_E', ($L_E' \approx p/2$ at sufficiently large x), an additional modulation with a period corresponding to the much longer elliptical beat length L_E ($L_E = 4\,L_B^2\,/\,p$ for $2L_B \gg p$) [249–251]. The oscillation amplitude is smaller for fibers with larger embedded birefringence and grows with tighter bending radii but can be suppressed to zero by an appropriate choice of the input polarization state [249]. Whereas the short period oscillations hardly affect the line integration of the magnetic field, the long period oscillations may result in an imperfect integration if they are not appropriately accommodated or suppressed [278].

7.5.5 Spun Fibers Operated with Broadband Light

In the previous sections, the light was assumed as monochromatic. However, most fiber-optic current sensors work with a broadband source such as a super-luminescent diode (SLED). Typically, the source spectra have center wavelengths l_0 of 1310 nm or 1550 nm, and a spectral width Dl (full width at half maximum, FWHM) of about 20–40 nm. The low coherence length (a few tens of micro-meters) helps to suppress disturbing interferences from parasitic light waves that result, for example, from polarization cross-coupling at imperfect fiber splices. Because the linear birefringence b_B is wavelength-dependent, the different spectral constituents accumulate different birefringent phase delays. Accordingly, at a given position z, the spectral components of an arbitrary input SOP have different ellipticities and azimuths, resulting in a reduced degree of polarization, DOP. The DOP ($0 \leq \text{DOP} \leq 1$) is defined as

$$\text{DOP} = I_p\,/\left(I_p + I_u\right), \tag{7.62}$$

where I_p and I_u are the intensity fractions of polarized and unpolarized light. The length of fiber at which the phase delays have spread out over a 2p-range, i.e., over a full circle on the Poincaré sphere, is referred to as depolarization length l_{dep} [234]. It should be noted that the underlying quasi eigenmodes themselves remain polarized (DOP = 1), since the spectral constituents of an eigenmode maintain common SOP ellipticities and azimuths. At $z \gg l_{\text{dep}}$, the DOP depends on how the input field distributes itself onto the two eigenmodes (amplitudes C_1, C_2 in [7.60]). The DOP becomes zero (fully depolarized) if the input light is linearly polarized at 45° to the slow axis ($C_1 = C_2$), and the DOP reaches a maximum, dependent on the spin ratio x, if the input light is polarized at 0° or 90°. For circular input light, the output DOP is independent of the input azimuth and only determined by x; see [255] for more details.

The description in terms of depolarization is equivalent to saying that, at $z > l_{\text{dep}}$, the two quasi eigenmodes have accumulated a group delay longer than the vacuum coherence length l_c of the light source. The fiber length $l_{c,f}$ corresponding to l_c equals $l_{c,f} = l_c/Dn_{g,E}$. The vacuum coherence length of a SLED with a near-Gaussian spectrum is $L_c = 0.88(\lambda_0^2/Dl)$; see Section 3.4. Neglecting the dispersion of Dn_B (i.e., assuming $Dn_{g,B} = Dn_B$), one can rewrite $Dn_{g,E}$ of (7.54) in agreement with [234] as follows:

$$\Delta n_{g,E} = \lambda_0 \frac{\xi}{\pi} \frac{x^{-2}}{\sqrt{x^{-2}+1}}. \tag{7.63}$$

The fiber length $l_{c,f}$ is then

$$l_{c,f} = 0.88 \frac{\pi}{\xi} \frac{\sqrt{x^{-2}+1}}{x^{-2}} \frac{\lambda_0}{\Delta\lambda}. \tag{7.64}$$

For small spin ratios ($x \ll 1$), $l_{c,f}$ is independent of the spin pitch p and approximately equals $l_{c,f} = 0.88L_B(\lambda_0/\Delta\lambda)$. For large x, $l_{c,f}$ increases in proportion to x: $l_{c,f} = 0.88xL_B(\lambda_0/Dl)$ [234]. As an example, $l_{c,f}$ is about 4.8 m for $p = 5$ mm and $L_B = 12.5$ mm ($x = 5$), assuming a 1310 nm-light-source with a FWHM of 30 nm. It is obvious that no Faraday polarization rotation occurs at fiber lengths much longer than $l_{c,f}$, where the quasi eigenmodes have lost their coherence. Therefore, spun fibers are often operated in reflection mode (the differential group delay is compensated on the return trip) or in a Sagnac interferometer with counter-propagating beams.

7.5.6 Photonic Crystal Spun Fibers

Several research groups have reported micro-structured elliptical or circular birefringent spun fiber [279–289]. Analog to conventional spun highly birefringent fiber, A. Michie et al. fabricated such fiber by spinning the preform of a micro-structured polarization-maintaining fiber during fiber drawing. Three rings of different-sized air holes around a solid elliptical core resulted in a (group) beat length of the embedded linear birefringence of 1 mm at 1550 nm. The spin pitch was 4.4 mm [280]. Since such a fiber is largely free of internal stress and consists of pure fused silica, the embedded linear birefringence hardly changes with temperature. Consequently, only the Verdet constant should contribute to the temperature dependence of a corresponding current sensor [282]. Chamorovskiy et al. demonstrated a miniature fiber coil with 100 turns and a diameter of only 5 mm made from micro-structured spun fiber. Its magneto-optic sensitivity reached 70% of the ideal value [283].

Whereas the fibers reported by Michie et al. and Chamorovskiy et al. were, like a conventional spun fiber, elliptically birefringent, R. Beravat et al. presented a photonic crystal spun fiber that was purely circularly birefringent [287, 288]. The fiber structure consisted of a hexagonal array of hollow air holes arranged around a central solid glass core. The sixfold symmetry of the air holes resulted in negligible linear birefringence. Here, the differing reflection coefficients for left and right circularly polarized light at the helically curved interfaces between the core and the air holes gave rise to circular birefringence. The birefringence could be tuned via the diameter-to-spacing ratio of the air holes. Owing to the fiber's circular eigenmodes, there was no reduction in the magneto-optic sensitivity as compared to a corresponding unspun fiber.

7.6 HELICALLY WOUND FIBERS

Helical fiber windings represent a further technique for supressing bending-induced linear birefringence in fiber-optic current sensors. Like spun highly birefringent fibers, helically wound fibers are elliptical birefringent. The circular component of the elliptical birefringence is the result of Berry's topological phase, i.e., is of geometrical origin. "Berry's phase" was initially introduced in a quantum mechanical context [290], but also plays a role in other areas of physics. A particular manifestation of Berry's phase is the optical activity of a fiber describing a nonplanar curve in space. In an early paper on the matter, R. N. Ross theoretically quantified and experimentally demonstrated the effect in single-mode fibers [291]. Numerous subsequent authors have further investigated the topological optical phase, particularly in helically wound fibers [189, 292–303].

The line a curved fiber describes in space is characterized by a moving trihedron of orthogonal unit vectors $\boldsymbol{t}$, $\boldsymbol{n}$, and $\boldsymbol{b}$, also known as Serret-Frenet coordinate system, where at any point z along the fiber, $\boldsymbol{t}$ is a tangential to the fiber, $\boldsymbol{n}$ is the normal vector that points to the center of curvature, and the binormal $\boldsymbol{b}$ is orthogonal to both, $\boldsymbol{t}$ and $\boldsymbol{n}$ (Figure 7.10a) [291]. The angle between $\boldsymbol{b}_1$ and $\boldsymbol{b}_2$ at two fiber points P_1 and P_2 separated by an infinitesimal distance dz is $t\,dz$, where $t = t(z)$ is the geometrical torsion of fiber:

$$d\boldsymbol{b} / dz = -\tau \boldsymbol{n}. \tag{7.72}$$

Light traveling from P_1 to P_2, being linearly polarized at an angle θ_1 with respect to $\boldsymbol{b}_1$ at point P_1, will be polarized at an angle θ_2 with respect to $\boldsymbol{b}_2$ at P_2 [291]:

$$\theta_2 = \theta_1 - \tau(z)dz. \tag{7.73}$$

Here, it is assumed that the fiber is so thin that linear birefringence b_c due to the fiber curvature is negligible and the fiber is free of torsional stress (fiber twist). The geometric polarization rotation $\theta(z)$ over a longer distance z is then obtained by integration:

$$\theta(z) = \theta(0) - \int_0^z \tau(z)dz. \tag{7.74}$$

The polarization evolution in the presence of both geometric polarization rotation and non-negligible linear birefringence as a result of the fiber curvature is again described by coupled mode equations [296]. The coupled mode equations for a fiber helix of constant pitch p and radius R are independent of z and obtained from the equations of a (straight) spun birefringent fiber by replacing 2ξ and b_B in (7.40) with $2t$ and b_c, respectively [189]. The principal axes of the linear birefringence b_c coincide with the $\boldsymbol{n}$ and $\boldsymbol{b}$ axes of the Serret-Frenet frame and hence rotate at a constant rate t along the fiber, equivalent to the axes rotation in a spun fiber; albeit in a spun fiber, the rotation rate is commonly much higher. It should be noted that the reference coordinate system for the dielectric perturbation tensor De of (7.8) and coupled mode equations of a helical fiber is Tang's coordinate frame [189, 304]. The z-axis of Tang's frame coincides with the $\boldsymbol{t}$-axis of the Serret-Frenet frame (fiber direction), whereas the x and y axes rotate with respect to the $\boldsymbol{n}$ and $\boldsymbol{b}$ axes of Serret-Frenet frame at a rate of $-t$. Tang's frame is thus equivalent to the laboratory frame in case of a spun fiber.

Consequently, a helically wound fiber supports, like a spun fiber, right-handed and left-handed elliptical polarization modes of constant ellipticity. Their azimuths coincide with the ***n*** and ***b*** axes of the Serret-Frenet frame. The elliptical birefringence (rad/m) in the Serret-Frenet frame is [296, 305]

$$\gamma = \left[(2\tau)^2 + \beta_c^2 \right]^{1/2}. \tag{7.75}$$

The geometrical geometric torsion t equals

$$\tau = \frac{2\pi p}{p^2 + (2\pi R)^2} = \frac{2\pi p}{s^2}, \tag{7.76}$$

with s being the arc length (fiber length per turn): $s = (p^2 + (2pR)^2)^{1/2}$. τ is positive for a right-handed helix, which means that a right circular SOP propagates faster than a left circular SOP. With the material parameters of fused silica fiber, the linear birefringence b_c equals [305]

$$\Delta\beta_c = 0.77 \frac{1}{\lambda} \frac{r^2}{R_k^2} = 0.77 \frac{1}{\lambda} \left(\frac{r\cos(\sigma)}{R} \right)^2. \tag{7.77}$$

Here, r is fiber radius, and R_k is the radius of fiber curvature, $R_k = R\left[1 + \left(\frac{p}{2\pi R}\right)^2\right]$, and s is the pitch angle, $s = \text{atan}(p/2pR)$. Analog to a spun fiber, the ellipticity of the modes is given by $e = \tan\psi$, $\psi = \pm(1/2)\text{atan}(2\tau/Db)$. The maximum torsion occurs at $s = 45°$ (i.e., if $p = 2\pi R$) and amounts to $t = 1/(2R)$ [189]. As an example, for a helix diameter $2R$ of 30 mm, a pitch p of 188.5 mm (at $s = 45°$), a fiber diameter $2r$ of 125 mm, and a wavelength of 1310 nm, one obtains a torsion (and geometrical polarization rotation) of $\tau = 33.3$ rad/m and linear birefringence of $Db_c = 2.55$ rad/m. The elliptical birefringence is $\gamma = 66.6$ rad/m ($\sim 2\tau$), and the corresponding beat length is 94 mm.

For current sensing, an integer number N of helical windings of a low birefringent single-mode fiber are wound with constant pitch and radius, either on a cylindric carrier body with the current conductor along the cylinder axis as shown (Figure 7.10a) [189, 306–308] or on a toroidal carrier with the conductor passing through the center of the torus (Figure 7.10b) [136, 309, 310]. The magneto-optic phase retardation of the two polarization modes is then again unaffected by conductor movements and, provided $Db_c \ll 2\tau$, given by [305]

$$\Delta\phi = (2\tau / \gamma)\Delta\phi_{\text{ideal}}, \tag{7.78}$$

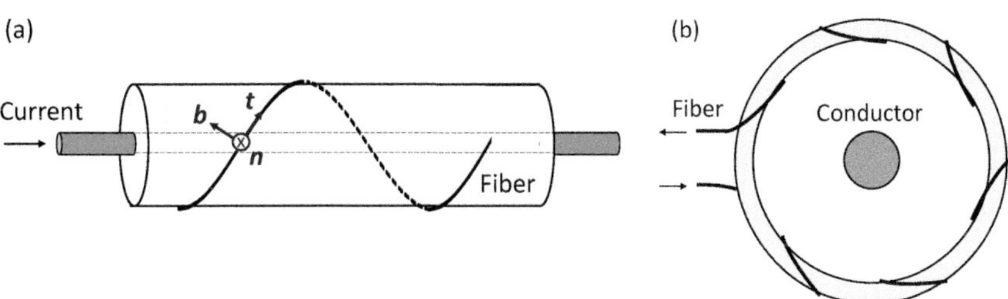

FIGURE 7.10 Helically wound fiber on a cylindrical carrier (a) and on a toroidal carrier (b). The fiber helix is described by a moving trihedron of orthogonal unit vectors ***t***, ***n***, and ***b*** (Serret-Frenet frame).

with $Df_{\text{ideal}} = 2NVH_z$, equivalent to (7.66) for a spun fiber. Here, a cylindric carrier body has been assumed. It is obvious that a toroidal helical fiber configuration results in additional bent-induced fiber birefringence b_b that influences the sensor scale factor in the same way as it does for a spun fiber [296].

Papp and Harms theoretically and experimentally characterized helically wound fibers for current sensing in form of liquid-core multimode fibers already in 1977, well before the above cited works on geometric polarization rotation in nonplanar solid fibers [306]. More than 10 years later, Maystre et al. reported what was presumably one of the first experimental demonstrations of a fused silica helical fiber current sensor [305]. That sensor consisted of a single turn of low birefringent fiber with a pitch angle of 30 degrees and an arc length of 17 cm on a cylindrical carrier. The fiber was operated as a Fabry-Pérot resonator which resulted in a fourfold enhancement of the magneto-optic sensitivity versus a single-pass sensor. The helix parameters and elliptical birefringence were chosen for a single-pass birefringent phase retardation g of the near-circular polarization modes equal to p. Short et al. [309] investigated a toroidal helical current sensor with a torus radius of R = 150 mm, a carrier body radius of r = 5 mm, and 12 fiber turns. The experiments showed that it was important to wind the fiber with slight axial strain for well-defined principal axes of the linear birefringence from the fiber curvature. Zhang et al. presented theoretical and experimental investigations of such sensors in further detail and showed how the geometrical parameters of the fiber helix affect the scale factor [310]. The final sensor had a toroidal quartz glass carrier (R = 90 mm, r = 6 mm) with a total of 35 helical windings that made six revolutions around the torus. The total fiber length was 21.3 m. The authors achieved scale factor stability to within ±0.1% between −40°C and 60°C at a current of 1200 A rms. It is interesting to note that already in 1978, the German company AEG filed a patent claiming straight and toroidal helical glass fiber coils. The patent did not consider the topological phase, however, but the helical geometry was to mimic a twisted fiber. To this end, the fiber ends were prevented from rotating, while the initially straight and torsion-free fiber was wound into a helix [311].

Advantages of helical fiber current sensors are that the sensing fiber is a simple low-birefringent single-mode fiber and is less affected by temperature than spun fibers with stress-induced linear birefringence. The geometrical circular birefringence is fully temperature-independent. On the other hand, helical fiber current sensors are more complicated to manufacture [310] and offer less application flexibility. As a result, they have found little, if any, practical use.

7.7 HELICAL-CORE FIBERS

Helical-core fibers [312–316] are fabricated by spinning the preform of a single-mode fiber with an off-center core. As a result, the core follows a helical path. An ideal helical-core fiber is purely circularly birefringent (assuming that the core is circular and the fiber free of internal stresses). The circular birefringence is again the result of Berry's topological phase, i.e., the equations of the previous section apply again but with vanishing linear birefringence. At a given pitch angle, the helical-core pitch can be made much smaller than the pitch of a helically wound fiber, which results in significantly stronger circular birefringence. Husey et al. reported the first helical core fibers in the mid-1980s [312]. The fibers had a cladding diameter of 110 mm and a core diameter of 20 mm and were fabricated with various core pitches. The circular beat lengths were as small as 4.3 mm at a pitch of 8.6 mm and a wavelength of 633 nm. One must assume, though, that the small curvature radii of the helical core resulted in considerable loss. Birch investigated helical-core fibers with larger diameters ranging from 450 mm to 800 mm and core offsets from the fiber axis between 170 mm and 340 mm, respectively. The 800-mm-diameter fiber had a core pitch of 2 mm with a circular beat length of 3.6 mm at 633 nm [315].

In principle, the high circular birefringence of helical-core fibers should make them good candidates for fiber-optic current sensors. Nevertheless, apart from some exploratory research, little is known about the use of helical-core fibers for current sensing. Obviously, the core offset makes optical coupling more cumbersome, and large fiber diameters exclude small fiber coil diameters. In more recent times, helical-core optical fibers have found interest for use in fiber lasers [317]. Also, photonic crystal fibers with chiral micro-structures have been the subject of active research, [318] and references therein.

7.8 SPUN LOW BIREFRINGENCE FIBERS

Early single-mode fibers still had considerable intrinsic linear birefringence as a result of imperfect circular core shapes and frozen-in stress. Birefringent phase retardations as high as several hundred degrees per meter were not unusual [222, 233]. It was found, however, that spinning the preform during the fiber drawing process restored an average circular symmetry and effectively eliminated the birefringence [222, 319–324]. To this end, the spin pitch should be smaller than the linear beat length of the unspun fiber by a factor of 10 or more. Barlow et al. reduced the linear birefringence from 60 deg/m at 633 nm (L_B = 6 m) to less than 1 deg/m by spinning with a pitch of 5 mm [222]. Already a few years earlier, in 1978, Schneider et al. at Siemens had reported unspun low-bi (low bi-refringence) fibers with retardations of less than 6°/m [200]. As the spinning occurs at a state of low glass viscosity, low birefringence fibers have, other than twisted fibers, negligible intrinsic circular birefringence. Fiber spinning has also been employed to reduce polarization mode dispersion in telecom fibers [222, 325].

A low birefringence fiber comes close to an ideal fiber for current sensing but remains sensitive to bending-induced fiber birefringence and other external perturbations.

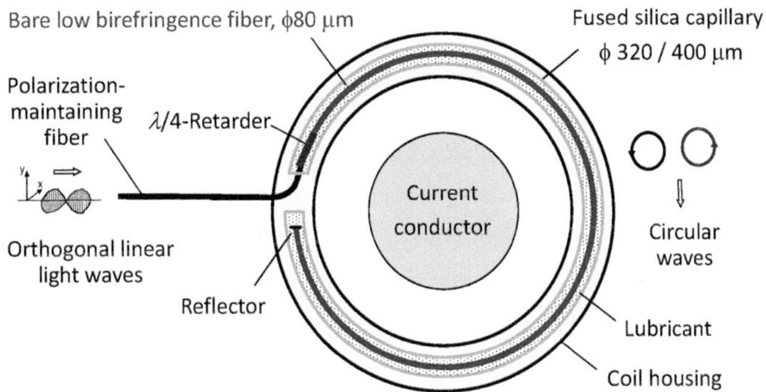

FIGURE 7.11 Bare low birefringence fiber operated in reflection mode and packaged in a thin capillary of fused silica; for simplicity, only one capillary winding is shown.

Bending-induced birefringence is well-defined and rather stable, however, and in many cases can be kept sufficiently small (e.g., $d_b < 20°$) by accordingly choosing the diameter of the fiber coil and number of fiber windings. Here, fibers with a diameter of 80 µm instead of standard 125 µm are beneficial, as the birefringence varies in inverse proportion to the square of the diameter; eq. (7.17).

By contrast, other sources of birefringence such as stresses from the fiber coating, particularly at low temperature, and coil packaging often have poor repeatability and may introduce intolerable temperature dependence. Bohnert et al. eliminated coating and packaging related stress by inserting a bare low birefringence fiber, i.e., with the coating removed, into a thin lubricant-filled capillary of fused silica (320-mm inner diameter). Depending on the sensor application, the capillary was either embedded in a flexible strip of fiber-reinforced epoxy [326] or wound up to a coil and placed into a rigid coil housing (Figure 7.11) [327, 328]. In combination with measures for intrinsic temperature compensation (Section 8.6), the sensors were accurate to within ±0.1% between −40°C and 85°C.

7.9 THERMALLY ANNEALED FIBER COILS

In the 1980s and early 1990s, G. W. Day and co-workers at the National Bureau of Standards and Technology (now NIST) in Boulder, Colorado, developed thermal annealing for the removal of bending-induced linear birefringence in fused silica fiber coils [7, 193, 233, 329–331]. The method was successfully applied to coils with diameters between 5 mm and 100 mm and up to 200 and more fiber turns. To this end, the coils were placed into a circular channel in a ceramic annealing fixture and annealed in air at 850°C for 24 hours. The temperature was then gradually lowered to room temperature at a rate of 0.2°C/min. A complete annealing cycle lasted 3–4 days [193]. Shorter cycle times were reported in later work [331, 334, 335]. The fiber coating burned off during the initial temperature up-ramp (5°C/min). It was important to first loosen the coating by means of a solvent in order to avoid fiber breakage. The annealed coils were placed into a plastic housing with a circular aperture for the current conductor. The interior of the housing featured a channel similar to the one of the annealing fixture, within which the fiber windings resided in a viscous lubricant. The fibers were various types of single-mode communication fiber with cut-off wavelengths between 600 and 900 nm. The annealing effectively removed the bending-induced birefringence and partially also the intrinsic birefringence due to imperfect circular symmetry and embedded stress. (The latter birefringence corresponded to some tens of degrees per meter.) Pre-twisting the fibers also helped to reduce the intrinsic birefringence to near zero [233].

The technique was commercialized by the 3M company in the 1990s. The commercial coils were fully packaged and had polarizing fiber leads for interfacing. Reports showed Verdet-constant-limited performance between −20°C and 80°C with an error bar of less than ±0.2% [332–334].

Bohnert et al. reported inherently temperature-compensated coils of low birefringence fiber packaged and annealed in a thin capillary of fused silica with residual temperature dependence of the current signal of -0.14×10^{-4} K^{-1} between −35°C and 85°C that was about five times smaller than the temperature dependence of the Verdet constant [335–337].

It should be noted that annealing can lead to changes in the fiber's spectral transmission and dispersion, shifts in the cut-off wavelength, and higher attenuation. The severity of the effects depends on the type and concentration of the fiber dopants. The modifications become more pronounced at annealing temperatures beyond 850°C [193]. Another phenomenon, especially above 1000°C, is devitrification, i.e., the formation of crystallites. Consequences are substantial scattering losses and depolarization resulting in a reduction of the measured Faraday rotation. Furthermore, devitrification may severely diminish the mechanical strength of the fiber [331].

Thermal annealing has resulted in fiber coils with excellent performance over wide temperature ranges and, with proper packaging, good reliability. Nevertheless, after spun highly birefringent fibers became commercially available, the method was largely abandoned. Downsides were the lengthy fabrication process and delicate handling and splicing of the brittle coils.

7.10 FLINT GLASS FIBERS

In the early 1990s, Kurosawa and co-workers developed flint glass fiber for current sensing in a collaboration between the Tokyo Electric Power Company and Hoya Corporation [31, 32, 338–341]. Flint glass contains a high fraction of lead oxide which results in relatively high refractive indices between 1.5 and 2.0. Fibers from flint glass are of interest to current sensing due to their small stress-optic coefficient that is about 740 times smaller than in fused silica fiber. At 633 nm, the corresponding numbers are 4.58×10^{-15} m^2/N [339] and 3.40×10^{-12} m^2/N [194], respectively. Hence, bending-induced birefringence in flint glass fiber is negligibly small. Circular birefringence due to fiber twist is about two orders of magnitude lower than in fused silica fiber [339], and the near-infrared Faraday polarization rotation is about six times stronger [31]. Its variation as function of temperature is 1.0×10^{-4} K^{-1} and only slightly larger than in fused silica (0.7×10^{-4} K^{-1}). The phase retardation of the fabricated fibers due to intrinsic fiber birefringence corresponded to a few degrees per meter.

However, the optical loss (about 1 dB/m at 850 nm [32], 1.3 dB/m at 1550 nm [341]) is much larger than in fused silica fiber but in many cases still tolerable due to the relatively short fiber lengths. It should also be mentioned that flint glass fiber cannot readily be spliced to fused silica fiber. Typically, bulk analyzer optics interfaces the flint glass fiber coil with fused silica leads [32, 341]. FOCS with flint glass fiber have been successfully deployed in numerous electric power facilities in Japan (see, for example, [341] and references therein). While flint-sensing fiber has had some success in Japan, commonly, other fiber solutions have been preferred elsewhere.

7.11 FIBERS WITH ENHANCED MAGNETO-OPTIC SENSITIVITY

The Faraday effect in fused silica fiber is relatively small. The polarization rotation is well sufficient at the high currents in electric power transmission and other high current applications, but precise measurement of small currents, e.g., in the milliampere range, would require an excessive number of fiber loops. Adding appropriate dopants to the fiber core can significantly enhance the Faraday effect. Suitable dopants include rare earth irons such as Eu3+, Ce3+, Tb3+, or Pr3+ [342–344] or nanoparticles, e.g., of cobalt [345], CdSe [346, 347], or Gd_2O_3 [348]. Further dopants may be added to control the refractive index difference between the fiber core and cladding or to prevent dopant clustering at high concentrations. Compound glass fibers also exhibit an enhanced Faraday effect [31, 32, 349–352]. Table 7.1 lists the Verdet constants and optical attenuation (if specified) for a selection of

TABLE 7.1 Doped Fibers and Compound Glass Fibers with Enhanced Magneto-Optic Sensitivity

Host Material / Glass Type	Dopants	Verdet Constant (rad T^{-1} m^{-1})	Optical Loss (dB/m)	Wavelength (nm)	References
Doped fibers					
Fused silica	Tb3+	−24.5	-	1053	[343]
Fused silica	Tb3+	−15.5	8	1300	[38]
Fused silica	Eu3+	−4.56	-	660	[344]
		−3.43		802	
		−1.37		1310	
		−0.98		1550	
Flint glass	CdSe quantum dots	7.2	-	633	[347]
Alumino-silicate	Gd_2O_3 nano particles	3.19	-	650	[348]
Compound glass fibers					
Schott F7 (93 mol% SiO_2)	-	~8.3	0.77	633	[349]
		~5.2	0.49	789	
Tellurite glass (multimode fiber)	-	−28	-3.5	633	[350]
		-	2.6	980	
				1550	
TZN glass ($75TeO_2$-$20ZnO$-$5Na_2O$)	-	26.1	-	633	[39]
Flint glass fiber (undoped)	-	~20	-11.3	633	[31, 32]
		~11		850	
		3.2		1550	
For comparison					
Fused silica fiber	-	3.25	negligible	633	[28, 30]
		1.83		825	
		0.54		1523	

doped fibers and compound glass fibers as well as of fused silica fiber for comparison. Note that doped fibers have negative Verdet constants due to the paramagnetic nature of the dopants. Commonly, the stronger Faraday effect is accompanied by significant optical loss. Hence, higher signal noise increasingly eats up the precision gains as the fiber length increases. Beyond a certain length, fused silica fiber will again be superior. For a Schott F7 glass fiber having relatively low attenuation (see Table 7.1), this length is about 50 m at 633 nm and a detection bandwidth of 1 Hz [349]. The corresponding lengths of doped fibers with higher attenuation can be much shorter, though.

7.12 MULTIMODE FIBERS

Early researchers have also explored the Faraday effect in graded-index multimode fibers [353, 354]. Multimode fiber sensors may incorporate other low-cost components such as an LED light source or thin film polarizers and allow for an economic sensor solution. On the other hand, internal birefringence, modal dispersion, and mode coupling due to internal and external perturbations tend to reduce the degree of polarization at the fiber end and make such sensors rather prone to environmental disturbances. Corresponding sensors can be adequate for some diagnostic and monitoring purposes but are commonly not an option for accurate and reliable current measurement. Wessel et al., for instance, used a straight piece of multimode fiber to monitor magnetic field pulses of a flux compression generator with peak fields of 160 T and rise times of 3 T/ns [355]. Shiyu et al. investigated the Faraday effect in a Tellurite glass multimode fiber [350].

7.13 LIQUID CORE FIBERS

Without going into detail, it is worth mentioning that early works also explored liquid core fibers for current measurement [356, 357]. To this end, Papp and Harms filled thin glass capillaries with hexachlorobutadiene (C_4Cl_6) as a magneto-optic liquid. The linear birefringence associated with bending the capillaries into a coil was suppressed by configuring the capillary along a helical path as mentioned further above or in form of orthogonal loops [201, 306]. Frenkel et al. arranged four straight tubes containing toluene or mineral oil along the sides of a square around the current conductor with ortho-conjugate prisms for beam steering at the corners [358].

7.14 MICROFIBER COILS

Microfibers and nanofibers (MNF) [359, 360] are typically produced by tapering a section of standard single-mode fiber via heating and stretching to a diameter of a few micrometers or less (Figure 7.12) [361]. In the process, the dopants of the original core diffuse out over the taper cross-section. The taper then represents the new core. A significant fraction of the optical power propagates as an evanescent wave outside of the taper. MNF have been demonstrated as sensors for refractive index, humidity, biochemical parameters, gases, and physical quantities such as temperature, acceleration, rotation, or acoustic waves [360, 361]. Also, current sensors based on the Faraday effect in micro-fiber coils have been investigated [362–365]. Micro-fibers enable millimeter-sized fiber coils with potential detection bandwidths in the GHz range due to short transit time of the light. Challenges have been

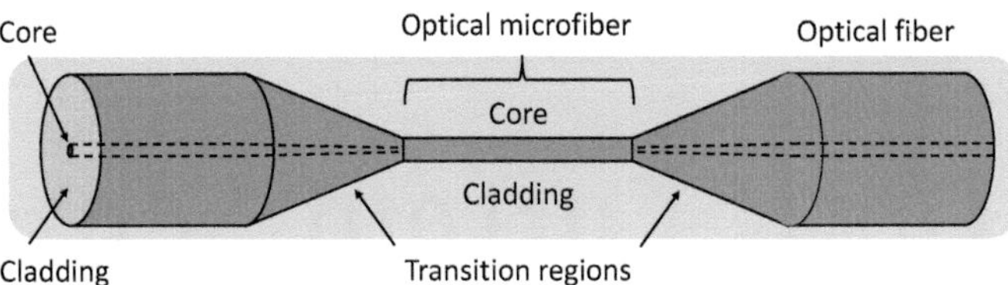

FIGURE 7.12 Optical micro-fiber with an external medium such as a liquid or gas (grey) representing the cladding (adapted from Chen et al. [361].)

the birefringence and stress-free packaging of the coils [361]. Chen et al. demonstrated a spun birefringent microfiber coil with a fiber diameter of 2 mm, a spin rate of 24 turns/cm, and embedded linear birefringence of 10^{-3}. The latter was produced by side polishing the original fiber. The coil consisted of 7.5 turns around a 1-mm-thick electric wire and had a sensitivity of 8.6 mrad/A [363]. In related work, microfiber coils were also demonstrated as indirect current sensors, where the current produced, via ohmic heating, thermal optical phase shifts in the fiber. Commonly, the fibers were operated as resonators or in an interferometer [366–372]. Given the thermal detection principle, the sensitivity decreased with increasing current frequency [366].

CHAPTER 8

Fiber-Optic Current Sensors

8.1 OVERVIEW

Over the years, researchers have explored many versions of all-fiber optical current sensors (FOCS) based on the Faraday effect. Naturally, there are different ways to categorize the sensors. Here, we group them as follows:

- Basic polarimetric FOCS
  - Transmissive configurations
  - Reflective configurations
- Polarimetric FOCS with biasing Faraday rotator
- Sagnac interferometer current sensors
  - Phase-modulated Sagnac interferometers
  - Sagnac interferometers with 3×3 fiber directional coupler
- FOCS with polarization-rotated reflection
  - Phase-modulated sensors
  - Sensors with static phase bias
- Fabry-Pérot and light recirculating current sensors
- Fiber laser current sensors
- Integrated optical current sensors
- Faraday effect in fiber Bragg gratings
- Multiplexed and distributed current sensing

DOI: 10.1201/9781003100324-11

8.2 BASIC POLARIMETRIC FOCS

8.2.1 Transmissive Operation Mode

The earliest all-fiber current sensors substituted the discrete Faraday sensor elements of Chapter 5 by an optical fiber coil, again employing polarimetric detection schemes such as illustrated in Figure 5.1. S. Yoshikawa and A. Ueki from the Nippon Selfoc Company were, in 1968, among the first to propose the concept [373]. Others followed in subsequent years [171, 196, 214, 226, 320, 374–376]. As already discussed, bend-induced fiber birefringence proved to be a serious obstacle to reliable current measurement. Furthermore, mechanical perturbations of the fiber coil and its fiber leads, e.g., by bending and vibration, caused prohibitively large signal perturbations. Similarly, as in the case of bulk-optic sensors [136, 143, 146], configurations with counter-propagating light helped to improve the situation, albeit at the cost of higher complexity (Figure 8.1). By taking the difference of the two signals, the non-reciprocal Faraday effect doubles, whereas reciprocal effects tend to cancel out [136, 377–380]. Note, however, that the light propagation in the sensing fiber is fully reciprocal only at zero current. Hence, the compensation becomes less effective with increasing Faraday phase shifts. Fang et al. reported detailed theoretical and experimental studies of the matter for different types of sensing fiber [378, 379]. Over time, transmissive polarimetric sensors were largely abandoned. Not least, accurate measurement of direct currents (dc) has remained elusive for simple designs. It is worth mentioning that closed-loop detection for polarimetric sensors was also explored, where an additional wire coil compensated the polarization rotation by the primary current, as in the case of the magnetic field sensor in Figure 5.2 [381]. Heterodyne detection schemes with a frequency or intensity modulated light sources were reported in [382, 383].

8.2.2 Reflective Operation Mode

A sensor coil operated in reflection mode (also called in-line configuration, Figure 8.2) simplifies the sensor setup, and the free end of the sensing fiber allows for winding the fiber around the conductor without opening the electric circuit [205, 227, 253, 257, 286, 384–390]. Often, the sensors employ a Faraday rotator mirror that rotates the polarization azimuth by 90°. As a result, phase shifts on the forward path due to linear birefringence cancel out on the return path (this is strictly true only at zero current, however), and the rotated polarization significantly reduces sensitivity to vibration [384, 385, 389]; see [205] for a theoretical analysis. Pistoni et al. measured 30 dB rejection of mechanical disturbances [384]. Note, however,

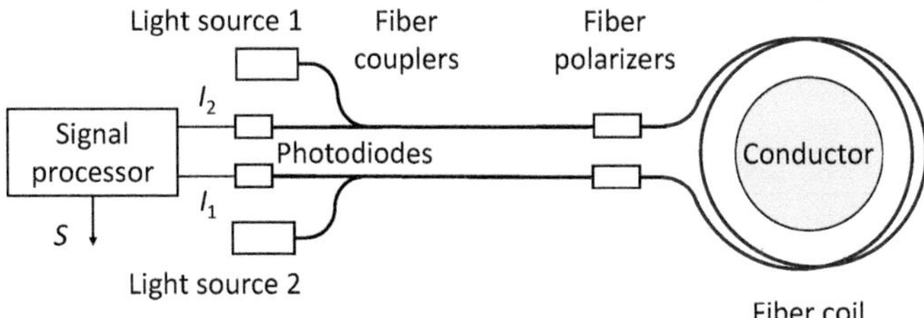

FIGURE 8.1 Polarimetric all-fiber current sensor with counter-propagating beams for vibration compensation [377–380].

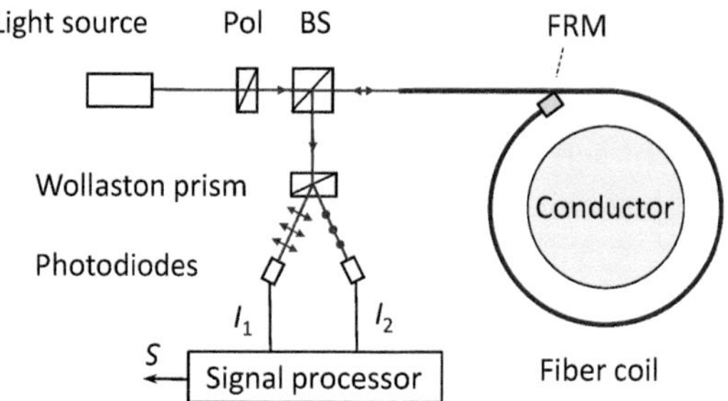

FIGURE 8.2 Polarimetric FOCS operated in refection mode; BS: beam splitter, FRM: Faraday rotator mirror [384, 385].

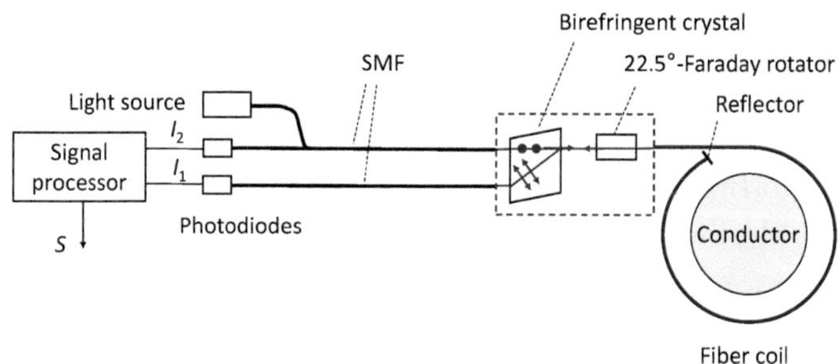

FIGURE 8.3 Polarimetric FOCS with 22.5°-Faraday rotator for phase biasing (lenses not shown) (adapted from US patent 7,176,671 [392].)

that quenching of the Faraday effect in standard single-mode fibers by bend-induced linear birefringence continues to persist.

8.3 POLARIMETRIC FOCS WITH BIASING FARADAY ROTATOR

The sensor configuration is similar as in the previous section with the difference that a 22.5°-Faraday rotator in front of the fiber coil rotates the polarization by 2 × 22.5° on a roundtrip of the light, that is, the rotator introduces a 90°-phase bias between the left and right circular constituents (Figure 8.3) [31, 261, 391–394]. The entry polarizer can then also serve as an exit polarizer (analyzer), as the rotator makes a 45°-relative analyzer orientation obsolete. The sensor in Figure 8.3 utilizes a birefringent prism instead of a polarizer to generate two anti-phase signals. Insensitive single-mode fibers guide the light to and from the sensor head. It is obvious that the concept results in a simple overall design. Kurosawa et al. at Tokyo Electric Power Co. developed a corresponding sensor with flint glass sensing fiber that was subsequently commercialized by Adamant Kogyo Co., Japan [31, 393].

8.4 SAGNAC INTERFEROMETER CURRENT SENSORS

8.4.1 Sagnac Interferometer with Non-Reciprocal Phase Modulation

The 1980s saw the development of the first optical fiber gyroscopes as alternatives to mechanical and laser gyroscopes for applications, e.g., in navigation systems. Fiber gyros utilize the Sagnac effect in a coil of polarization maintaining fiber (PMF) to measure rotation velocities. The two counter-propagating light waves in the coil accumulate a differential

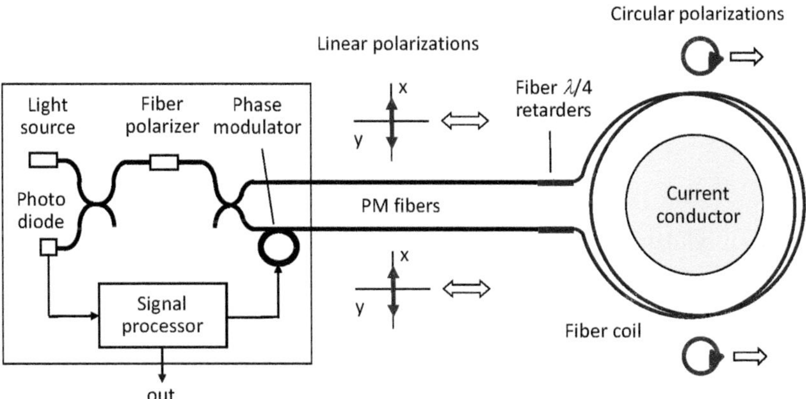

FIGURE 8.4 Sagnac interferometer current sensor (© 2002 IEEE. Reprinted, with permission from [336].)

phase shift in proportion to the rotation velocity about the coil normal. The fundamentals of fiber gyros are described in [395, 396] and references therein. It was soon recognized that fiber gyroscopes could easily be adapted to current sensing by replacing the PMF with a circularly birefringent or low birefringence fiber and additional quarter-wave retarders, as illustrated in Figure 8.4 [336, 398–409].

Typically, the light source is a temperature-stabilized 1310-nm or 1550-nm superluminescent diode (SLED) with a spectral width of 30–40 nm, FWHM. Two linearly polarized light waves (here parallel to x) travel through PM fiber leads to the fiber coil. (In this context, the term *wave* represents the whole spectrum of waves emitted by the source.) Quarter-wave retarders at the entrances to the coil convert the linear polarization states into circular polarizations of equal handedness. Upon leaving the coil, the retarders convert the circular polarizations back to linear states, again parallel to x. The linear SOP interfere in the coil-side 1×2 fiber coupler. The interference intensity from the reciprocal port of that coupler represents the optical signal. Initially, loops of single-mode fiber, the loop normal at 45° to x, served as all-fiber retarders. With appropriate loop diameter, the bend-induced birefringence gave 90°-phase retardation [410, 411, 402]. Later, short sections of PM fiber were introduced as more convenient and reliable solutions [404, 336].

8.4.1.1 Open-Loop Operation

The two counter-propagating light waves accumulate a differential magneto-optic phase shift given by

$$\Delta\phi = 2VNI. \tag{8.1}$$

Here, the fiber coil is assumed as ideal, i.e., with a sensing fiber free of linear birefringence and perfect quarter-wave retarders. The Verdet constant V is about 1.0 mrad/A at 1310 nm and 0.7 mrad/A at 1550 nm [7]. Without further measures, the interference signal is

$$I = \left(I_0/2\right)\left[1+\cos\left(\Delta\phi\right)\right]. \tag{8.2}$$

To generate an output that varies (at small signals) linearly with $\Delta\phi$, a modulator introduces an additional differential phase modulation $\phi(t)$ of the two waves with amplitude ϕ_m and angular frequency $\varpi_m = 2\text{p}\nu_m$ (Figure 8.5) [395]:

$$\phi(t) = \phi_m \sin(\varpi_m t). \tag{8.3}$$

Simple phase modulators consist of a piezoceramic disk (PZT) with a few fiber loops attached to it. The modulator is located at one end of the Sagnac loop, and, ideally, half the modulation period corresponds to the transit time t of the light through the loop, i.e., $\nu_m = 1/(2t)$.

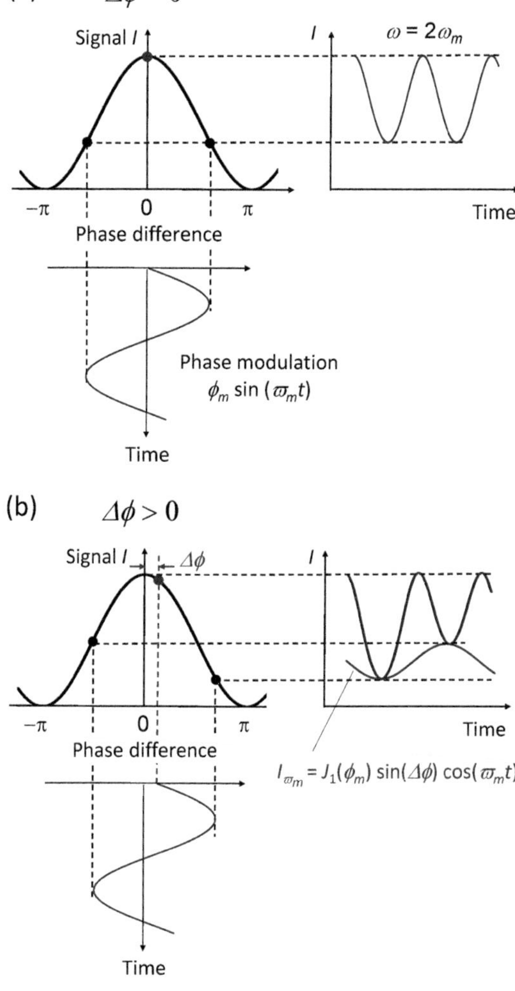

FIGURE 8.5 Sagnac sensor with non-reciprocal phase modulation (open-loop operation): (a) Without applied current ($Df = 0$), the operating point of the interferometer (red) is at the maximum of the cosine interference function, and the output signal is modulated at frequency $2\varpi_m$; (b) With applied current ($Df > 0$), the 1st harmonic signal in the sensor output (red curve) is proportional to $\sin(Df)$.

The modulation of the cw and ccw waves is then anti-phase, which maximizes the net modulation amplitude ϕ_m. This kind of phase bias modulation is commonly referred to as non-reciprocal. The interferometer output is now

$$I = (I_0/2)\left[1 + \cos\left(\Delta\phi + \phi_m \varpi_m t\right)\right], \tag{8.4}$$

which can be expanded in a series of Bessel functions:

$$\begin{aligned} I = \frac{I_0}{2} \Bigg\{ 1 + & \left[J_0(\phi_m) + 2\sum_{k=1}^{\infty} J_{2k}(\phi_m)\cos(2k\varpi_m t) \right] \cos(\Delta\phi) \\ & - \left[2\sum_{k=1}^{\infty} J_{2k-1}(\phi_m)\cos\left((2k-1)\varpi_m t\right) \right] \sin(\Delta\phi) \Bigg\}. \end{aligned} \tag{8.5}$$

The magneto-optic phase shift *Df* then follows from the ratio of the signals at frequencies ϖ_m and $2\varpi_m$:

$$\Delta\phi = a\tan\left(\frac{J_2(\phi_m)}{J_1(\phi_m)} \frac{I(\varpi_m)}{I(2\varpi_m)} \right). \tag{8.6}$$

At small currents ($Df \ll 1$), the first harmonic signal $I(\varpi_m)$ varies in proportion to the current. Typically, ϕ_m is chosen at the first maximum of $J_1(\phi_m)$, i.e., $\phi_m = 1.84$ rad, which ascertains the best signal-to-noise ratio. As the modulation amplitude of PZT modulators can significantly vary with temperature, the ratio of the signals at the 2nd and 4th harmonic is often utilized as feedback to stabilize the amplitude ϕ_m [412]:

$$\frac{I(2\varpi_m)}{I(4\varpi_m)} = \frac{J_2(\phi_m)}{J_4(\phi_m)}. \tag{8.7}$$

If the magneto-optic phase shift is alternating at frequency ν_{ac}, the first harmonic signal splits in two side bands at frequencies $\nu_m \pm \nu_{ac}$. Open-loop operation is straightforward to implement, but the sinusoidal response limits, like in polarimetric sensors, the usable measurement range. Moreover, the piezoelectric modulators are commonly operated on a resonance much lower than the proper frequency $\nu_p = 1/(2t)$ of the Sagnac loop. For example, the PZT frequency may be on the order of 100 kHz, whereas ν_p is 1 MHz for a Sagnac loop length of 100 m and correspondingly higher at shorter lengths. For compensation, the modulators must be driven at correspondingly higher amplitude to obtain the desired modulation depth of $\phi_m = 1.84$ rad. For these reasons and others, high-end gyros and corresponding current sensors work with electro-optic modulators, in particular integrated-optic lithium niobate modulators and closed loop detection circuits. Still, PZT-based gyros with closed-detection are also known [413].

8.4.1.2 Closed-Loop Operation

Electro-optic modulators provide much larger bandwidths than the PZT modulators of open-loop detection circuits. (Details on the lithium niobate electro-optic modulators will follow further below.) Typically, the modulators are combined with all-digital signal processing and generate a square-wave phase bias modulation with amplitude of ±p/2 at the proper frequency ν_p (Figure 8.6) [395, 414]. Hence, at zero current (*Df* = 0), the output signal exhibits a flat plateau with a series of spikes at intervals of τ (Figure 8.6a). At finite current (*Df* > 0) and open-loop operation, the plateau splits into lower and upper plateaus with a separation proportional to J_1sin(*Df*) (Figure 8.6b). At closed-loop operation, the signal processor generates, in addition to the bias modulation, a step-wise phase ramp synchronous with the bias modulation that compensates the current-induced phase shift *Df* (Figure 8.6c). At constant *Df*, the average slope of the phase ramp is $\partial\phi_{\text{ramp}}/\partial t = -(1/t)Df$, i.e., a step of the ramp having a width *t* equals −*Df*. Hence, the ramp slope is a measure for

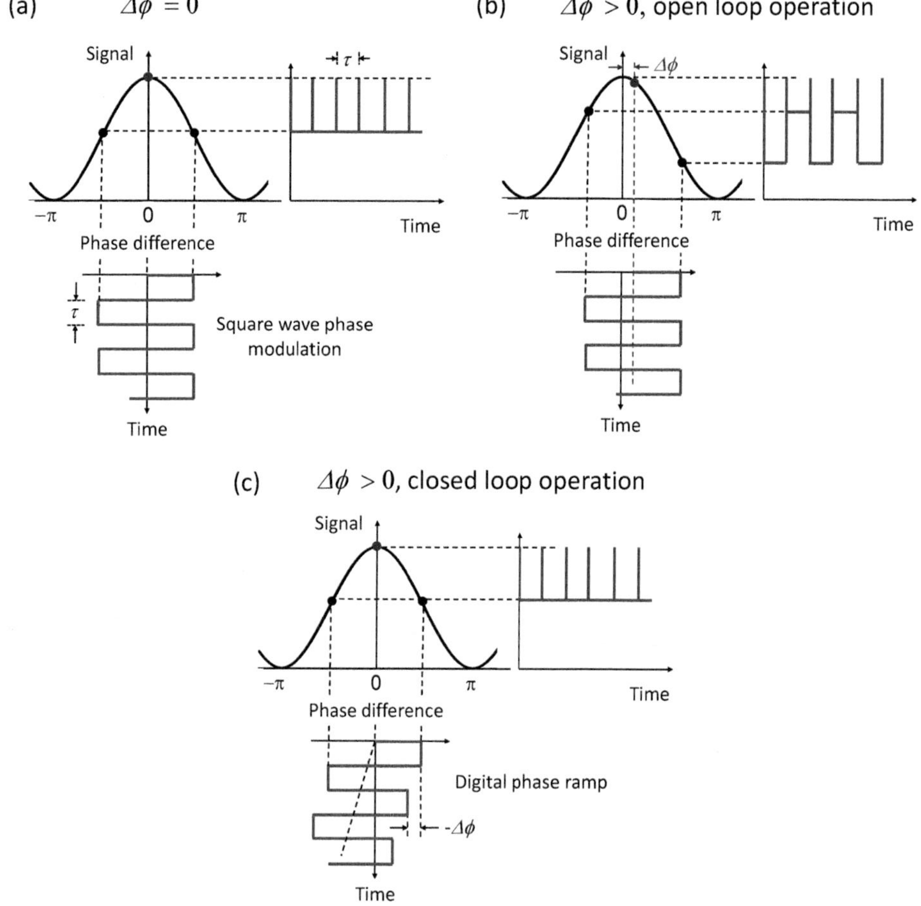

FIGURE 8.6 Sagnac sensor with digital signal processing and square-wave phase modulation at frequency $v_p = 1/(2\tau)$: (a) no applied current, *Df* = 0; (b) *Df* > 0, open-loop operation; (c) *Df* > 0, closed-loop operation: a digital phase ramp with steps of -*Df* compensates the current-induced phase shift *Df*.

the current. Once the ramp voltage reaches its positive or negative limits, the processor initiates a 2p-phase reset. Note that Figure 8.6 only illustrates the very basics of the method. Numerous modifications and upgrades have been reported, such as multiple level modulation and additional feedback loops, most of which aim to reduce potential errors, e.g., due to thermal drifts, dead zones, and others; see [415–418] for a few examples.

As mentioned, eq. (8.2) assumes ideal sensor components. The influence of imperfections such as non-ideal retarders and sensing fibers has been investigated in [336, 406, 411, 419] and will be discussed in context of a sensor version operated in reflective mode further below. An advantage of a Sagnac interferometer is that, other than in polarimetric sensors, changes in any reciprocal circular birefringence does not affect the signal. Likewise, relative polarizer/analyzer alignment drifts cannot occur. In combination with non-reciprocal phase modulation, Sagnac current sensors are therefore well suited for dc. On the other hand, Sagnac sensors still have significant sensitivity to dynamic mechanical perturbations such as vibrations, particularly if the perturbations occur near the fiber loop ends where the time delay between the counter-propagating waves is largest. Moreover, the Sagnac effect translates any rotational movement of the sensor into a signal perturbation. Special fiber winding can eliminate the effect, however [276]. Sagnac current sensors with non-reciprocal phase modulation became largely obsolete after the invention of a more attractive sensor version with polarization-rotated reflection in the early 1990s.

8.4.2 Sagnac Interferometer with 3×3 Fiber Coupler

Sagnac interferometers with a 3×3 fiber coupler instead of a 2×2 coupler provide three signals with mutual phase differences of 120°. Appropriate processing results in a small signal output that varies linearly with the applied current. Hence, a 3×3 coupler is a simple passive alternative to an active phase modulator. In the 1980s, Sheem and others demonstrated Sagnac fiber gyros with 3×3 couplers [420–424]. Corresponding Sagnac current sensors soon followed [425–432]. Figure 8.7 illustrates an implementation with a polarization-maintaining 3×3 coupler that otherwise stays close to the configuration of Figure 8.6 [432]. A 3×3 matrix of coupling coefficients, $\kappa_{m,n}$, n, m = 1,2,3, relates the optical field $E_{o,m}$ at output port m of the coupler to the input fields $E_{i,n}$. The coefficients of an ideal lossless coupler with (1/3)-power-splitting-ratios are [420–423]

$$\kappa_{mn} = \left(1/\sqrt{3}\right)\exp\left(-i\pi/2\right) m = n, \tag{8.8}$$

$$\kappa_{mn} = \left(1/\sqrt{3}\right)\exp\left(i\pi/6\right) m \neq n. \tag{8.9}$$

If light is sent into one of the three input ports, the two cross-coupled transmitted fields ($m \neq n$) are phase shifted with respect to the directly transmitted field ($m = n$) by 120°. It is then straightforward to show that for input light intensity I_0 (I_0 = 1, middle port), the three interference signals I_i, i = 1,2,3, are (Figure 8.8) [425, 426]

$$I_1 = \left(2/9\right)\left(1+\cos\Delta\phi\right), \tag{8.10}$$

$$I_2 = (2/9)\left[1 + \cos\left(120^\circ - \Delta\phi\right)\right], \tag{8.11}$$

$$I_3 = (2/9)\left[1 + \cos\left(120^\circ + \Delta\phi\right)\right]. \tag{8.12}$$

The difference of I_2 and I_3 is proportional to sin *Df*:

$$I_2 - I_3 = (2/9)\sqrt{3}\sin\Delta\phi. \tag{8.13}$$

It is obvious that the method only works if the two counter-propagating waves have circular (or near circular) polarization in the sensing fiber. The effect of non-circular states has been investigated by Rochford et al. [426]. As an alternative to the quarter-wave retarders in Figure 8.7 for generating the circular states, Haywood et al. at the University of Sydney reported a polarizing spun high birefringence fiber [429, 431]. In the window between the two cut-off wavelengths of the first order higher mode, the fiber guides only one of the two circular polarization states, while the other state leaks out. Hence, apart from the sensing fiber, all-fiber components, including the 3×3 coupler, can be from standard single-mode fiber.

It should be noted that the optical paths of the two counter-propagating waves in the 3×3 coupler are not reciprocal. Since the amplitudes and phases of the transmitted fields of a practical coupler vary with temperature, the non-reciprocity can lead to prohibitive errors [429]. Basset, Hayward, et al. have addressed the problem by sending light into all three input ports of the coupler and recording all together nine interference signals. Appropriate signal processing then cancelled coupler imperfections to first order (Figure 8.9) [428–431]. To this end, the light source (a 1540 nm SLD or 1532 nm superfluorescent fiber source) was pulsed with a pulse width of 200 ns and a repetition period of 5 ms, and its light coupled into a 1×3 fiber coupler followed by fiber delay coils (of 70 m and 140 m lengths). Then, derived from each initial emitted pulse, three pulses were generated, separated in time due to the delay cause by the fiber coils, were then fed into the

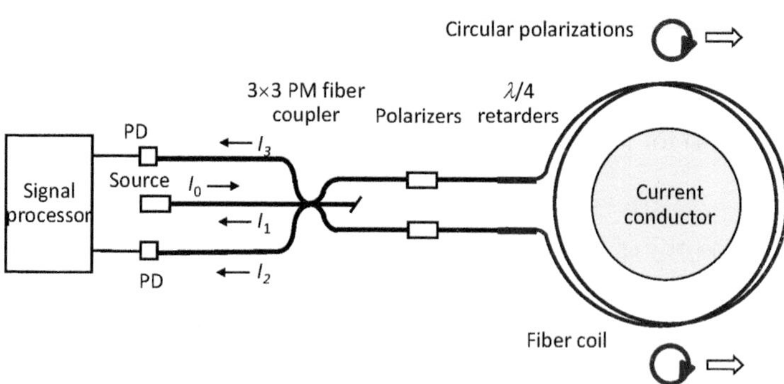

FIGURE 8.7 Sagnac interferometer current sensor with 3×3 PM fiber coupler (adapted from US patent 7,492,977B2 [432].)

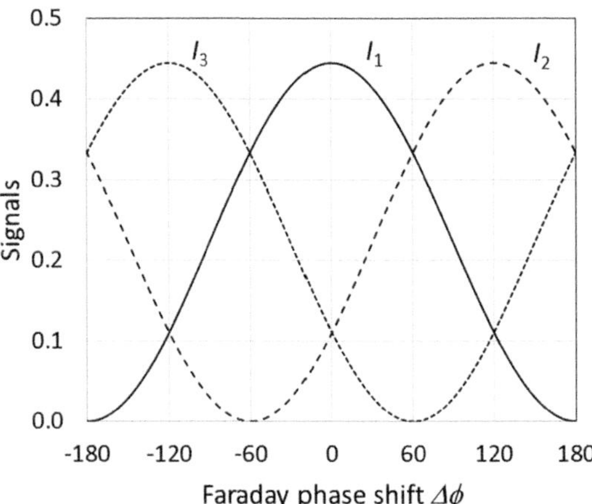

FIGURE 8.8 Signals I_1, I_2, and I_3 as a function of the Faraday phase shift $\Delta\phi$.

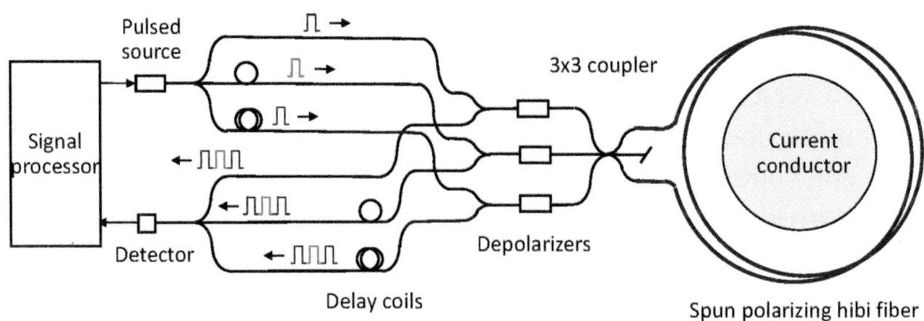

FIGURE 8.9 Sagnac interferometer current sensor with non-reciprocity compensation of the 3×3 fiber coupler (according to Matar et al. [431].)

source-side ports of the 3×3 coupler. Hence, each of the three returning interference signals consisted of a train of three pulses. Further delay lines (210 m and 420 m) separated the trains, and a single avalanche photodiode recorded the nine time-separated pulses. At a current of 400 A, the measurement error was less than ±0.2% between −25°C and 8°C [429, 431]. Sensing fiber coils with 100 and 300 windings were mentioned [429]. The temperature dependence of the Verdet constant was removed via a temperature measurement. The sensor was commercialized by a University of Sydney spin-off (Smart Digital Optics, SDO, founded in 2004) for applications in electric power transmission and industry.

8.5 FIBER-OPTIC CURRENT SENSORS WITH POLARIZATION-ROTATED REFLECTION

8.5.1 RFOCS with Non-Reciprocal Phase Modulation

Reflective fiber-optic current sensors (RFOCS) with non-reciprocal phase modulation, also referred to as in-line Sagnac interferometer current sensors, were derived from the phase-modulated Sagnac current sensor and arguably have become the most successful type of FOCS. At least half a dozen companies have developed corresponding commercial products. The underlying technique was first demonstrated by K. Goto in 1972 in the context of traveling wave light intensity modulators [433]. In 1987 A. Enokihara demonstrated a corresponding magnetic-optic field sensor, albeit with static phase biasing and a 50-mm long glass rod as Faraday medium [434]. In the early 1990s Frosio et al. at the University of Neuchatel, Switzerland [435, 436] and, independently, Blake et al. [437], then at Texas A&M University, demonstrated the first fiber-optic current sensors of this type, with both static and dynamic phase biasing. Further development by Blake et al. [309, 404–406, 438] and Bohnert et al. at ABB [326, 336, 439, 440] led to the first commercial applications in the 2000s. Since then, numerous further works on various sensor aspects have been published, e.g., [251, 254, 265, 266, 441–457].

Figure 8.10 illustrates a typical sensor setup [266]. Light from a broadband source first passes a depolarizer (not shown) and a fiber polarizer and then excites at a 45°-fiber splice the two orthogonal polarization modes of a PM fiber. A fiber-optic quarter-wave retarder at the near end of the sensing fiber coil converts the linear SOP into left and right circular polarizations. Upon reflection at the coil's far end, the circular waves swap their polarization states, i.e., left circular becomes right circular, and vice versa. The fiber retarder converts the returning circular waves back to orthogonal linear SOP, the polarization directions of which are also swapped versus the ones of the forward traveling waves. Finally, the orthogonal waves interfere at the fiber polarizer, and the interference signal is recorded at a photodiode. An integrated-optic birefringence modulator in a closed-loop circuit modulates the differential phase of the orthogonal waves, analog to the modulation of a Sagnac interferometer. The roundtrip Faraday phase shift is twice as large as in a Sagnac interferometer:

$$\Delta\phi = 4VNI. \tag{8.14}$$

Typically, the modulator is an integrated-optic lithium niobate modulator [458, 459]. Piezoelectric modulators in combination with open-loop detection have also been used,

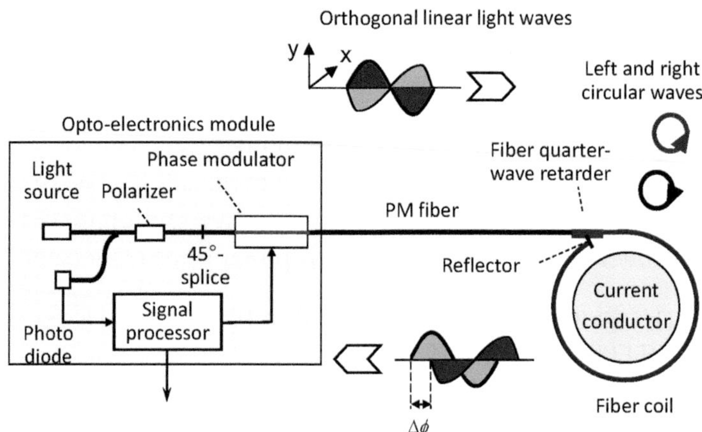

FIGURE 8.10 Fiber-optic current sensor with polarization-rotated reflection (© 2019 IEEE. Reprinted with permission from [266].)

however, e.g., in [436, 405, 336], but for the same modulation depth the modulator must provide much stronger actuation than in a Sagnac sensor. The retarder is commonly a short piece of appropriately oriented PM fiber with a length corresponding to a quarter of a beat length. Retarders from elliptical-core fiber are preferred due to their comparatively low temperature dependence. Retarders from polarization-transforming fibers have also been demonstrated [254, 276]. It should be noted that the orthogonal modes of the PM fiber link accumulate a differential optical path difference much longer than the source coherence length, i.e., the modes are incoherent when they arrive at the fiber coil, but they regain their coherence on the return trip due to the swapped polarization states. The differential group delay per meter of PM fiber is typically several hundred mm, whereas the source coherence length is about 30 mm.

It is obvious that the integrated-optic modulator is a critical component of the optical circuit. The waveguide is produced by in-diffusion of titanium into a lithium niobate substrate, which increases both the material's ordinary and extraordinary indices of refraction. The substrate is either x-cut or z-cut (that is, the crystallographic x-axis or z-axis is perpendicular to the plane of the substrate), and the waveguide is in the y-direction. Naturally, the crystal cut determines the arrangement of the electrodes relative to the waveguide. Particularly for z-cut substrates, the electrodes reside on transparent and conducting buffer layers to prevent optical loss and buildup of pyroelectric charges on the z-faces. The half-wave voltage, i.e., the voltage to produce a p-phase shift between the orthogonal polarization modes, commonly amounts to a few volts; for details, see [458, 459] and references therein. A parameter important to the FOCS signal stability is the polarization extinction ratio (PER) of the modulator and its stability versus temperature and time. The PER is defined as

$$\mathrm{PER}(\mathrm{dB}) = 10\log_{10}\left(P_0/P_c\right), \tag{8.15}$$

where P_0 is the optical power injected into a given polarization mode, and P_c is the power cross-coupled into the orthogonal mode. Cross-coupling mainly occurs at the fiber-to-waveguide junctions. By comparison, cross-coupling within the waveguide is negligible. It can be shown that crosstalk at the source-side junction is largely uncritical, whereas crosstalk at the coil-side junction enhances the apparent magneto-optic phase shift but also generates incoherent light that does not contribute to the signal [449, 454, 460]. If the sensor signal is to remain stable to within 0.1%, the modulator PER must remain above 32 dB within the optoelectronic unit's temperature range of operation, e.g., from –25°C to 65°C. Off-the-shelf modulators from the telecom industry were found to have an extinction ratio of only around 23–26 dB at room temperature. Design adaptions and improved assembly procedures consistently increased the PER of the fiber-to-waveguide junctions to values above 35 dB between –25°C and 65°C [460]. In contrast to PER changes, variations in the modulator's half-wave voltage with temperature do not affect the signal, as the closed-loop signal processing continuously recalibrates the voltage in terms of optical phase via the 2p-resets of the phase ramp.

FOCS with polarization-rotated reflection have resolved persistent shortcomings of earlier sensor types:

- The sensor is highly immune to mechanical shock and vibration [336, 405]. Compared to the Sagnac sensor of Figure 8.4, for instance, the sensitivity of the PM lead fiber is inherently lower by factors of 100 to 1000 depending on the type of PM fiber, as the differential phase of the orthogonal waves is much harder to disturb than the phases of the two individual waves of the Sagnac sensor. Moreover, in both cases counter-propagating beams help to cancel phase perturbations down to a level determined by the location and frequency spectrum of the disturbance.
- The signal processing technique derived from fiber gyros provides high accuracy not only for ac but also dc due to superior scale factor and zero-point stability.
- Closed-loop detection results in a linear signal up to magneto-optic phase shifts of at least ±p, corresponding to about ±780 kA at 1310 nm for single fiber loop.
- The fiber retarder can be adjusted for balancing the temperature dependence of the fiber coil. This way, sensor accuracy to within ±0.1% between −40°C and 85°C has been achieved without an extra temperature sensor (Section 8.6) [265, 266, 336].
- A minimum of optical components makes possible economic sensor fabrication.

Figure 8.11 shows an alternative implementation of the optical circuit [326, 439]. Here, a MIOC (multi-functional integrated optical chip) lithium niobate modulator known from fiber gyroscopes replaces the birefringence modulator of Figure 8.10. The modulator waveguides are produced by proton exchange with subsequent high temperature annealing [458, 459]. Replacement of lithium ions by protons increases the extraordinary index and leaves the ordinary index essentially unchanged. As a result, the waveguides are polarizing

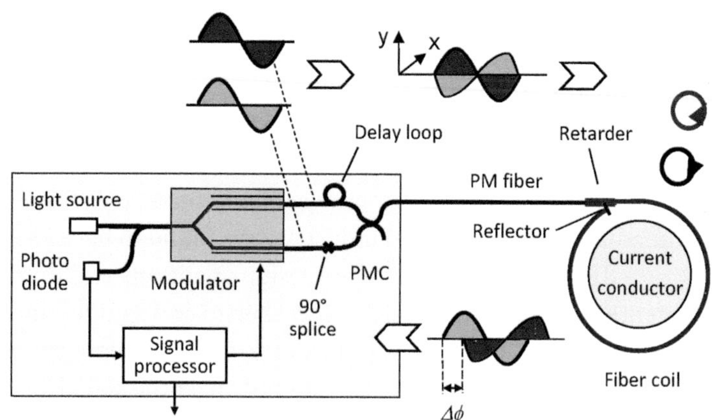

FIGURE 8.11 Reflective fiber-optic current sensor with gyro-type modulator (MIOC); PMC: polarization-maintaining fiber coupler (© 2007 IEEE. Reprinted with permission from [326].)

with a PER typically above 50 dB. A fiber polarizer is no longer needed. A polarization-maintaining fiber coupler, PMC (alternatively a 2×1 integrated-optic splitter), with a 90°-splice in one of the two fiber branches after the MIOC combines the two parallel linear SOP to two orthogonal SOP in the PM fiber link to the coil. Light waves that have traveled on reciprocal paths upon their return to the modulator interfere at the waveguide junction. By contrast, light waves on non-reciprocal paths return with SOP that are blocked by the modulator. Moreover, they accumulate a relative path delay that renders them incoherent [326]. The configuration makes it possible to operate a FOCS with a virtually unmodified fiber gyro module. Other than with a birefringence modulator, there is no risk of signal deterioration by insufficient modulator PER, as the modulator blocks returning parasite waves or they become incoherent. For the same reasons, the configuration also enables the use of PM fiber connectors with modest PER in the PM fiber arms after the MIOC without negative effect on the signal stability [454].

A certain drawback of the above sensor configurations is the need for a PM fiber between the opto-electronic unit and the fiber coil. Distances well above 100 m are preferably avoided. Also, a PM fiber makes fiber splicing in the field or the use of fiber connectors more demanding or may exclude connectors altogether in high-end applications [454, 460]. The sensor version of the next section is an alternative that works with a static phase bias and employs a standard single-mode fiber link.

8.5.2 RFOCS with Static Phase Bias

Static phase biasing by means of a quarter-wave retarder is an economic alternative to an integrated-optic phase modulator. In the basic setup, the orthogonal polarization modes returning from the fiber coil reflect off a beam splitter and pass the biasing retarder platelet before they interfere at a polarizer [435]. Commonly, sensors with static phase biasing are less accurate, though, as packaging stress and its variation with temperature and time may result in bias drift and changes in fringe visibility. This particularly concerns dc measurement but, to a lesser degree, is also true for ac. In a sensor with one fiber loop, a bias uncertainty of 1°, for instance, corresponds to a dc current uncertainty of 4.3 kA at 1310 nm.

To address the issue, Bohnert et al. have developed a sensor that employs an integrated-optic 1×3-polarization-splitter (IOPS) for passive interrogation of the coil and demonstrated accuracy for AC between −40°C and 85°C on par with corresponding phase modulated sensors [266, 328]. Figure 8.12 depicts the sensor setup. The depolarized light from a SLED travels through a single-mode fiber to the IOPS, which is part of the sensor head assembly. A thin polarizer platelet at the source-side facet of the IOPS (P1, 25-m thick) polarizes the light at 45° to the principal axes of a short PM fiber link to the coil, i.e., both orthogonal polarization modes of the fiber are again excited with equal amplitude. A principal fiber axis is orthogonal to the splitter plane. After a roundtrip through the coil, the two orthogonal modes again return with swapped polarization states and interfere at polarizers P1 and P2. The two interference signals travel through single-mode fibers to the opto-electronics unit. A quarter-wave retarder platelet (QWR) in the source channel of the IOPS introduces the p/2-phase-bias between the interfering waves.

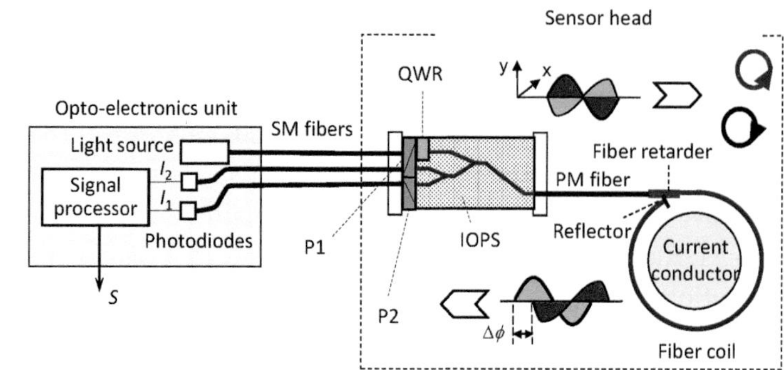

FIGURE 8.12 Reflective fiber-optic current sensor with integrated-optic polarization splitter (IOPS); P: polarizer, QWR: quarter-wave retarder (© 2019 IEEE. Reprinted with permission from [266].)

The IOPS is derived from a commercial 1×3-planar-lightwave-circuit (PLC) splitter from the telecom industry with waveguides produced in a glass substrate by means of electric field-assisted ion exchange [461]. Even though the waveguides are weakly birefringent, the roundtrip differential phase retardation between the orthogonal polarization states in the IOPS is small (less than 2° – 3°) due to the polarization-rotated reflection at the coil end. The quartz QWR platelet is of zeroth or first order (the corresponding thicknesses are 38.7 m and 116 m, respectively) and resides in a recess in the source-side IOPS facet. Embedding in a soft gel prevents stress from affecting the bias; see [328] for further details.

The difference of the two (pre-normalized) photodiode signals I_1, I_2 divided by their sum gives a normalized final signal from which the magneto-optic AC phase shift *Df* is derived [328]:

$$S = kK\cos\left(\Delta\theta/2\right)\sin\left(\Delta\phi+\theta\right), \tag{8.16}$$

with $\theta = (\theta_1 + \theta_2)/2$ and $\Delta\,\theta = \theta_1 - \theta_2$. Here, θ_1, θ_2 are the deviations from a perfect p/2-phase bias in the two channels. Deviations originate in the QWR temperature dependence $((\partial\theta_i/\partial T) = -0.0274$ deg/°C for a first-order retarder) and residual IOPS waveguide contributions. The parameter k is given as

$$k = \left[1 + K\cos\left(4VNI+\theta\right)\sin\left(\Delta\theta/2\right)\right]^{-1}, \tag{8.17}$$

where $K = K_s K_c$ is the fringe contrast with contributions K_s and K_c from the IOPS and fiber coil, respectively ($K_c = 1$ for an ideal coil). For small magneto-optic phase shifts ($Df \ll 1$) and $\theta_1 = \theta_2$, (8.16) reduces to

$$S = K\left(\Delta\phi+\theta\right). \tag{8.18}$$

The phase angle θ and hence the dc offset of S vary linearly with temperature. (The variation of K_s is negligible.) In case of AC currents, the dc offset can be utilized to compensate any thermal scale factor variation associated with the IOPS [328]. An important benefit of the sensor is the standard single-mode fiber link instead of a PM fiber between the opto-electronics and sensor head. Single-mode fibers facilitate splicing during field installation, are less sensitive to external stresses, and allow for longer distances.

In similar work, Min-Cheol Oh and co-workers at Pusan National University in Korea developed a polymer integrated-optic chip for sensor interrogation. The chip unites the functions of directional couplers, polarizers, and quasi-static thermo-optic phase biasing [462–464]. The group also reported a chip with an additional multimode interference coupler that generated two transfer functions at quadrature [465, 466]. H. Lin et al. at the National Sun Yat-Sen University, Taiwan, reported a polarimetric sensor with polarization-rotated reflection, where a 22.5°-Faraday rotator after the coil's quarter-wave retarder generated the static 90°-phase bias [467–469]. The group also demonstrated time-division multiplexing of several coils [470]. And in a 2005 patent application, J. Blake positioned the rotator at the source-side end of the PM fiber link [471]; a theoretical analysis followed in [472].

8.6 TEMPERATURE COMPENSATION OF FOCS WITH POLARIZATION-ROTATED REFLECTION

The above equations for the magneto-optic phase shift *Df* or sensor signal *S* assume an ideal fiber coil, i.e., the circular birefringence of the Faraday effect is the only birefringence, and the fiber retarder at the coil entrance is a perfect quarter-wave retarder. In practice, linear birefringence from coiling a nominally low birefringence fiber or the embedded linear birefringence of a spun fiber modify the detected phase shift. Likewise, a retarder that deviates from perfect p/2-retardation (or is imperfectly oriented with the respect the PM fiber axes) also modifies the signal. Moreover, the variation of the retarder and sensing fiber parameters with temperature may cause considerable temperature dependence of the signal adding to the contribution of the Verdet constant. The following shows how the retarder can be employed to balance the overall temperature dependence of the fiber coil. The coils include both, coils made from low birefringence fiber with bend-induced birefringence and coils from spun fiber. We will see that there are significant differences in this regard between phase-modulated FOCS and FOCS with static phase bias. For simplicity, we call the two sensor types according to their detection techniques *Interferometric FOCS* and *Polarimetric FOCS*, respectively, also since the interferometric FOCS roots in the interferometric Sagnac sensor.

8.6.1 Compensation of Low Birefringence Fiber Coils

The Jones matrix calculus gives for the *interferometric FOCS* of Figures 8.10 and 8.11 and a coil of low birefringence fiber a magneto-optic phase shift [326]

$$\Delta\phi = \arctan\frac{2T}{1-T^2}. \tag{8.19}$$

Here, the parameter T stands for the ratio $T = j'_{12}/j''_{12}$, where j'_{12} and j''_{12} are the real and imaginary parts, respectively, of the matrix element j_{12} $\left(j_{12} = -j^*_{21}\right)$ of the system Jones matrix J (in a linear base) that describes a roundtrip of the light through the fiber coil including the fiber retarder; stated differently, $\Delta\phi$ corresponds to the difference in the arguments of the output optical fields $E_x^{(\mathrm{out})}$, $E_y^{(\mathrm{out})}$ parallel to the birefringent axes, x and y, of the PM fiber lead. T is obtained as

$$T = \frac{2VNI\dfrac{\tan\gamma}{\gamma}}{\cos\varepsilon - \delta_b \sin\varepsilon \cos(2\alpha)\dfrac{\tan\gamma}{\gamma}}, \tag{8.20}$$

$$\gamma = \left[(\delta_b)^2 + (2VNI)^2\right]^{1/2}. \tag{8.21}$$

Here, ε is the deviation of the fiber retarder with retardation r from perfect p/2-retardation ($\varepsilon = r - \mathrm{p}/2$), δ_b is the (one way) bend-induced birefringent phase retardation of the coil, and a is the angle between the slow retarder axis and slow axis of the bend-induced birefringence (coil normal). (At this point, all other parameters of the optical circuit are assumed as ideal). For small linear birefringence and small magneto-optic phase shifts ($g \ll 1$) and a perfect p/2-retarder ($\varepsilon = 0$), (8.19) reduces to [309, 326]

$$\Delta\phi = 4VMI\left(1 + (1/3)\delta_b^2\right). \tag{8.22}$$

We now first consider temperature compensation for the simple case of a fiber coil with *negligible* linear birefringence. An example is a single large-diameter fiber loop for the measurement of high dc at aluminum smelters [326]. Hence, only the Verdet constant and the retarder change with temperature. With $4VNI \ll 1$, $\Delta\phi$ is then given by [309, 326, 336, 435]

$$\Delta\phi = 4VNI/\cos\varepsilon. \tag{8.23}$$

Figure 8.13a shows the small signal scale factor SF normalized to the scale factor $4VNI$ of an ideal coil as function of the retarder retardation r:

$$\mathrm{SF} = \frac{\Delta\phi}{4VNI} = \frac{1}{\cos\varepsilon} = \frac{1}{\cos(\rho - \pi/2)}. \tag{8.24}$$

At sufficiently small ε, the scale factor follows a parabola centered at $r = 90°$ ($e = 0°$). We assume an elliptical-core fiber retarder of zero order with a temperature coefficient $(1/r)\,\partial r/\partial T = -2.2 \times 10^{-4}\ \mathrm{K}^{-1}$ [326] and set the room temperature retardation r_0 to 100.4°. The red curve segment then shows the variation of the scale factor by the retarder between −40°C and 90°C. This variation balances to first order the linear increase of the Verdet constant with temperature; $(1/V)\,\partial V/\partial T = 0.7 \times 10^{-4}\ \mathrm{K}^{-1}$ [28] (Figure 8.13b). The residual scale factor change due to the retarder's second-order contribution is only 0.06% over the given temperature range compared to 0.91% without compensation.

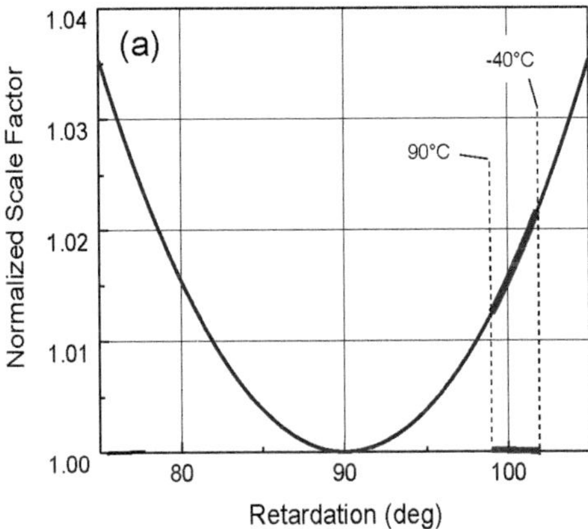

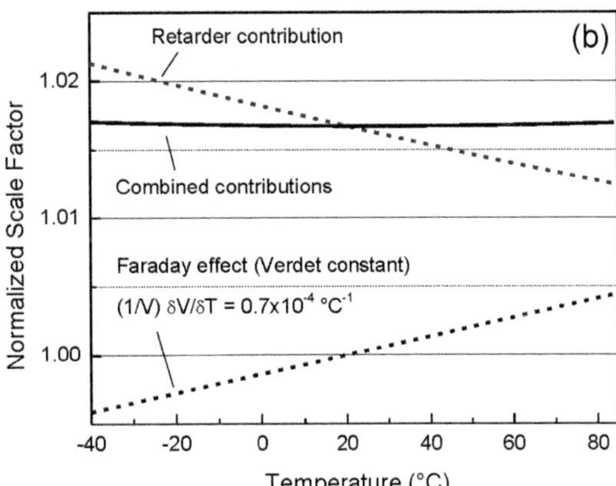

FIGURE 8.13 (a) Normalized scale factor versus retardation r of fiber retarder for a sensing fiber free of linear birefringence. A retarder with a room temperature retardation of 100.4° decreases the scale factor between −40°C and 90°C along the red curve segment. (b) The retarder thus balances to first order the increase of the Verdet constant with temperature (© 2002 IEEE. Reprinted with permission from [336].)

In case of *non-negligible* bend-induced birefringence d_b, the scale factor becomes a function of the retarder orientation a; eq. (8.19). Figure 8.14a shows the scale factor versus r for a low-bi fiber coil with d_b = 16° and three characteristic retarder orientations. d_b = 16° corresponds to a coil with four 172-mm-diameter windings and a fiber diameter of 80 m [328]. The scale factor for d_b = 0 is indicated for comparison (black dotted curve). At retarder orientations of a = 0°, 90°, bend-induced birefringence shifts the parabola vertices to r = 90° − d_b and r = 90° + d_b, respectively, i.e., the coil birefringence adds to or subtracts from the retarder birefringence. The scale factors at the vertices are below unity which reflects the quenching of the Faraday effect by the coil linear birefringence. The curves for a = ±45° coincide and are centered about r = 90° for reasons of symmetry. It is obvious that all parabolas must intersect at r = 90°, where the injected SOP are circular. Temperature compensation works like for d_b = 0 if the room temperature retardation is adjusted accordingly, as the red curve segment indicates for a = 0°; here with $(1/r)\ \partial r/\partial T = -3.2 \times 10^{-4}\ \mathrm{K}^{-1}$ [328]. The coil birefringence $d_b(T)$ also contributes to the overall temperature dependence but only weakly [180], and is included in the compensation.

The corresponding normalized scale factor of the *polarimetric FOCS* is (Figure 8.14b) [328]

$$\mathrm{SF} = \frac{1}{2}\left(\cos\varepsilon\,\frac{\sin(2\gamma)}{\gamma} - 2\delta_b \sin\varepsilon\,\cos(2\alpha)\,\frac{\sin\gamma^2}{\gamma^2}\right). \tag{8.25}$$

For d_b = 0°, SF reduces to

$$\mathrm{SF} = \cos\varepsilon. \tag{8.26}$$

Note that the scale factor parabolas of the polarimetric sensor are essentially inverted versions of the parabolas of the interferometric sensor. The red curve segment in Figure 8.14b

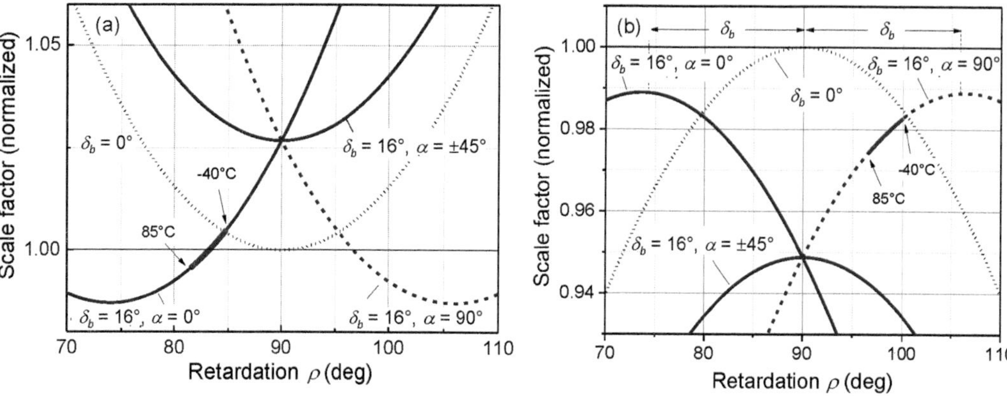

FIGURE 8.14 Normalized scale factor versus retardation of fiber retarder for a low-bi fiber coil with bend-induced linear birefringence of δ_β = 16° and three characteristic angles a between the slow retarder axis and fiber coil normal: (a) interferometric FOCS, (b) polarimetric metric FOCS. The solid (red) curve segments indicate the scale factor variation between −40°C and 85°C due to the temperature dependence of the fiber retarder which balances the temperature dependence of the Faraday effect in temperature-compensated sensors. For comparison, the scale factors for δ_β = 0° are also indicated (© 2019 IEEE. Reprinted with permission from [328].)

indicates temperature compensation for a = 90° with the same retarder and fiber parameters as in Figure 8.14a.

The curves of Figure 8.14 reflect the interference of several pairs of orthogonal waves (again, the term *wave* is to include the whole source spectrum). A fiber retarder deviating from 90° partially cross-couples the orthogonal polarization states both on the forward and return path. As a result, not only one pair of orthogonal SOP returns from the coil but altogether four pairs, as Figure 8.15 illustrates [328, 336, 454]. Two pairs return with swapped polarization states and are coherent, whereas the other two pairs return with unswapped polarizations and therefore have become incoherent. The primary waves (A_x, A_y) have not cross-coupled at the retarder but return with reduced amplitude of $\cos^2(\varepsilon/2)$ (at δ_b = 0); assuming an amplitude of unity for the original parent waves and neglecting other optical loss. The pair of secondary waves (B_x, B_y) with amplitude $\sin^2(\varepsilon/2)$ results from twofold cross-coupling at the retarder during a roundtrip and is 180° out of phase with the primary waves (A_x, A_y). *A*-waves and *B*-waves experience Faraday phase shifts $2\varphi_F$ ($\varphi_F = VNI$) of opposite direction as indicated by the arrows. This, in combination with the 180°-phase difference, enhances the apparent magneto-optic phase shift $\Delta\Phi$. Hence, the scale factor increases in the interferometric sensor as ρ moves away from a parabola vertex (Figure 8.14a).

By contrast, in the polarimetric sensor, the same effects, in combination with the static p/2-phase bias, reduce the slope of the signal *S*(*I*), that is, the scale factor drops as ρ moves away from a parabola vertex, as seen in Figure 8.14b. The two pairs of *C*-waves (C_x, C_y) with amplitude $\cos(\varepsilon/2)\sin(\varepsilon/2)$ represent light that has cross-coupled only once, either on the forward or backward path. The *C*-waves accumulate group delays of $DL_g = \pm lDn_g$ with respect to the *A* and *B* waves, where l is the PM fiber length and Dn_g the group linear birefringence; in addition, there may be the delays by the modulator or IOPS. The *C*-waves

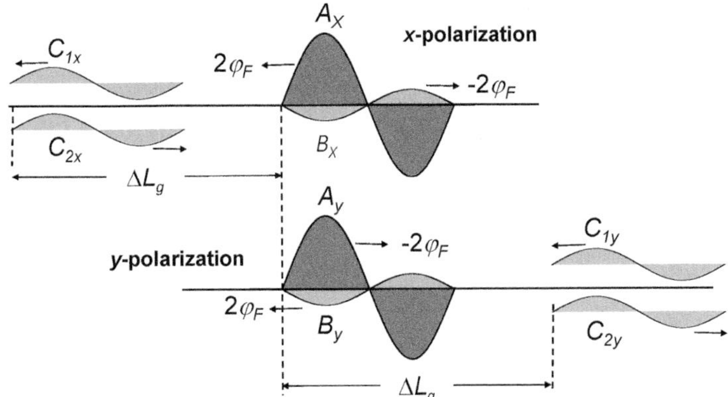

FIGURE 8.15 Interfering light waves at zero phase bias and zero current in case of a fiber retarder deviating from p/2. The arrows indicate the magneto-optic phase shift directions at applied current with $j_F = VNI$ (© 2019 IEEE. Reprinted with permission from [328].)

represent incoherent background light that reduces in both sensor versions the fringe visibility [328, 336].

Therefore, the enhancement of the apparent magneto-optic phase shift of the interferometric FOCS by a non-p/2-retarder does not improve but does somewhat reduce the signal-to-noise ratio. As noted, the curves in Figure 8.14 depict the small signal scale factor ($4VNI \ll 1$). Sensors in electric power transmission typically operate in this regime. At larger magneto-optic phase shifts the temperature compensation effect weakens — the parabolas flatten out — and vanishes at $4VNI = \text{p}$ [326, 441, 473]. Moreover, it was assumed that the fiber retarder axes are perfectly aligned at 45° to the PM fiber axes. Misalignment by an angle Da, e.g., due to fabrication tolerances, changes the normalized scale factor in proportion to 1/cos(2Da) in interferometric FOCS and cos(2Da) in polarimetric FOCS [436, 445]. This scale factor change is independent of operation conditions, however, and inherently accounted for by the sensor calibration.

Figure 8.16a shows the measured signal as a function of the coil temperature of a temperature compensated *polarimetric FOCS* (low-bi fiber coil with four 172-mm-diameter fiber windings, 80-m-fiber, packaged in a fused-silica capillary, as illustrated in Figure 7.11 [328]. The signal is independent of temperature within about ±0.03% between −40°C and 85°C and agrees well with the theoretical signal; solid (red) line. Figure 8.16b shows the signal versus time during an endurance test. Here, the polarization splitter and fiber coil were jointly exposed to temperature cycles between −45°C and 85°C over more than 200 hours, including 48-hour dwelling periods at the extreme temperatures. The signal remains within ±0.05% without any noticeable drift.

Figure 8.17 shows for the same sensor the measurement error versus current up to about 1400 A. Each data point represents the average of 50 measurements, each measurement over ten 60-Hz current cycles, that is, ten 167-ms time periods. The error bars indicate the standard deviation ($\pm 1\sigma$) of the 50 measurements. For four fiber windings, the corresponding noise-equivalent current is about 10 mA/$\sqrt{\text{Hz}}$. This corresponds to a minimum detectable magneto-optic phase shift of 0.16 μ rad/$\sqrt{\text{Hz}}$.

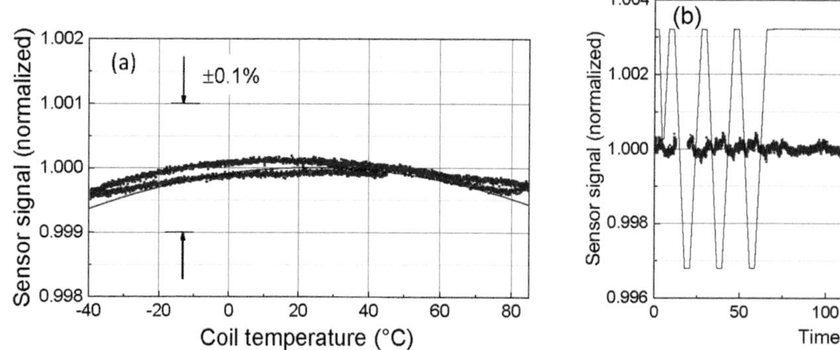

FIGURE 8.16 Polarimetric FOCS with a temperature-compensated low-bi fiber coil: (a) Signal vs. coil temperature at a constant alternating current (1000 A, 60 Hz). The solid (red) line represents the calculated signal; (b) Signal vs. time during repeated temperature cycling of sensor head including polarization splitter and fiber coil between −45°C and 85°C (© 2019 IEEE. Reprinted with permission from [328].)

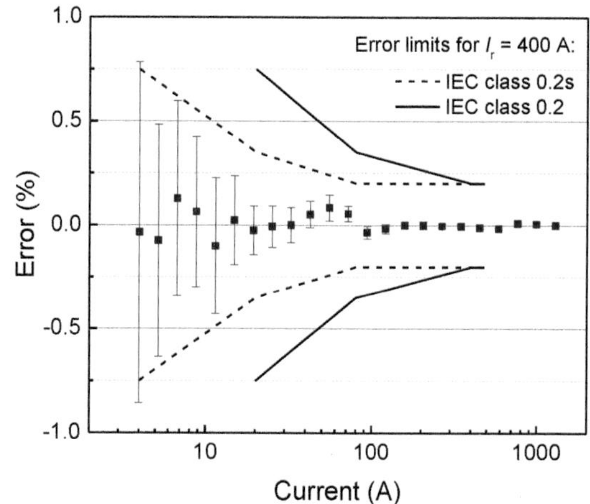

FIGURE 8.17 Measurement error vs. current and error limits according to IEC metering classes 0.2s and 0.2. The error bars represent the standard deviation (±1σ) of 50 individual measurements of ten 60-Hz current cycles (© 2019 IEEE. Reprinted with permission from [328].)

Typical rated currents in electric power transmission are between 400 A and 6000 A. Electricity metering according to IEC metering class 0.2s [474] requires that the sensor is accurate to within ±0.2% between 20% and 120% of the rated current. Allowed phase errors are ±10 min. Somewhat larger errors are allowed between 1% and 20% of the rated current. The dashed and solid lines in Figure 8.17 indicate the amplitude error limits of class 0.2s and the less strict class 0.2, respectively, for a rated current of 400 A. The error limits are the same at higher rated currents but with the current range shifted in proportion.

At normal operating conditions, the magneto-optic phase shift *Df* remains small, and linearization of the sinusoidal response is not necessary. By contrast, fault currents, e.g., in case of a short circuit, may well exceed 100 kA and require linearization of the sine function. This, however, is not particularly critical, since typical error limits in that regime are ±5% (IEC protection class 5P [474]). Tests showed that with linearization the sensor stays within the ±0.2%-accuracy limits up to magneto-optic phase shifts of ±(35°–45°). The ±5%-error margins are reached only beyond ±75° (corresponding to 327 kA turns). Equivalent results for temperature-compensated interferometric FOCS can be found in [326, 327].

8.6.2 Compensation of Spun Highly Birefringent Fiber Coils

Interferometric and polarimetric FOCS with spun fiber exhibit the same basic phenomena as with low-bi fiber. However, polarimetric sensors are more sensitive to variations of the fiber's linear birefringence with temperature. Whereas interferometric sensors can be prepared such that the temperature dependence of the birefringence does not appear in the signal, this is not possible for polarimetric sensors. Therefore, a careful choice of adequate fiber parameters is particularly critical for polarimetric sensors.

At a given spin pitch p, the scale factors of both sensor versions and their variation with temperature are governed by the Verdet constant $V(T)$, the embedded linear beat length $L_B(T)$, and the orientation a of the slow retarder axis relative to the slow spun fiber axis at the retarder-side fiber facet. (We assume that bend-induced birefringence is sufficiently quenched and can be disregarded; see Figure 7.9.) Assuming $4VNI \ll 1$, the magneto-optic phase shift of the *interferometric FOCS* is [266]

$$\Delta\phi = \frac{4VNI}{\sqrt{1+\left(\sin\left(2\alpha/x\right)\right)^2}\cos\left(\epsilon+\tan^{-1}\left[\sin\left(2\alpha/x\right)\right]\right)}, \tag{8.27}$$

where $x = 2L_B/p$ is the spin ratio introduced in Section 7.5. Furthermore, we assume that the spun fiber's quasi eigenmodes accumulate a group delay much larger than the source coherence length in order to suppress disturbing interference effects from secondary polarization states generated upon reflection of the modes at the far end of the fiber [256]. For spun fiber orientations $\alpha = \pm 45°$, (8.27) reduces to

$$\Delta\phi_{\alpha=\pm45°} = \frac{4VNI}{\sqrt{1+x^{-2}}\cos\left(\epsilon\pm\cot^{-1}x\right)} \tag{8.28}$$

and for $a = 0°$ and $a = 90°$ to

$$\Delta\phi_{\alpha=0°,90°} = \frac{4VNI}{\cos\varepsilon}. \tag{8.29}$$

Figure 8.18a shows again the normalized scale factor, SF = $\Delta\phi/4VNI$, as a function of the retarder deviation e from p/2 for the four characteristic spun fiber orientations a. The assumed spin ratio of $x = 4$ provides good robustness against fiber bending with only modest loss of sensitivity (see Figures 7.8 and 7.9). It is interesting to note that at $a = 0°$, 90° (8.29),

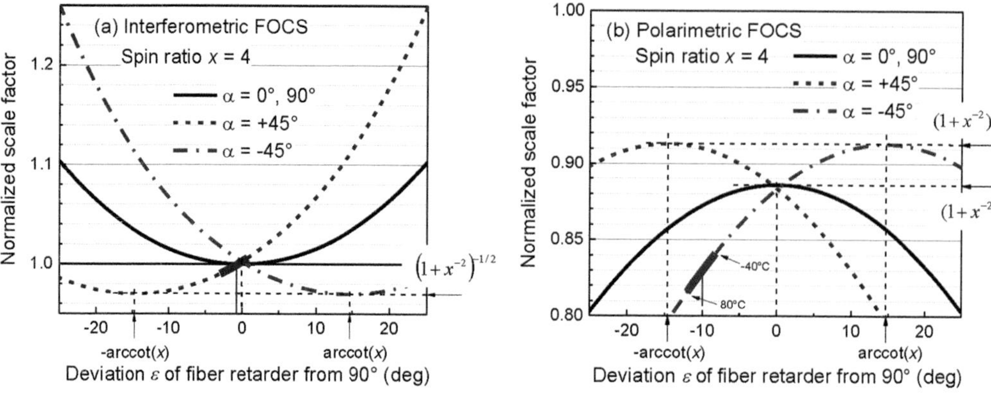

FIGURE 8.18 Scale factor vs. ε for a spun fiber coil with a spin ratio $x = 4$ and different angles α between the retarder and spun fiber slow axes: (a) Interferometric FOCS, (b) Polarimetric FOCS. The thick curve segments indicate the scale factor variation between −40°C and 80°C, which balances the variations of the Verdet constant and embedded linear fiber birefringence with temperature (© 2019 IEEE. Reprinted with permission from [266].)

a spun fiber provides the same magneto-optic phase shift as an ideal non-birefringent fiber. In particular, the scale factor corresponds to unity at $r = 90°$ ($e = 0$), and the fiber's embedded linear birefringence and its variation with temperature does not affect the signal. The similarity can be understood from the fact that (8.27) reduces to (8.29) as the spin ratio approaches infinity, which represents a non-birefringent fiber. However, a difference to an ideal fiber is that the retarder/spun fiber combination generates additional secondary light states that return with unswapped polarization directions and reduce the fringe contrast.

If the fiber orientation changes from $a = 0°, 90°$ to $a = \pm 45°$, the parabola minima shift from $\varepsilon = 0$ to $\varepsilon = \cot^{-1}x$. At those retardations, the left and right elliptical SOP emerging from the fiber retarder map one-to-one onto the quasieigenmodes of the spun fiber. Note, that the scale factor at the minima corresponds to the intrinsic magneto-optic sensitivity of a spun fiber according to (7.66): $SF_{\min,\,\alpha=\pm 45°} = (1 + x^{-2})^{-1/2}$. At all other α and ε, the SOP emerging from the retarder excite, besides the primary pair of eigenmodes, a secondary pair of modes with respective opposite handedness. The amplitude and phases of the secondary pair depend on α and ε [256]. The secondary pair again enhances the recovered magneto-optic phase shift through the mechanisms illustrated in Figure 8.15.

The small signal scale factor ($4VNI \ll 1$) of the *polarimetric sensor* with spun fiber reads [266]

$$\mathrm{SF} = \frac{1}{\left(1+x^{-2}\right)^{2}} \sqrt{1+\left(\sin\left(2\alpha/x\right)\right)^{2}} \cos\left(\epsilon + \tan^{-1}\left[\sin\left(2\alpha/x\right)\right]\right). \tag{8.30}$$

At spun fiber orientations $\alpha = \pm 45°$, this expression becomes

$$SF_{\alpha=\pm 45°} = \frac{1}{\left(1+x^{-2}\right)^{3/2}} \cos\left(\epsilon \pm \cot^{-1} x\right), \tag{8.31}$$

and at $\alpha = 0, 90°$, one obtains

$$\mathrm{SF}_{\alpha=0°,90°} = \frac{1}{\left(1+x^{-2}\right)^2}\cos(\epsilon). \tag{8.32}$$

In the limit of an infinite spin ratio x, the scale factor reduces to cos(ε), that is, the scale factor of a polarimetric FOCS with low-bi fiber. Compared to the interferometric FOCS, the scale factor parabolas of the polarimetric version (Figure 8.18b) are again inverted. But in contrast to the interferometric sensor, the scale factor depends on the spin ratio x at all fiber orientations a, i.e., the intrinsic linear birefringence always contributes to the temperature dependence of the signal. The vertices at $\alpha = \pm 45°$ are again located at $\varepsilon = \cot^{-1}x$ and now correspond to a scale factor of $\mathrm{SF}_{\max,\alpha = \pm 45°} = (1 + x^{-2})^{-3/2}$, which is the product of two terms, namely the intrinsic magneto-optic sensitivity of the spun fiber, that is $(1 + x^{-2})^{-1/2}$, and the reduced fringe contrast, $(1 + x^{-2})^{-1}$. Other values of α and ε (for which the elliptical SOP from the retarder do not match the quasi eigenmodes of the spun fiber) further reduce the fringe contrast and scale factor. All curves intersect at the vertex at $\varepsilon = 0$ ($\alpha = 0°, 90°$) with a scale factor of $\mathrm{SF}_{\max,\,\alpha = 0°,90°} = (1 + x^{-2})^{-2}$.

Figure 8.19 shows the calculated contribution of the fiber's embedded linear birefringence to the temperature dependence of the two sensor versions for $\alpha = 0°, 90°$ and $\alpha = \pm 45°$. The beat length and spin pitch are $L_B = 10$ mm (at 20°C) and $p = 5$ mm, respectively, which gives a spin ratio of $x = 4$. The temperature coefficient of the beat length is 1.0×10^{-3}°C^{-1} (which is typical for spun fiber with stress-induced birefringence such as bowtie spun fiber; typically, L_B increases with temperature). As mentioned, the signal of the interferometric sensor is independent of the beat length at $\alpha = 0°$ and 90° (Figure 8.19a, solid line), and changes at a rate of 0.6×10^{-4}°C^{-1} at $\alpha = \pm 45°$ (dashed line). The corresponding slopes for the polarimetric sensor are 2.4×10^{-4}°C^{-1} and 1.8×10^{-4}°C^{-1}, respectively (Figure 8.19b).

Since the temperature coefficients of the beat length contribution (if any) and Verdet constant (0.7×10^{-4}°C^{-1}) are both of positive sign, the embedded fiber birefringence

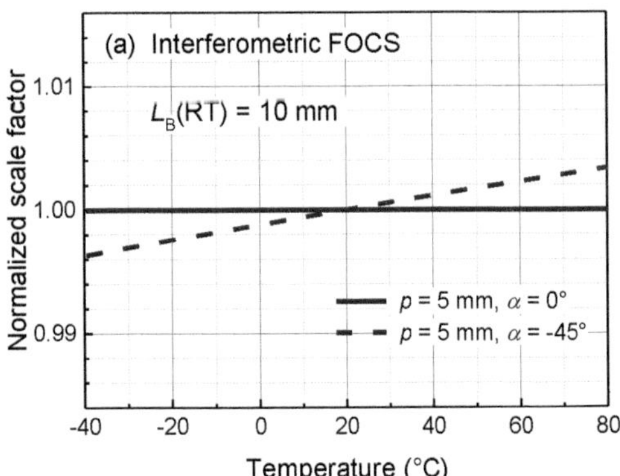

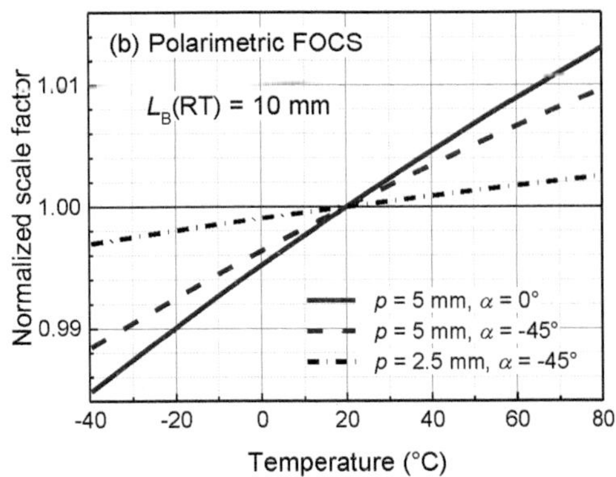

FIGURE 8.19 Contribution of the spun fiber linear beat length $L_B(T)$ to the temperature dependence of the scale factor: (a) interferometric and (b) polarimetric FOCS; RT: room temperature, 20°C (© 2019 IEEE. Reprinted with permission from [266].)

enhances the temperature dependence. It is obvious that thermal beat length variations affect the polarimetric sensor much more than the interferometric sensor. The temperature dependence of the polarimetric sensor can be minimized, however, by orienting the fiber at $\alpha = \pm 45°$. Reducing the spin pitch p at a given beat length reduces the temperature dependence, e.g., from $1.8 \times 10^{-4}°C^{-1}$ at $p = 5$ mm to $0.5 \times 10^{-4}°C^{-1}$ at $p = 2.5$ mm (dash-dotted line). Doubling the beat length from 10 mm to 20 mm and leaving the pitch at 5 mm has the same effect, but the fiber becomes more prone to bending; see [266] for experimental data. It is obvious that polarimetric sensors are preferably operated with spun fiber types such as elliptical-core fiber that have smaller temperature coefficients of the birefringence.

In spite of their differences, interferometric and polarimetric sensors with spun fiber can be temperature-compensated analog to the sensors with low-bi fiber. For $x = 4$, temperature coefficients for L_B and ρ of $1.0 \times 10^{-3}°C^{-1}$ and $-3.2 \times 10^{-4}°C^{-1}$, respectively, and fiber retarders of zero order, ε_{RT} must be set to $\varepsilon_{RT} = -1°$ ($\rho_{RT} = 89°$) in the interferometric sensor (with $\alpha = +45°$) and to $\varepsilon_{RT} = -10°$ ($\rho_{RT} = 80°$) in the polarimetric sensor (with $\alpha = -45°$). The heavy line segments in Figure 8.18 indicate again the corresponding scale factor variations between −40°C and 80°C due to the retarder that offset the thermal changes of the Verdet constant and embedded linear birefringence. Experimental results demonstrate temperature compensation for both sensors to within about ±0.1% between −40°C and 85°C [266].

8.7 FABRY-PÉROT RESONATOR AND LIGHT RECIRCULATING CURRENT SENSORS

Placing a magneto-optic material into a resonant Fabry-Pérot cavity can greatly enhance the Faraday polarization rotation [475–479]. Cavity-enhanced polarization rotation has been of interest, e.g., in the context of optical isolators [476, 477], but was also explored for magnetic field and current sensing [305, 479]. Already in 1964, Rosenberg et al. demonstrated 180-fold rotation increase in Freon gas (CCl_2F_2) at 662.8 nm in an 83-cm-long cavity with spherical mirrors of 98.9% reflectivity [475]. In an applied magnetic field, the Airy function of the cavity splits into two functions corresponding to the left and right circular constituents of linearly polarized input light. On resonance (i.e., the single pass phase shift in the absence of a magnetic field equals an integer multiple of p), the transmitted circular constituents again add up to a linear polarization state that is rotated against the input state by an angle [475, 478]

$$\varphi_F^{(\mathrm{FP})} = \arctan\left(\frac{1+R}{1-R}\tan\varphi_F\right), \tag{8.33}$$

where R is the power reflectivity of the cavity mirrors, and φ_F is the single-pass Faraday polarization rotation of (4.1). Hence, at small φ_F, the cavity enhances the Faraday rotation by a factor $(1 + R)/(1 - R)$. In the high reflectivity limit (R approaches $R = 1$), the accumulated small signal Faraday rotation approaches 45°. Without an analyzer, the cavity transmission is [478]

$$\frac{I_t}{I_0} = \frac{(1-R)^2}{(1-R)^2 + 4R\sin^2\varphi_F}, \tag{8.34}$$

and with an analyzer oriented at 0° to the entry polarizer, the transmission becomes

$$\frac{I_t}{I_0} = \frac{\cos\left(\varphi_F\right)^2}{\left(1 + F'^2 \sin^2 \varphi_F\right)^2}, \tag{8.35}$$

with $F' = 2\sqrt{R}/(1-R)$. As already noted in context of helically wound fibers, Maystre et al. reported a Fabry-Pérot current sensor consisting of a single helical fiber loop with an arc length of 17 cm, as shown in Figure 7.10a. Reflective coatings with reflectivity of 58% and 97% resulted in a measured cavity finesse of $F' = 6$ and a fourfold enhancement of the Faraday effect [305].

Wang et al. presented a fiber cavity formed by 100 m of standard single-mode fiber, birefringence-compensating Faraday rotator mirrors (FRM) at both ends, and an in-cavity fiber coupler for light incoupling and outcoupling. The cavity resulted in 6 dB signal gain. (Note that this cavity is not strictly a Fabry-Pérot resonator, however [480].) Other researchers aimed to increase the sensitivity by recirculating continuous wave (cw) or pulsed light multiple times through the fiber coil [481–485].

8.8 FIBER LASER CURRENT SENSORS

Polarimetric fiber laser sensors have been explored for the measurement of physical parameters such as lateral force, strain, temperature, torsion, or hydrostatic pressure [486–489]. The laser cavity may comprise separate fiber sections for gain and sensing, or the two functions may be combined in a single fiber. The laser emission is made up of two trains of longitudinal modes associated with the two orthogonal linear polarization eigenmodes. At zero net cavity birefringence, the two mode trains are degenerate. Birefringence induced by an external disturbance lifts the degeneracy and frequency-shifts the mode trains relative to each other. The frequency shift is detected as a polarization mode beat (PMB) frequency proportional to the magnitude of the disturbance. Preferably, at least one of the cavity reflectors is a narrow-band fiber Bragg grating in order to limit the number of active modes and improve the lasing stability. Distributed feedback (DFB) fiber laser sensors support only a single longitudinal mode at each polarization [490]. The easy-to-monitor frequency encoded output is an attractive feature of fiber laser sensors.

Fiber laser magnetic field and current sensors must operate with left and right circular eigenpolarizations rather than linear polarization states. To this end, B. Y. Kim and co-workers replaced one of the cavity reflectors by a Faraday rotator mirror (FRM) (Figure 8.20) [491–493]. With (single pass) FRM circular birefringence of p/2 and a cavity free of linear birefringence, the modes are LHC and RHC polarized throughout the cavity. In the presence of linear birefringence, the modes assume various degrees of elliptical polarization with correspondingly reduced magnetic field sensitivity; see [491] for a Jones matrix analysis. An Er^{3+}-doped fiber section before the FRM acts as a gain medium and is pumped by a 980-nm DFB laser via a wavelength division multiplexer (WDM, 980/1550 nm). A second, unpumped Er^{3+}-doped fiber section at the other side of the WDM acts as saturable absorber and serves for magnetic field sensing [492]. The standing wave patterns in the cavity lead to gain and absorption gratings in the respective doped fiber sections.

According to the authors, nonlinear gain and loss at those gratings reduce the number of active longitudinal modes to one per circular SOP. Due to the relative phase shift at the FRM, the LHC and RHC longitudinal modes trains are displaced against each other by half a free spectral range at zero magnetic field. The corresponding frequency difference is $f_{SFR}/2 = c/4nL$ and constitutes the zero-field PMB frequency; see bottom part of Figure 8.20; (c is the speed of light, n is the effective refractive index, and L is geometrical length of the cavity). An applied magnetic field increases the FSR for one circular SOP and decreases the FSR for the other and, depending on the field direction, accordingly increases or decreases the PMB frequency. The PMB frequency is monitored after a polarizer outside the cavity. The PMB frequency shift as a function of the current is $Df_{PMB} = [(NVc)/(nLp)]I$, assuming that the sensing fiber section forms an N-windings coil around the conductor (V is again the Verdet constant of the fiber). A laser sensor setup as shown in Figure 8.20 with a cavity length of 9 m resulted in a free spectral range of 11 MHz and a PMB frequency shift of 8 kHz/$A_{p\text{-}p}$ with a solenoid of 1528 turns. The noise equivalent current was 460 $mA_{rms}/\sqrt{Hz}$ at 60 Hz [492]. In other works, Takahashi et al. demonstrated a flint glass fiber ring laser for current sensing [494], and Cheng et al. reported a DFB fiber laser for magnetic field sensing [495, 496]. In spite of the attractive frequency-encoded output, laser current sensors have had little practical impact. Particularly, long-term stable operation under field conditions has remained unproven.

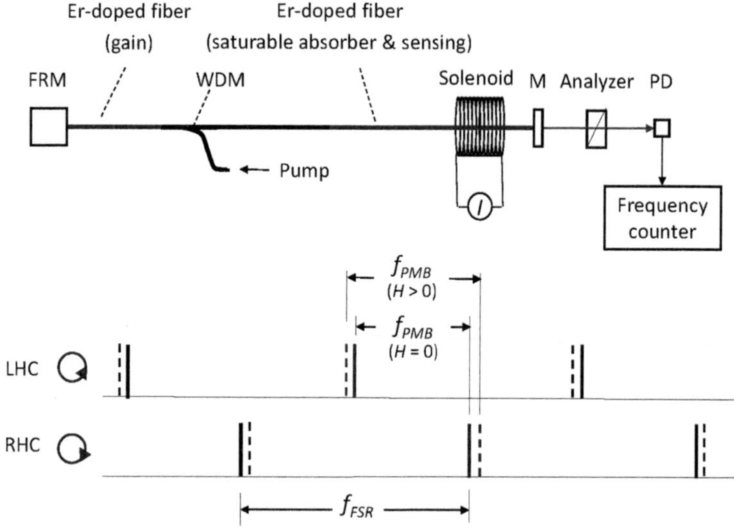

FIGURE 8.20 Fiber laser current sensor (top) and *LHC* and *RHC* polarized longitudinal laser modes (bottom) with and without applied magnetic field (dashed and solid lines, respectively); active lasing modes are indicated in red. FRM: Faraday rotator mirror, M: mirror, PD: photodiode, WDM: wavelength division multiplexer, f_{FSR}: free spectral range, f_{PMB}: polarization mode beat frequency (adapted with permission from [491, 492] © The Optical Society.)

8.9 INTEGRATED-OPTIC CURRENT SENSORS

Integrated optical current sensors found some interest in the early and mid-1990s as an alternative to twisted fiber coils [497–501]. The single-mode waveguides were formed in a glass substrate (e.g., from soda-lime silicate or borosilicate glass) by means of electric field assisted ion exchange, a technique also used for the fabrication of PLC splitters [461, 502]. For current sensing, the waveguide encircles a central aperture in the substrate for the current conductor. Straight waveguide sections to the substrate edges allow for fiber coupling. The architecture may include a waveguide spitter and polarizers for the generation of two polarimetric anti-phase signals [497]. Again, sufficiently low linear birefringence of the waveguides ($Dn < 10^{-6}$) constituted a practical challenge. Minier et al. reduced the initial birefringence (around 10^{-5}) to about 0.5×10^{-6} by thermal annealing and achieved 90% sensitivity of an ideal waveguide. That sensor had a single waveguide loop with a radius of 20 mm [500]. Further development reduced the birefringence of multi-loop sensors to $Dn < 0.5 \times 10^{-7}$ at waveguide losses of <0.15 dB/cm [501]. But with the advent of good quality spun fibers, interest in integrated optical sensors soon waned.

8.10 FARADAY EFFECT IN FIBER BRAGG GRATINGS

In a magnetic field, the Bragg wavelengths λ_B of a fiber grating slightly differ for left and right circularly polarized light as a result of the circular birefringence, $Dn = n_R - n_L$, induced by the Faraday effect [503]:

$$\lambda_{B,R} = 2n_R\Lambda, \tag{8.36}$$

$$\lambda_{B,L} = 2n_L\Lambda, \tag{8.37}$$

where Λ is the grating period. With Dn from (4.2), the Bragg wavelength split $Dl_B = l_{B,L} - l_{B,R}$ is

$$\Delta\lambda_B = V'B\left(\lambda/\pi\right)\Lambda. \tag{8.38}$$

It is obvious, that even at magnetic flux densities as high as 1 T, the wavelength split is less than 1 pm for a 1550-nm grating in fused silica fiber and therefore hardly observable in the grating's reflection spectra. (The spectral widths [FWHM] of typical FBG are 100 pm and more.) Instead, researchers have tried to extract the Faraday polarization rotation interferometrically [503] or from the grating's spectra of polarization dependent loss (PDL) [504, 505, 506], circular polarization dependent loss (CDL) [507, 508], and the differential group delay (DGD) [506]. Using an unbalanced fiber Mach-Zehnder interferometer for measuring the Bragg wavelength shifts of a 1300-nm grating, Kersey et al. achieved a resolution of a 400-Hz ac magnetic field of 4×10^{-4} T at a detection bandwidth of 0.15 Hz. Wu et al. measured a dc magnetic field-induced CDL of 1 dB at 1.15 T for a 1550-nm grating in an Er-doped fiber with strongly enhanced Faraday effect ($V = -11$ rad/T/m) [508]. But again, intrinsic linear birefringence of the host fiber and photo-induced birefringence of the grating itself compromised the achievable field resolution. For improvement, Descamps et al.

computed from the Mueller matrix of the FBG the field-induced rotation of the diattenuation vector [509] in order to discriminate between linear and field-induced circular birefringence [506]. They achieved a resolution of 0.1 T at dc fields up to 1 T with a 5-mm long 1550-nm grating written in hydrogen-loaded standard fused silica fiber. Orr et al. investigated compound distributed feedback grating structures with resonantly enhanced Faraday effect and predicted an achievable resolution of 0.1 mT/$\sqrt{\text{Hz}}$ [510, 511].

8.11 MULTIPLEXED AND DISTRIBUTED CURRENT SENSING

Various reports have concerned multiplexed fiber coils and distributed current sensing. As already noted further above, Lin et al. reported on time-division multiplexed sensors with polarization-rotated reflection [470]. Yong et al. investigated frequency-division-multiplexed fiber laser current sensors that operated at different wavelengths [512]. Other multiplexing architectures were presented in [485, 513]. Palmeri et al. measured spatially resolved dc current in electric power cables by extracting the Faraday polarization rotation from the polarization analysis of backscattered light [514]. A loose-tube packaged standard single-mode fiber was helically wound with a pitch of 0.1 m and a radius of 45 mm on two parallel 20-m-long cables. The total length of the wound fiber was 130 m. Using polarization optical frequency domain reflectometry (POFDR), the team achieved under stationary conditions a current resolution of 100 A and a spatial resolution of 4 m at currents up to 2.5 kA.

CHAPTER 9

Hybrid Current Transformers

Hybrid current transformers utilize conventional means to convert a high voltage current into a small electric signal and then encode that signal into optical information. An optical fiber transmits the optical signal from high voltage to ground potential. Compared to traditional current transformers, the hybrid approach drastically reduces the expenditures for electric insulation. The following presents three classes of hybrid current transformers.

9.1 ELECTRONIC CURRENT TRANSFORMERS (ECT) WITH OPTICAL COMMUNICATION

Electronic current transformers (ECT) employ a conventional transformer core (CT core) with a secondary winding, a Rogowski coil, or a precision shunt to derive from the primary current a small secondary current [515–531]. A burden resistor R converts that current into a voltage proportional to the primary current or, in the case of a Rogowski coil, a voltage proportional to the time-derivative of the current. For the case of direct current (DC), a precision shunt serves as the primary transducer. Low power electronics encode that voltage signal, and an optical transmitter, commonly an LED, sends the encoded signal via fiber optics to a photodetector on ground. After amplification and processing, the electronics located on the ground system, would send a standardized-format signal to substation control equipment. Hence, ECTs eliminate the need for massive oil- or gas-filled insulators of traditional current transformers. Already in 1965, E. R. Perry reported one of the first transformers of this type [515]. In that case, the electronics on high voltage was powered by a second CT core with energy from the power line. An obvious drawback of a powerline supply is that the electronics is not operational at zero current flow, i.e., a defective transducer and a status of zero current are not distinguishable. Therefore, subsequent work focused on optically powered ECTs, where a laser diode on ground potential energized dedicated low power electronics, either via a second fiber link or via the communication fiber [520–523, 528–531]. M. Adolfsson et al. at the Swedish company Asea (later ABB) developed an optically powered ECT, particularly for use at HVDC converter stations, series capacitor banks for grid control, and line filters (Figure 9.1) [520, 532]. Since the mid-1990s, ABB has deployed numerous dc and ac systems for various rated currents

DOI: 10.1201/9781003100324-12

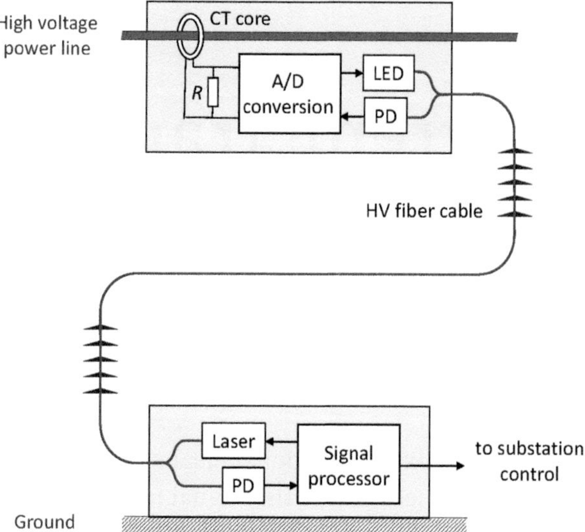

FIGURE 9.1 Electronic current transformer with optical powered electronics (according to [532]).

and accuracy classes. Bandwidths reach up to 100 kHz. Commonly, the transducers are mounted on existing high voltage infrastructure, so that an extra support insulator is not necessary. The fibers are protected in a thin suspension-mounted insulator with silicone sheds for sufficient electric creep distance. The transformer can also be employed for voltage sensing by combining it with a capacitive voltage divider.

More recently, Bassan et al. reported a clamp-on transformer for the measurement of current, voltage, and temperature in medium voltage distribution grids [531]. The voltage signal was obtained via a capacitive divider. The sensor unit on high voltage was powered by a 976-nm, 2-W laser via a multimode fiber. Two further multimode fibers transmitted the digitized sensor data and corresponding clock data to a processing unit on ground. Two low power vertical cavity surface emitting lasers (VCSEL, 850 mn) served as emitters on high voltage. At room temperature, the voltage sensor met the requirements of IEC61869-11 accuracy class 0.5 (accuracy to within ±0.5% between 80% and 120% of the rated voltage) [533] at 11.9 kV, 13.8 kV and 23.0 kV (phase-to-phase voltages), and the current sensor met IEC61869-10 accuracy class 0.5 [534] at a rated current of 25 A and class 0.2 at a rated current of 750 A. Tests versus temperature (0–70°C) showed that both sensors maintained their accuracy after temperature compensation. (The errors were significantly larger, however, before compensation). The sensors were successfully tested in the field.

Whereas the sensors above use active electronics with power supply for signal conversion, simpler techniques have used passive electronics for direct LED modulation [521, 535–537]. If the LED is operated in the linear range of its response curve, the intensity modulation is largely proportional to the primary current (but strongly dependent on temperature) [521]. In [536], a spectral edge filter converted the shifts of the LED spectrum at changing drive voltage into an intensity modulation, and in [537] the sinusoidal current to be measured was encoded in the temporal width of light pulses emitted by the source by setting a current threshold below which the source was switched off.

9.2 HYBRID OCT WITH PIEZOELECTRIC TRANSDUCERS

Like ECT, optical current transformers (OCT) with piezoelectric transducers again employ a CT core, a Rogowski coil, or a shunt for first converting the primary alternating current into a low voltage signal. A piezoelectric transducer converts that voltage in a periodic mechanical deformation (converse piezoelectric effect), and an optical fiber sensor measures the deformation (Figure 9.2). The transducers are typically made from poled piezoelectric ceramics (lead zirconium titanate, $PbZr_xTi_{1-x}O_3$, PZT) [538] that exhibit a large piezoelectric effect. The first OCT of this type utilized optical fiber interferometers for PZT interrogation [134, 539–542]. Ning et al. reported an OCT consisting of a CT core (with a 3000:1 current conversion ratio) as a primary converter, burden resistors of 10 Ω and 100 Ω, and a tube-type PZT with 12 fiber loops operated in a fiber Michelson interferometer at 780 nm (Figure 9.2b) [539]. With the 10-Ω resistor, currents up to 140 A were measured in a frequency range from 30 Hz to 10 kHz and a resolution of $0.1\text{A}/\sqrt{\text{Hz}}$ at 50 Hz. A clamp-on sensor for currents up to 43 kA was demonstrated in [540].

Fiber interferometers for PZT readout were soon replaced by more attractive fiber Bragg gratings (FBG), Figure 9.2c [543–549]. Here, a fiber grating is attached to a stack of PZT elements. Interlaced electrodes expose each PZT element to the full derived voltage for sufficiently large piezoelectric strain $\varepsilon = Dd/d$:

$$\varepsilon = d_{33}\left(V/d\right). \tag{9.1}$$

Here, Dd is the voltage-induced thickness change of a piezoelectric element with thickness d, and d_{33} is the relevant piezoelectric strain coefficient (piezoelectric charge constant). Broadband light, e.g., from a super-fluorescent doped fiber source, illuminates the grating. Out of the broad spectrum (some tens of nanometers), the grating reflects a narrow

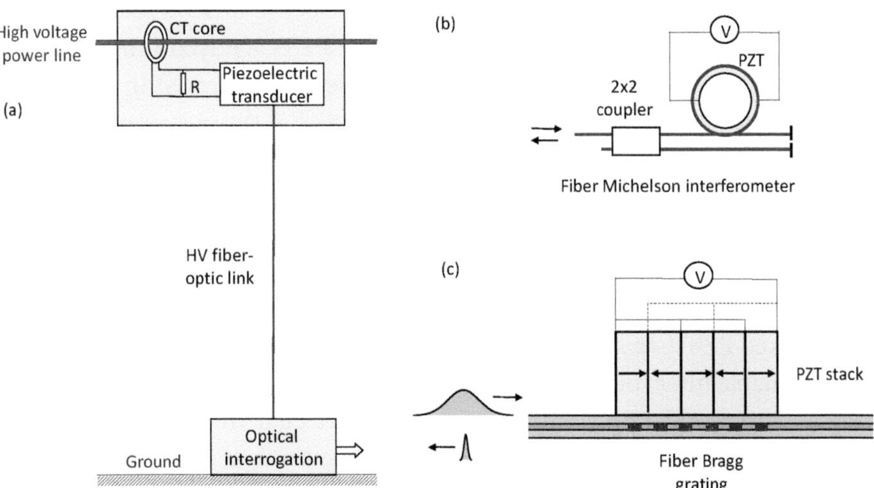

FIGURE 9.2 Hybrid optical current transformer with piezoelectric transducers: (a) System set-up; (b) interferometric interrogation of a PZT tube; (c) fiber-Bragg-grating-based interrogation of a PZT stack; the arrows indicate the alternating poling directions.

wavelength band with a center wavelength $l_B = 2nL$, where n is the average refractive, and L is the grating period. A typical spectral width is 200 pm (FWHM) for gratings with Bragg wavelengths around 1550 nm. A change in voltage DV shifts the Bragg wavelength by [545]

$$\Delta\lambda_B = kd_{33}\left(V/d\right), \tag{9.2}$$

where k is the strain sensitivity of the grating (1.2 pm for strain change of 10^{-6} at 1550 nm) [550]. A spectrometer such as a scanning Fabry-Pérot filter measures the wavelength shift. For example, a voltage of 200 V, 1-mm thick PZT elements, and d_{33} = 200 pm/V result in a wavelength shift of 48 pm. Typically, this shift is measured with a resolution of about 1 pm. A hybrid OCT that converted the CT current into a Bragg wavelength shift by heating the grating was demonstrated in [551].

Hybrid sensors with FBG-based readout are attractive in several aspects:

- A significant number of sensors can be multiplexed along the same fiber for distributed current measurement (e.g., distributed fault monitoring) [552]; the cost per measurement point reduces accordingly.
- In contrast to interferometric interrogation, the wavelength encoded signal is largely insensitive to mechanical fiber perturbations.
- The distances between the interrogator and the sensors can amount to tens of km [552], enabling current measurement at remote locations without electric infrastructure.

On the other hand, there is a relatively long chain of conversion stages from the primary current to the final signal with corresponding sources of error. In particular, the proper choice of the PZT material and its pre-aging are critical. PZTs are known for non-linear response, hysteresis, and dielectric loss, and their dielectric and piezoelectric properties exhibit relatively large temperature dependence. Also, PZTs age according to a logarithmic function of time. These negative effects tend to be less pronounced in so-called "hard" ceramics and more pronounced in "soft" ceramics [538]. But the piezoelectric effect is typically stronger in soft ceramics. More information on PZT materials is given in Chapter 16 on piezo-optic voltage transducers.

Hendersen et al. demonstrated one of the first hybrid sensors with FBG-based readout, but still utilizing a tube-shaped PZT, and achieved a current resolution of 0.7 A/$\sqrt{\text{Hz}}$ [543, 544]. Dziuda et al., aiming at applications in the oil industry (downhole), introduced a sensor with a stacked PZT (six stack layers, each with a thickness of 1.5 mm) and a scanning Fabry-Pérot tunable filter for FBG interrogation. A specially designed CT core with 2000 secondary windings and a 6 kΩ burden resistor generated a voltage of 150 V at the rated current of 50 A. The signal-to-noise ratio at that current was of 36 dB. Between 20°C and 90°C, the signal rose by 4.6% without temperature compensation and was within ±2% after compensation [545]. Numerical techniques for PZT hysteresis compensation were reported in [546, 547]. The use of a Rogowski coil as a primary converter instead of CT was investigated in [553]. Various examples of the application of multiplexed sensors to power grid monitoring and control, fault localization, etc., were reported in [554–556]. A dedicated FBG interrogator for multiplexed sensors combining a switchable wavelength division multiplexer and a passive unbalanced Mach-Zehnder interferometer with a 3×3 output coupler was presented in [557]. Corresponding sensors have been commercialized by the UK company Synaptec Ltd.

9.3 INTENSITY MODULATING AND MICRO-MACHINED OCT

P. Wei et al. reported hybrid OCT which incorporated a light intensity modulator. Here, the voltage signal from the primary converter controlled, by means of an optically powered electronic circuit, a variable optical attenuator or a reflective MEMS device (micro electromechanical system). The attenuator or MEMS modulated the intensity of a light beam in proportion to the primary current [558–560]. MEMS directly modulated by the current magnetic field were investigated in [561, 562]. Duplessis et al. proposed a sensor where a small current derived from the primary current caused a Lorentz-force-induced vibration of a reflecting diaphragm. The diaphragm movement was interrogated with a multimode fiber bundle vibration sensor. The central fiber of the bundle illuminated the diaphragm; the other fibers collected the reflected light. The returning light was modulated according to the current-dependent gap between the fiber facets and diaphragm [563]. A hybrid OCT powered by the line and working with liquid crystal-based chromatic light modulation was reported in [564]. And Bull et al. presented a hybrid OCT consisting of a Rogowski coil, a passive integrator, and an integrated electro-optic modulator. At currents between 30 A and 30 kA, the reported measurement error was less than 0.3% at temperatures between −30°C and 70°C [565].

CHAPTER 10

Further Magnetic Field and Current Sensors

10.1 MAGNETOSTRICTIVE SENSORS

Magnetostriction, discovered in 1842 by J. P. Joule at a sample of iron [566], is the phenomenon that a ferromagnetic material changes its dimensions when exposed to a magnetic field H. Magnetostriction results from the reorientation of ferromagnetic domains in the material, which minimizes the free energy of the system. Today, magnetostriction is employed in sensors and a wide variety of transducers including motors, acoustic noise and vibration cancelling, ultrasonic cleaning, sonar communication, etc. [567, 568]. Magnetostriction is also the cause of the well-known humming noise of transformers at twice the line frequency. Like magnetization, magnetostriction is a nonlinear function of the magnetic field, goes into saturation above a material-dependent saturation field H_s, and exhibits hysteresis. Well below $H_s/2$, the magneto-strictive strain parallel to the field, often denominated as l, is approximately proportional to the square of the field strength [569]:

$$\lambda = \Delta l / l = CH^2. \tag{10.1}$$

Here, l is the sample length, Dl the length change, and C is a material-dependent constant, which can be positive or negative. More sophisticated models can be found, for example, in [570] and references therein. Accordingly, l can be expressed in a 3-parameter model (omitting hysteresis) as (Figure 10.1)

$$\lambda = \lambda_s \left(1 - \frac{3}{\cosh(\kappa H) - 2} \right), \tag{10.2}$$

where l_s is the saturation magnetostriction, and $\kappa = m_0 A_s H_s$; A_s is a parameter determined by the material's susceptibility and saturation magnetization. The magneto-strictive strain perpendicular to the field direction is $-l/2$, since the sample volume remains constant.

DOI: 10.1201/9781003100324-13

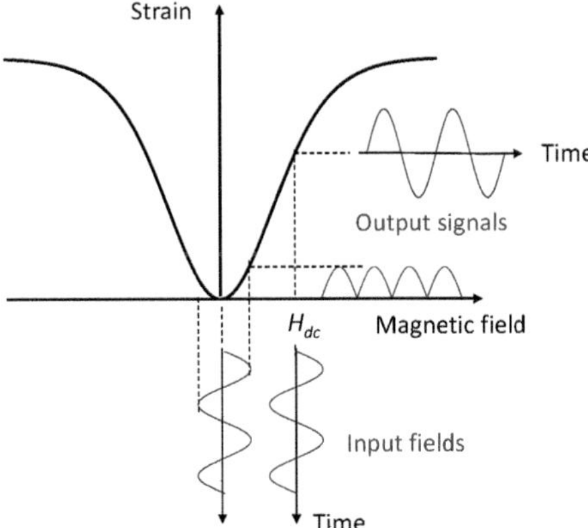

FIGURE 10.1 Magneto-strictive strain l as a function of applied magnetic field H; the blue traces schematically represent alternating input magnetic fields and corresponding strain signals with and without a bias field H_{dc}.

Magneto-strictive materials include monoatomic ferromagnetic metals (such as iron, cobalt, or nickel) and ferromagnetic alloys. Well-known examples of the latter are Galfenol (Fe_xGa_{1-x}) and Terfenol-D ($Tb_xDy_{1-x}Fe_2$). Another frequently used material is the amorphous alloy $Fe_{0.81}B_{13.5}Si_{3.5}C_2$ (trade name Metglass 2605SC) and related metallic glasses. Table 10.1 compares l_s and the magnetic flux densities B_s at saturation for some selected materials [568], and references therein.

In 1980, A. Yariv and H. V. Windsor proposed to employ magnetostriction for optical fiber magnetic field measurement [571]. In the same year, A. D. Dandridge et al. experimentally demonstrated a corresponding sensor [572]. Since then, magneto-strictive optical fiber sensors for magnetic fields and currents have been an active area of research [573–629]. Typically, the fiber is coated with the magneto-strictive material [573, 585, 597], bonded to a magneto-strictive slab or strip [575, 579], or wrapped on a magneto-strictive cylinder [576, 580]. The longitudinal strain transmitted from the magneto-strictive material onto the fiber is measured in a fiber interferometer [572–585] or by a Bragg grating written into the fiber [589–629]. Fabry-Pérot-cavity-based sensors were also reported [587, 588]. In the 1980s and early 1990s, much of the pioneering work was done at the Naval Research Laboratory in

TABLE 10.1 Saturation Magnetostriction l_s and Saturation Magnetic Flux Density B_s of Selected Materials; [568] and References Therein

Material	**$(3/2)l_s$(mm/m)**	**B_s(T)**
Fe	−14	2.15
Ni	−50	0.61
Co	−93	1.79
Co(50%)Fe(50%)	87	2.45
$TbFe_2$	2630	1.1
$Tb_{0.3}Dy_{0.7}Fe_2$ (Terfenol-D)	1620	1.0
$Fe_{0.81}B_{13.5}Si_{3.5}C_2$ (Metglass 2605SC)	60	1.65

Washington, D.C. A focus was on sensors for small and slowly varying magnetic fields; such sensors can be utilized for the tracking of vessels and submarines via their perturbations of the earth's magnetic field [572–583]. The magneto-strictive material (frequently a strip of metallic glass, about 10 cm long with 1 m of fiber attached to it) modulated the optical phase in an arm of a Mach-Zehnder or Michelson interferometer. In the low frequency regime, however, environmental and thermal noises tend to mask small magneto-strictive strain signals. Here, Koo et al. [576] and Kersey et al. [577] made use of the quadratic response at small fields and applied, in addition to the field H_{dc} to be measured, an ac dither field H_{ac} of constant amplitude. The frequency ν_{d} of the dither field was above the cut-off frequency of the servo loop that kept the interferometer at quadrature. The strain signal of (10.1) is then

$$\lambda = CH^2 = C\left(H_{\mathrm{dc}} + H_{\mathrm{ac}}\right)^2 = C\left(H_{\mathrm{dc}}^2 + H_{\mathrm{ac}}^2 + 2H_{\mathrm{dc}}H_{\mathrm{ac}}\right). \tag{10.3}$$

Hence, the interferometer output $CH_{\mathrm{dc}}H_{\mathrm{ac}}$ at frequency ν_{d} is directly proportional to H_{dc} and unaffected by low frequency noise. Closed loop operation (a second servo loop adds a compensating field $-H_{\mathrm{dc}}$) resulted in a hysteresis-free linear range of more than five orders of magnitude with a detection sensitivity of 10^{-10} T at frequencies below 2 Hz [578, 579]. For comparison, the earth's magnetic field at the equator is about 30 mT.

It is obvious that unbiased ac magnetic fields with frequency ν_{ac} result in a rectified strain response at frequency $2\nu_{\mathrm{ac}}$ (Figure 10.1). Adding an appropriate dc bias field H_{dc}, e.g., by means of a solenoid, provides an output at the fundamental frequency that varies over a limited range largely proportional to the input ac field. F. Bucholtz et al. reported a corresponding interferometric magneto-strictive magnetometer with a resolution of 70 fT/$\sqrt{\mathrm{Hz}}$ near 34 kHz [582]. Sedlar et al. demonstrated for a ceramic-jacketed fiber (cobalt-doped nickel ferrite coating over 1m) operated in a Mach-Zehnder interferometer a field resolution of 2.5×10^{-3} A/m/$\sqrt{\mathrm{Hz}}$ (equivalent to 3.1×10^{-9} T/$\sqrt{\mathrm{Hz}}$ in air) at frequencies below 2 kHz. With a bias field of 13.5 kA/m, the signal was free of hysteresis up to 20 kA/m. The jacketed fiber exhibited mechanical resonances between 0.5 and 1.5 kHz [585].

Most subsequent magneto-strictive sensors employed fiber Bragg gratings for interrogation [589–629]. FBG-based sensors commonly do not achieve the high field resolution of their interferometric counterparts but have the benefit of simplicity. Nazare et al., for example, measured a Bragg wavelength shift at 1540 nm of 1.45 nm/T (induced strain of 1.2×10^{-3}/T) for a FBG on a Terfenol-D rod [622]. Assuming a wavelength resolution of 5 pm, the corresponding magnetic flux resolution is about 3.5 mT. Distributed and quasi-distributed magneto-strictive field sensors were reported in [600, 601, 603, 604, 617].

Whereas most FBG-based sensors focused on the measurement of local magnetic fields, several publications have been dedicated to ac current measurement [618–629]. Typically, the magneto-strictive material with an attached FBG is part of a toroidal assembly around the current conductor. The required dc bias magnetic field is generated by permanent magnets [618, 620, 623], mechanical stress applied to the magneto-strictive material [618, 621, 626], or a combination of both. Note that mechanical stress generates a bias field through the inverse magneto-strictive effect (Villari effect). As an example, Figure 10.2 shows a sensor developed by A. Dante et al. [624, 625]. Magnets generated the dc bias field; steel ring

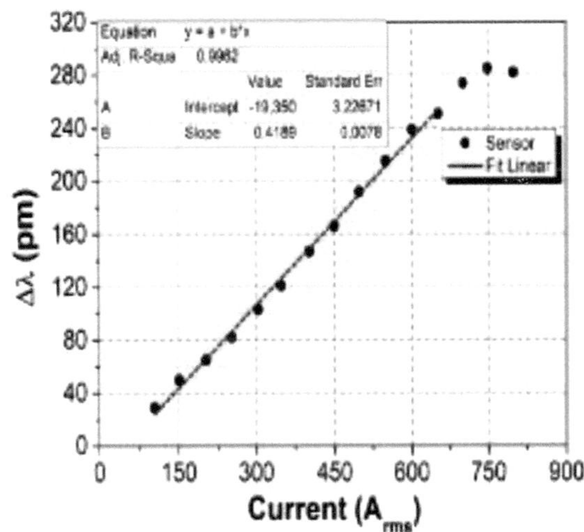

FIGURE 10.2 Magneto-strictive current sensor (left) and peak-to-peak Bragg wavelength shift vs. applied ac current (right) (reprinted from Dante et al. [624], with permission.)

segments concentrated the bias field as well as the ac field of the current to be measured onto a Terfenol-D (TD) rod with an attached FBG. The magnetic flux concentration allowed the authors to reduce the mass of the expensive TD to 5.6 g, and after further optimization, to 2.0 g. The Bragg wavelength shift as a function of an applied ac current (60 Hz) showed good linearity with a slope of 0.42 pm/A up to about 650 A (Figure 10.2). At higher currents, the TD rod went into saturation. The Bragg wavelength resolution of ±10 pm corresponded to a current resolution of 24 A. Lopez et al. reported similar designs with the TD masses as small as 0.42 g [627, 628]. At ac up to 800 A, nonlinearity errors were less than 1% of the full-scale current. In that case, the magneto-strictive transducer was a composite with TD particles dispersed in epoxy resin. A certain drawback of magneto-strictive current transducers is the inherent phase delay of the output signal with respect to applied current (e.g., 4.8° in [627]). For a review on magneto-strictive optical current sensors and the preparation of magneto-strictive composites and their characterization, see [629].

10.2 MAGNETIC FLUID-BASED SENSORS

Magnetic fluids (MF), or ferrofluids, are colloidal dispersions of ferromagnetic nanoparticles in a liquid carrier [630]. Carrier liquids include water and oils. The particles consist of materials such as Fe, Co, or magnetite (Fe_3O_4) and are commonly coated with a surfactant like hydroxyl or carboxyl to prevent clustering. Particularly, Fe_3O_4-based fluids are frequently used for sensing. Typical particle diameters are about 10 nm; particle volume fractions in the carrier liquid are on the order of 1%. Without an applied magnetic field, the particles are essentially uniformly distributed in the carrier liquid, whereas in a field, they tend to form chain-like structures parallel to the field [631]. The particle rearrangement is accompanied by changes in the refractive index and absorption spectra of the fluid [631, 632]. Also, induced birefringence and dichroism have been observed [633–635].

Like magnetostriction, the effects are nonlinear and saturate above a certain field strength. Naturally, response times are relatively slow (on the order of 10 s). With 1.8 vol% of Fe_3O_4-particles dispersed in water, Zhao et al. measured a refractive index change of about 2.0×10^{-2} between 0 mT and 70 mT for fields parallel to the optical path and -1.0×10^{-2} for fields perpendicular to the optical path [631]. The index changed roughly linearly with the field strength between 20 mT and 50 mT before reaching saturation. The field-induced refractive index change has been utilized in numerous magnetic field sensors. The sensor concepts include MF-filled Fabry-Pérot cavities [636–638], etched, tilted, or tapered fiber Bragg gratings [639–641], long period gratings [642, 643], fiber mode interferences [644–650], micro-fiber devices [651–653], silicon micro-ring resonators [654], light refraction [655, 656], fiber Mach-Zehnder and Sagnac interferometers [657, 658], fiber ring-down cavities [659], and others. Typically, the sensitive fiber section resides in a MF-filled capillary and interacts with the fluid via the evanescent field. As a result, compact and simple sensors become possible. Also, sensors based on transmission changes of photonic crystal fiber with MF infiltrated air holes were reported [660, 661]. Current measurement with MF-based sensors was reported in [636, 646, 654, 662]. It is worth mentioning that magnetic fluids also exhibit the Faraday effect. Martinez et al. reported a very high Verdet constant of 1.2×10^5 rad/T/m at 633 nm for a homemade magnetite-based fluid [663]. For reviews of MF H-field sensors and further references, see [638, 664].

10.3 NON-OPTICAL NON-CONVENTIONAL SENSORS

Optical current sensors are particularly advantageous at high voltages, as they inherently provide electric insulation for ground-based equipment. Also, optical sensors are beneficial at high currents (tens or hundreds of kiloamps), as they are free of saturation and can easily be adapted to application-specific conductor geometry. On the other hand, optical sensors are commonly not the solution of choice for lower voltage systems (e.g., in factories or power electronics) and small currents, not least for economic reasons. Besides more traditional current transducers based on shunts, Rogowski coils, or Hall effect devices, other non-conventional techniques have become available, which include fluxgate sensors, GMR sensors (gigantic magneto resistance), and AMR sensors (anisotropic magneto resistance). Often, the actual sensor element is within a gap of a magnetic core enclosing the current conductor and operated in a closed-loop circuit for linear response. These sensors are outside the scope of this book but have been reviewed, for example, by Lenz et al. [665] and S. Ziegler et al. [666].

Moreover, several institutions and companies have developed in the context of "smart grids" line-powered, wireless instrumentation for monitoring of medium and high voltage power lines. Typically, these devices are attached to the transmission line using a clamp-on style device; and may monitor a multitude of parameters like current and voltage waveforms including high frequency content, line temperature, line inclination angle, and sag. While these sensors do not replace conventional measurement and protective instrument transformers or corresponding optical sensors, they are complementary and do help improve the electric grid's performance and efficiency by allowing a wide-area monitoring of possible faults, disturbances, or abnormal operating conditions (such as voltage volatility or unbalanced load between phases), as well as facilitating the integration of renewable energy sources, while reducing operating expenses [667–671].

Other non-conventional alternatives for monitoring power line currents and voltages are low-cost magnetic and electric field sensors located on ground beneath the power lines. Sensors include inductive sensors, Hall effect sensors, magneto-resistive sensors, and capacitive probes. The number of sensors is equal (or larger) than the number of power lines to be monitored. A matrix of conversion coefficients, obtained by simulation or calibration, relates the sensor signals to the line currents or voltages [672–677].

CHAPTER 11

Accuracy, Precision, and Reliability of Optical Current Sensors

Applications of optical current sensors in the electric power industry demand accuracy as high as ±0.1% at temperatures that can range, for example, from −40°C to 85°C. The sensors are expected to maintain their accuracy over extended periods of service time. (Traditional instrument current transformers service times are more than 30 years.) The accuracy specification of a sensor indicates how close the mean of repeated measurements (at constant current) is to the true current. By contrast, the precision is a measure for the scattering of repeated measurements about their mean value [678]. Ideally, a sensor is both accurate and precise. However, a sensor can be accurate (the mean of repeated measurements is close to the true current) but unprecise (individual measurements strongly scatter) and, vice versa, a sensor can be inaccurate but precise. Furthermore, the sensors are expected to be reliable, i.e., they should be free of deficiencies and failures that could cause accuracy loss or signal loss. The following sections consider accuracy, precision, and long-term reliability for the example of FOCS with polarization-rotated reflection (Figure 8.10), but many of the findings also apply to the other sensor configurations discussed above and the voltage sensors of Part III.

11.1 FOCS ACCURACY

Manufacturing tolerances, inadequate sensor components, and poor packaging may lead to systematic scale factor instability and signal errors. Error causes include:

11.1.1 Manufacturing Tolerances

Manufacturing tolerances most notably concern the relative orientations of the polarizer, principal axes of the modulator, if applicable, PM fiber segments including the fiber retarder, and the sensing fiber (see Figure 8.10). There is also some tolerance in the length

DOI: 10.1201/9781003100324-14

and retardation of the retarder. Imperfect axes alignment at PM fiber splices or fiber-to-modulator interfaces results in cross-coupling between the orthogonal polarization modes and hence parasitic light waves propagating through the optical circuit. Deviations of the polarizer from perfect 45°-orientation reduce the interference fringe contrast but leave the scale factor unchanged [445, 449]. Polarization cross-coupling at a point between the polarizer and the modulator, particularly at the source-side modulator facet, also does not affect the scale factor. Cross-coupling at the coil-side modulator facet and at splices along the PM fiber link to the sensing coil enhances the scale factor — in proportion to $1/\cos(2Da)$, where Da is the misalignment angle — and reduces the fringe contrast, similarly as an imperfect 90°-retarder does, as shown in Figure 8.13 [406, 436, 445, 449, 454]. Imperfect 45°-orientation of the retarder with respect to PM fiber also enhances the scale factor. Scale factor changes resulting from a single misaligned PM fiber splice or imperfect retarder orientation are stable and accounted for by the sensor calibration. Therefore, they do not impact the sensor accuracy. On the other hand, multiple cross-coupling at serial PM fiber splices can seriously degrade a sensor, as will be explained below. The retardation and orientation of the fiber retarder affects the temperature dependence of the sensor, as we have seen further above (Figures 8.13, 8.14, 8.18).

11.1.2 Temperature Dependence of Optical Component Parameters

Typically, the sensing optical fiber contributes the most to the overall temperature dependence of a current sensor (Verdet constant, plus embedded linear birefringence in the particular case of a spun fiber). The corresponding scale factor changes are repeatable (provided the fiber is not exposed to excessive stress) and can be intrinsically compensated by means of the retarder as described in Section 8.6 or a temperature sensor. Residual errors may remain, e.g., due to temperature gradients. It is obvious that the remaining error can be more significant with the larger overall temperature dependence of the sensing fiber. Ideally, the signal variation to be compensated should not exceed 1–2% for a sensor with specified accuracy of ±0.2%. As Figure 8.16 has illustrated, residual errors can be kept below 0.1% for a properly designed and fabricated sensor, even over a temperature range as wide as 125°C.

The modulator's response (p-voltage) changes by several percent per 100°C but does not enter into the signal for reasons already discussed. The center wavelength of an uncooled 1310 nm SLED varies by about 0.6 nm/°C, which would result in an (unacceptable) scale factor change of about 0.09%/°C due to the wavelength dependence of the Verdet constant. Temperature stabilization of the SLED chip by a thermo-electric cooler (TEC) reduces the wavelength excursions to just a few hundred picometers between −25°C and 65°C [460]. The corresponding scale factor changes (within about ±0.025%) can be easily calibrated out.

As already indicated, multiple polarization cross-coupling at subsequent PM fiber splices or other coupling points may represent a more serious problem if not handled properly, even though principal axes misalignment is commonly less than 1–2° with state-of-the-art PM fiber splicers. In case of a single cross-coupling point, the pair of cross-coupled orthogonal waves becomes incoherent with respect to their parent waves after a distance

$l \gg l_c/Dng$ and will not affect the recovered magneto-optic phase shift. (l_c is the source coherence length and Dn_g the group index difference of the orthogonal modes.) If a second cross-coupling event occurs within a distance l_c/Dn_g from the first coupling point, the pair of double cross-coupled waves is at least partially coherent with the primary wave pair. At changing temperature, the phase between the primary and double cross-coupled waves varies. As a result, the sensor signal as a function of temperature exhibits oscillations with a peak-to-peak amplitude that can easily reach some tenths of a percent of the true current signal and a period (often a few °C) determined by the separation of the coupling points. Separating the coupling points by a distance $l \gg l_c/Dn_g$ removes the oscillations. Assuming a typical 1310-nm SLED and Panda PM fiber, the term l_c/Dn_g is on the order of 10 cm. Ideally, l should then amount to several tens of centimeters. Whereas this is easily implemented for two coupling points, the task becomes more challenging for multiple coupling points, as their separations must be properly staggered to exclude mutual coherence of a multitude of parasitic wave pairs. Moreover, the group delay of a spun birefringent sensing fiber must also be considered.

11.1.3 Stress Acting on Optical Components

Polarimetry is an excellent method to measure stress. A current sensor should measure exclusively the actual electric current flowing in the conductor, but not any "apparent current signal" caused by stress-induced effects on the device. Therefore, a significant part of the development of high-end FOCS is the identification and elimination of potential sources of random stress, as stress inevitably leads to unstable performance as a function of temperature and signal drift. Among the most common sources of possible stress acting on these devices are improper packaging of the sensing fiber, fiber feedthroughs (e.g., into the modulator housing), inadequate fiber coatings (especially at low temperature), splice protecting sleeves, poorly fabricated PM fiber cables, fiber connectors. Commercially available PM fiber connectors, for instance, even if upgraded for use in FOCS, were found to be incompatible with class 0.2 accuracy [452, 460].

11.1.4 Component Aging

Component aging is another potential cause for accuracy loss or even sensor failure and will be addressed separately further below.

11.2 FOCS PRECISION

11.2.1 Noise Sources

The precision of a sensor is limited by signal noise. Noise sources in FOCS are thermal noise (also known as Johnson-Nyquist noise), photon shot noise (Poisson noise), and light source intensity noise. Thermal noise reflects the thermal movement of the conduction electrons in the electric circuit and is independent of the received light power. Shot noise relates to the fundamental statistical fluctuations of the number of photons per time interval in a light beam and is proportional to the square root of the light power P (or number of photons per time interval). Hence, at increasing optical power the relative shot noise decreases in proportion to $1/\sqrt{P}$. Source intensity noise has its origin in the beating of spectral components at random phases and increases in proportion to the light power. Higher power will therefore not reduce the relative source intensity noise. Most commonly, a light source's intensity noise is higher the narrower the spectral width of the source's emitted light (see [679] for details on RIN of broadband sources). Since the three noise sources are uncorrelated, the total noise power is the sum of the individual powers (rather than the square of the sum of the noise rms amplitudes) [679, 680]:

$$\langle i_n^2 \rangle = \langle i_{\text{th}}^2 \rangle + \langle i_{\text{sh}}^2 \rangle + \langle i_{\text{in}}^2 \rangle. \tag{11.1}$$

Here, $\langle i_{\text{th}}^2 \rangle, \langle i_{\text{sh}}^2 \rangle, \langle i_{\text{in}}^2 \rangle$ are the time-averaged contributions from thermal, shot, and source intensity noise, respectively. The rms amplitudes of the three noise currents in the photodetector signal are

$$i_{\text{th}} = \sqrt{\frac{4kT}{R_{\text{L}}}\Delta f}, \tag{11.2}$$

$$i_{\text{sh}} = \sqrt{2e\langle i \rangle \Delta f}, \tag{11.3}$$

$$i_{\text{in}} = \sqrt{\frac{\langle i \rangle^2}{\Delta \nu_{\text{e}}}\Delta f}, \tag{11.4}$$

where k is the Boltzmann constant, T the absolute temperature, R_{L} the load resistance that the photocurrent experiences, e the electron charge, $\langle i \rangle$ the mean photodetector current, Dn_{e} the effective source spectral width, and Df the detection bandwidth.

An important characteristic of a fiber-optic current sensor is the noise equivalent current (NEC), that is, the current that produces a signal equal to the noise rms amplitude and represents the low end of the measurement range. Assuming a current sensor with polarization-rotated reflection and 90°-phase bias, the optical power arriving at the photodetector generates a photocurrent

$$i_{\text{d}} = r\left(P_0/2\right)\left(1 + \sin \Delta\phi\right), \tag{11.5}$$

where r is the power-to-photocurrent conversion coefficient (InGasAs photodiodes that are typically used at 1310 nm or 1550 nm have conversion coefficients close to $r = 1$ A/W), P_0 is the optical power at constructive interference, and $\Delta\phi = 4VNI$; the fringe visibility is assumed as $K = 1$. Considering only shot noise for a moment, the corresponding noise equivalent current $I_{\text{NE}}^{(\text{sh})}$ follows from (11.3) and (11.5) by setting the average photocurrent $\langle i \rangle$ equal to $rP_0/2$ and the current-induced signal change $Di_{\text{d}} = r(P_0/2)\Delta\phi$ equal to the shot noise i_{sh} (with $\Delta\phi \ll 1$):

$$I_{\text{NE}}^{(\text{sh})} = \frac{1}{2VN}\sqrt{\frac{e}{rP_0}\Delta f}. \tag{11.6}$$

The noise equivalent currents associated with thermal noise and intensity noise, $I_{NE}^{(th)}$ and $I_{NE}^{(in)}$, follow analogously. The total noise equivalent current is then

$$I_{NE} = \sqrt{\left(I_{NE}^{(th)}\right)^2 + \left(I_{NE}^{(sh)}\right)^2 + \left(I_{NE}^{(in)}\right)^2}, \tag{11.7}$$

and the signal-to-noise ratio is SNR = I/I_{NE}. Figure 11.1 shows the individual noise equivalent currents as well as the total as a function of the optical power P_0 for a fused silica fiber coil with a single loop (N = 1), a detection bandwidth of 1 Hz, a broadband source centered at 1310 nm, $Dn_e = 5.2 \times 10^{-12}$ Hz (Dl_e = 30 nm), and r = 1 A/W. It is obvious that thermal noise (here for R_L = 1 MΩ) plays a role only at optical powers well below 1 mW. Shot noise dominates up to about 5 mW. Above that level, intensity noise takes over as the main contribution. Note that optical power increase will improve the signal-to-noise ratio only up to about 5–10 mW, i.e., in the regime where shot noise dominates. Naturally, the noise equivalent current decreases with increasing number of fiber loops. Increasing the signal averaging time $t = 1/Df$ reduces the noise equivalent current in proportion to $1/\sqrt{t}$. The noise-equivalent current of a well-designed sensor with passive polarimetric detection is close to the theoretical limit. In sensors with active open-loop or closed loop detection circuits, the signal processing typically adds some extra noise and correspondingly increases the NEC.

11.2.2 Controlling Precision

At m repeated measurements of the current signal, the individual measurement samples x_i form a Gaussian distribution about their mean value m. The half-width at half maximum is the standard deviation σ of the samples:

$$\sigma = \sqrt{\frac{1}{m}\sum_{i=1}^{m}\left(x_i - \mu\right)^2}, \tag{11.8}$$

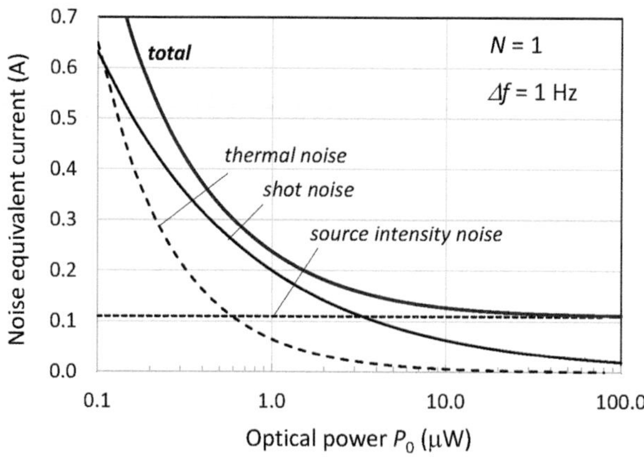

FIGURE 11.1 Contributions to noise equivalent current of thermal noise (R_L = 1 MΩ), shot noise, and source intensity noise vs. optical power P_0 for a FOCS with polarization-rotated reflection having a single fused silica fiber loop; l = 1310 nm, $Dn_e = 5.2 \times 10^{-12}$ Hz, r = 1. The detection bandwidth is Df = 1 Hz.

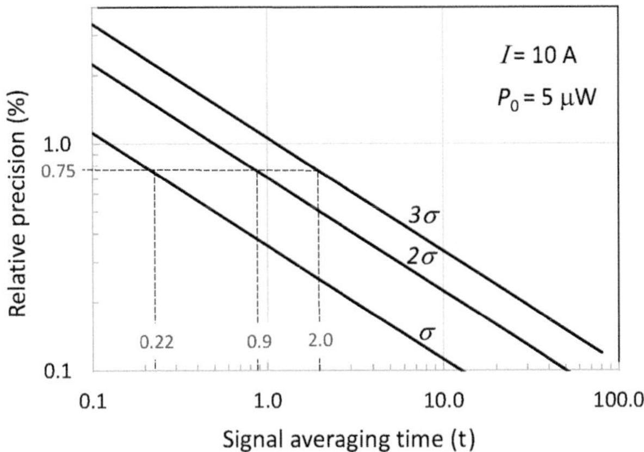

FIGURE 11.2 Relative sensor precision (σ, 2σ, and 3σ) vs. signal averaging time at a current I of 10 A with $N = 4$ and $P_0 = 5$ mW (see text for details).

$$\grave{\imath} = \frac{1}{m}\sum_{i=1}^{m} x_i \tag{11.9}$$

s is equal to the noise equivalent current of (11.7): $s = I_{NE}$. Hence, 68.3% of the samples are within a range of $m \pm\sigma$, 94.4% within $m \pm 2\sigma$, and 99.7% within $m \pm 3\sigma$. For further illustration, we consider the IEC accuracy class 0.2s for current transformers, which requires accuracy to within ±0.2% at currents between 20% and 120% of the rated primary current, ±0.35% at 5% and ±0.75% at 1% of the rated current [474]. Naturally, systematic sensor errors should be well within those limits. Furthermore, one may desire a precision corresponding to $\pm 1\sigma$, $\pm 2\sigma$, or even $\pm 3\sigma$, that is, 68.3%, 94.4%, or 99.7% of all samples should lie within the error limits. At a given optical power P_0, the precision can be controlled via the number N of fiber loops and bandwidth Df, i.e., the signal averaging time t, $t = 1/Df$; Eq. (11.7). As an example, we assume a rated primary current of 1 kA. The low end of the specified current range is then 10 A (1% of 1 kA) with error limits of ±0.75%. Figure 11.2 shows for the three cases the relative precision, defined as iI_{NE}/I, $i = 1, 2, 3$, at $I = 10$ A as function of t for $P_0 = 5$ mW and $N = 4$. It is seen that the signal samples must be time-averaged over at least over 0.22 s, 0.9 s, and 2.0 s to limit their scatter to ±0.75% of 10 A (75 mA) with σ, 2σ, and 3σ-precision, respectively. Whereas averaging times of seconds and longer are easily acceptable for measuring and metering purposes, protective relaying (recognition and interruption of faults) requires response times on a millisecond time scale. On the other hand, accuracy demands for protective current transformers are more relaxed. The frequently requested accuracy class 5P requires accuracy to within ±1% at the rated primary current; in our example, that accuracy is reached with 2σ-precision already after 50 ms.

11.3 LONG-TERM RELIABILITY

Most applications of fiber-optic current sensors are in the harsh environments of the electric power industry. Reliable operation is of utmost importance to the industry. Fiber-optic current sensors represent a new technology and, naturally, experience about their long-term reliability and performance is still limited. Over a period of several years, the author's team at ABB has carried out extensive accelerated aging tests on the individual optical FOCS components and field tests of the full system for phase-modulated FOCS with polarization-rotated reflection as depicted in Figure 8.10; see [460] and references therein. Typically, the component manufacturers qualify the components for the telecom industry and other uses [681]. Most of those tests only partially address the needs of FOCS. For example, long-term drift of the source wavelength may not be particularly critical in applications like optical coherence tomography but could lead to inacceptable accuracy loss in FOCS. Likewise, a polarization extinction (PER) of the integrated-optic modulator of 20–25 dB in a limited temperature range may suffice for cable television distribution but is again inadequate for FOCS and requires device improvements and subsequent long-term tests.

Prior to the tests, an extensive Failure Mode and Effects Analysis (FMEA) [682] of the sensor system as such and its essential components helped to define the tests. The components were exposed to temperature cycling (up to 15,000 cycles between –25°C and 65°C) and damp heat (85% relative humidity at 85°C for up to 7900 hours). Thermal cycling serves to discover potential failure modes associated with thermo-mechanical stress or fatigue at repeated differential thermal expansion of the involved materials. Damp heat accelerates diffusion and oxidation processes and is known to trigger packaging and material related failure modes [683–688]. Various PM fiber samples were exposed to constant high temperature at dry conditions (up to 115°C for up to 20,000 hours) to determine potential coating degradation and PER reduction [689, 690].

The tested optical components included SLED light sources in a butterfly package with integrated thermo-electric cooler, integrated-optic lithium niobate birefringence modulators, fiber polarizers, fiber couplers, photodiodes, custom-made PM fiber connectors, and fiber coatings. All tests involved multiple samples of each component, often from several suppliers. During the accelerated aging process, the components were repeatedly recharacterized at defined intervals to determine drifts, if any, or, in the worst case, component failure. Overall, the tests showed high maturity of the components, but also some differences among suppliers and a few component failures (that led to follow-up measures). The PER stability of PM fiber connectors, even with custom-adaptation, was found to be insufficient for 0.2%-accuracy, however; also see [454]. Hence, fiber splicing was determined as the more appropriate solution.

As an example, Figure 11.3 shows the evolution of the center wavelength of six 5-mW, 1310-nm SLDs from two suppliers during long-term temperature cycling. The symbols indicate the center wavelength measured at a constant drive current and a TEC temperature of 25°C. Note that the vertical lines are not error bars but indicate the residual wavelength excursions during temperature cycling (which in a sensor are accounted for by calibration). The long-term variations in the center wavelength were from ±190 pm to ±380 pm and well within the tolerances for accuracy class 0.2. Changes in the output power remained uncritical.

Figure 11.4 depicts the evolution of the polarization extinction ratio vs. temperature over 15,000 temperature cycles of one of the two fiber-to-waveguide junctions of a modulator. (Recall that only PER variations at the coil-side junction affect the sensor scale factor).

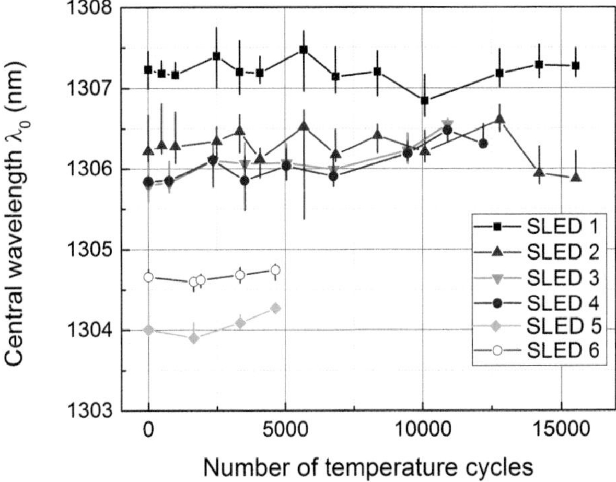

FIGURE 11.3 SLED center wavelengths vs. number of temperature cycles (between −25°C and 65°C). The wavelengths were measured at a constant drive current and a TEC temperature of 25°C. The symbols represent the center wavelengths at an ambient temperature of 25°C. The vertical bands indicate the residual wavelength excursions during the temperature cycles. SLED 1–4: Supplier 1. SLED 5, 6: Supplier 2 (© 2020 IEEE. Reprinted with permission from [460].)

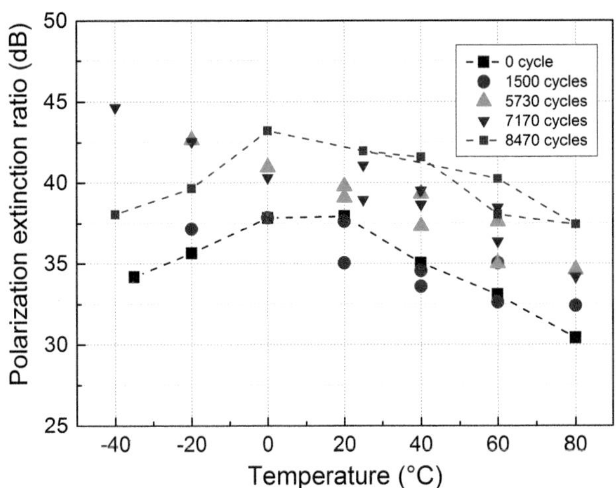

FIGURE 11.4 Birefringence modulator: Polarization extinction ratio at a fiber-to-waveguide junction vs. modulator temperature at increasing number of temperature cycles (© 2020 IEEE. Reprinted with permission from [460].)

The already high PER at the beginning of the test (black squares) further improved with increasing number of cycles. The corresponding scale factor changes would be well below ±0.1%.

Finally, Figure 11.5 shows data from a 3-year field trial of a 3-phase FOCS system integrated in 420-kV high voltage circuit breakers at a substation in Sweden (details are described in Section 12.2 below). The deviations of the optical sensors from corresponding conventional current transformers remained well within ±0.2% during the entire test (upper panel), independent of the ambient temperature (lower panel). Supplementary

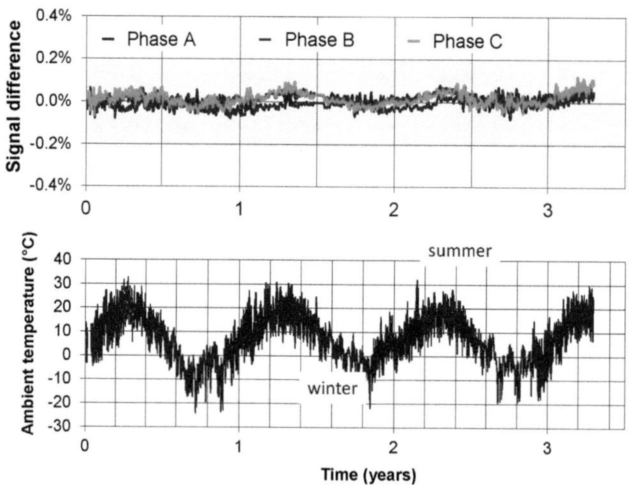

FIGURE 11.5 Field test of 3-phase FOCS integrated in high voltage circuit breakers: Difference vs. time between optical and conventional current measurement signals for the three phases (top); ambient temperature vs. time (bottom) (© 2020 IEEE. Reprinted with permission from [460].)

recordings of light source and modulator parameters also did not indicate any significant degradation in the power source, the modulator p-voltage, or insertion loss. Accelerated aging test results for magneto-optic current sensors with a glass rod sensing element were reported in [691]. And a review on measurement errors in optical current transducers was given in [692].

CHAPTER 12

Applications

The following sections describe a few applications of all-fiber current sensors and illustrate some of their benefits versus conventional solutions.

12.1 ELECTRO-WINNING OF METALS

One of the first successful commercial applications of fiber-optic current sensors was in the electro-winning of metals, most importantly, aluminum. First introduced in 2005 by ABB, FOCS rapidly gained a significant market share in that industry. Other FOCS manufacturers including DynAmp/NxtPhase (USA) and Smart Digital Optics (Australia) followed soon afterwards. Aluminum is won by electrolysis (Hall-Héroult process) from aluminum oxide (alumina) dissolved in molten cryolite (Na_3AlF_6) according to $2Al_2O_3 + 3C \rightarrow 4Al + 3CO_2$. Al_2O_3 is chemically extracted from Bauxite ore in a preceding process, and the carbon stems from the carbon anodes of the Hall-Héroult process [693, 694]. The electrolysis takes place in a chain of smelter pots at dc currents as high as 600 kA. The process consumes large amounts of energy — about 15.7 kWh per kg Al. Accordingly, large factories require several 100 MW of power. The current flows from groups of rectifiers through multiple parallel busbars into the serial smelter pots. Often, the busbars occupy an overall cross-section on the order of square meters (Figure 12.1). For process efficiency control, the total current must be measured with accuracy to within ±0.1%. Traditionally, Hall effect current transducers with magnetic flux compensation have done the task. The transducers have a segmented magnetic core surrounding the busbars with magnetic-field sensitive Hall elements positioned in air gaps of the core segments. Solenoids in closed-loop feedback circuits nullify the magnetic fields in the core segments. The total of the solenoid currents is then a measure for the smelter current.

Additionally, simpler open-loop Hall transducers with somewhat lower accuracy often serve to measure the currents of the individual rectifier groups. Hall transducers with flux compensation, although very precise, are rather complex devices, can weigh more than 2000 kg and may consume some kW of electric power. Installation and commissioning are time consuming and often require magnetic centering to prevent magnetic overload of

DOI: 10.1201/9781003100324-15

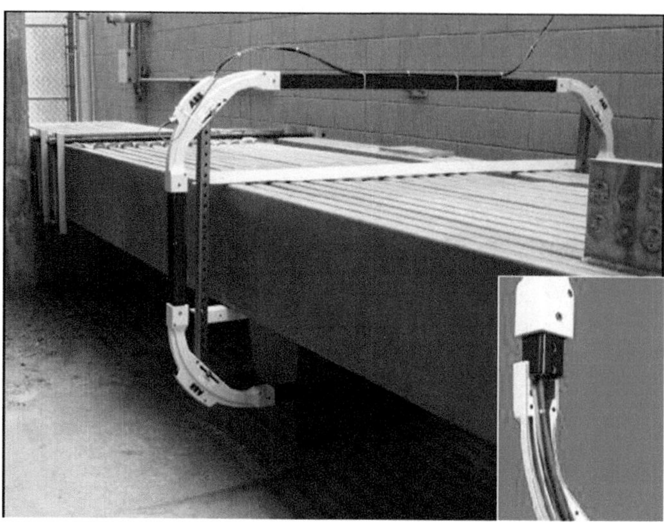

FIGURE 12.1 Fiber-optic current sensor mounted to the current-carrying busbars of an aluminum smelter. The rated current is 260 kA. The width of the sensor head housing is 2.25 m. Inset: The sensing fiber resides in a flexible strip of fiber reinforced epoxy (© 2007 IEEE. Reprinted with permission from [326]).

individual transducer segments. By contrast, a fiber-optic current sensor, as shown in Figure 12.1, measures the current with a single loop of fiber around the busbars [326, 440]. Here, the sensor is configured, as shown in Figure 8.11, and the sensing fiber resides in a fused-silica capillary embedded in a flexible strip of fiber reinforced epoxy (inset in Figure 12.1). A modular frame composed of hollow segments of fiber reinforced epoxy and mounted to the busbars accommodates the sensing strip. By adjusting the lengths of the fiber and straight frame segments, the sensor can be easily adapted to different conductor cross-sections (Figure 12.2). Magnetic centering is no longer necessary; the sensor correctly measures the current with the required accuracy even in strongly inhomogeneous fields, e.g., at angled conductors or in the presence of neighbor currents. Other important industrial FOCS applications include the electro-refining of copper and electrolytic production of chlorine.

12.2 ELECTRIC POWER TRANSMISSION

Conventional current transformers (CT) in air-insulated HV substations are typically stand-alone devices, as shown in the introduction. In a corresponding stand-alone FOCS, the fiber coil is mounted on top of a support insulator, similarly like the bulk-optic sensor of Figure 6.6. However, the flexible and compact design of FOCS sensing coils also allow for new installation options. As an example, Figure 12.3 illustrates the integration of FOCS into a double chamber high voltage circuit breaker [327]. The sensor head housing with two fiber coils for redundancy sits at the top end of the breaker's support insulator and inside the breaker's insulation gas volume. The current path is modified such that the current makes a detour through the aperture of the coil housing. Two three-phase opto-electronic

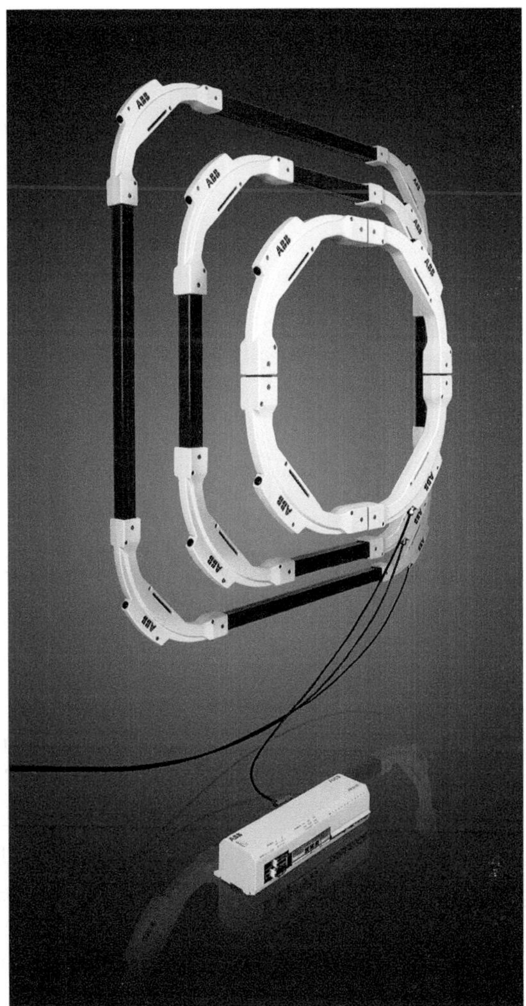

FIGURE 12.2 Sensor heads for different busbar cross-sections and opto-electronics module with analog and digital interfaces. The inner diameter of the smallest head is about 73 cm (© 2007 IEEE. Reprinted with permission from [440]).

units reside in a nearby cubicle and interrogate all six fiber coils of a three-phase breaker system. The fiber cable of a given sensor head runs through the hollow insulator column to ground and leaves the breaker through a gastight feed-through — the insulator also contains the breaker drive rod. The opto-electronic units send the six digital sensor signals (sampled values) via an optical process bus according to the IEC61850-9-2LE protocol to secondary electronics for metering or relaying in the substation control house. The integration of FOCS into breakers saves the space for conventional CT and results in a substantial footprint reduction. Moreover, there is less installation work in the field. The sensor heads are already mounted in the factory; the same applies to the opto-electronic units in the cubicle. Figure 12.4 shows a photograph of 420 kV circuit breakers in the field with integrated FOCS. As an alternative to full integration, others have mounted the FOCS

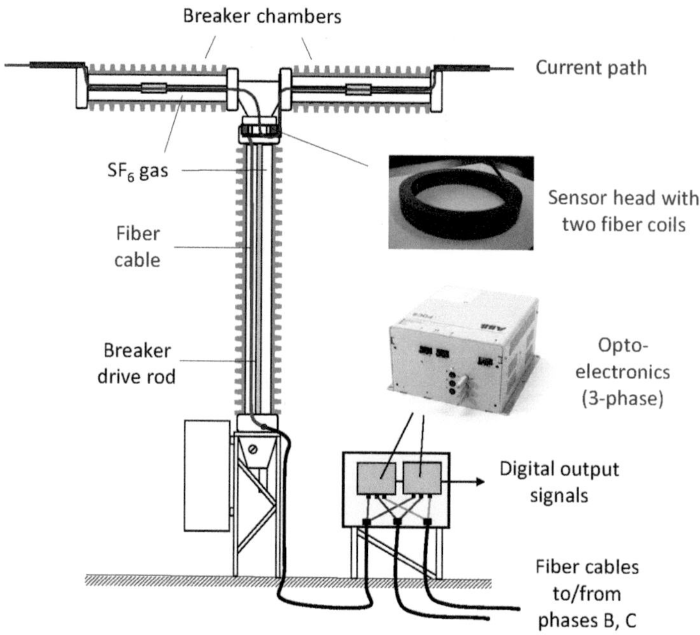

FIGURE 12.3 Integration of redundant FOCS in double chamber HV circuit breaker [327].

FIGURE 12.4 Live tank circuit breakers (420 kV) with integrated FOCS in HV substation [327].

fiber coil outside the breaker chambers at one of the terminals and guided the fiber lead to ground in a separate insulator column [695].

FOCS with large-diameter coils are not only of interest to the industrial applications of the previous section, but also to current measurement at generator circuit breakers, gas-insulated

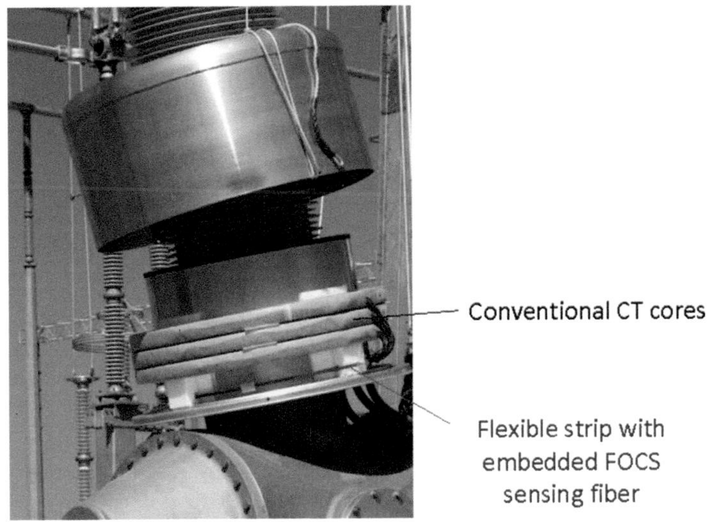

FIGURE 12.5 FOCS at a bushing of a dead tank circuit breaker [697].

switchgear (GIS) [227], or the base of high voltage bushings, e.g., of power transformers, dead tank circuit breakers (Figure 12.5), or HVDC converter stations [443, 696, 697]. Note that in the latter cases, current measurement occurs at ground potential. Another FOCS application that has gained particular interest is the protection of mixed overhead/underground power-lines, since the problem is difficult to solve with conventional CT [341, 698]. The identification of a potential fault on the underground cable section requires current measurement at both ends of the cable. Often, those locations are several km away from the nearest substations and without electric infrastructure. Other than conventional CT, FOCS fiber coils at those locations can be remotely interrogated. To this end, existing fibers in the optical ground wire cable of the power line may be used as optical links. Obviously, corresponding FOCS must work with non-PM fiber leads such as the ones in Figure 8.9 or Figure 8.12. More application examples can be found in the references of the previous chapters.

12.3 NUCLEAR FUSION RESEARCH

Fiber-optic current sensors with large-diameter fiber loops have also been developed for use in plasma physics, particularly at tokamaks for nuclear fusion research such as ITER (International Thermonuclear Experimental Reactor) in France. Here, the sensors are to be optimized for pulsed currents with peaks that can reach into the MA-range. Like in FOCS for electric power, much of the work has been dedicated to the exploration and mitigation of limiting effects due to birefringence, vibration, imperfect components, noise, etc. [225, 390, 699–703].

FURTHER READING

Since the 1980s, many papers have reviewed the status of optical current sensors at various stages of progress [134, 662, 704–718].

References

1. M. Faraday, "I. Experimental researches in electricity – nineteenth series," *Phil. Trans. R. Soc.* 136, 1–20, 1846, doi: 10.1098/rstl.1846.0001
2. E. Verdet, "On the optical properties developed in transparent bodies by the action of magnetism," *Ann. Chim. Phys.* 41, 570 (1854).
3. H. Becquerel, "Sur une interprétation applicable au phénomène de Faraday et au phénomène de Zeeman," *Comptes Rendus* 125, 679–685 (1897).
4. Y. Tamaru, H. Chen, A. Fuchimukai, et al., "Wavelength dependence of the Verdet constant in synthetic quartz glass for deep-ultraviolet light sources," *Opt. Mater. Express* 11, 814–820 (2021).
5. T. Yoshino, "Theory for the Faraday magneto-optic effect in optical fiber," *J. Opt. Soc. Am. B.* 22(9), 1856–1860 (2005).
6. J. Noda, T. Hosaka, Y. Sasaki, and R. Ulrich, "Dispersion of Verdet constant in stress-birefringent silica fibre," *Electron. Lett.* 20, 906–908 (1984).
7. A. H. Rose, S. M. Etzel, and C. M. Wang, "Verdet constant dispersion in annealed optical fiber current sensors," *J. Lightwave Technol.*, 15(5), 803–807 (1997), doi: 10.1109/50.580818
8. S. Donati, V. Annovazzi-Lodi, and T. Tambosso, "Magneto-optical fibre sensors for electrical industry: analysis of performances," *IEE Proc. Optoelectron.* 135(5), 372–382 (1988), doi: 10.1049/ip-j.1988.0069
9. D. Vojna, O. Slezák, A. Lucianetti, and T. Mocek, "Verdet constant of magneto-active materials developed for high-power Faraday devices," *Appl. Sci.* 9, 3160–3180 (2019), doi:10.3390/app9153160
10. L. I. Schiff, *Quantum Mechanics*, 3rd ed. (McGraw-Hill, Tokyo, 1968).
11. M. Battiato, G. Barbalinardo, and P. M. Oppeneer, "Quantum theory of the inverse Faraday effect," *Phys. Rev.* B89, 014413 (2014).
12. A. K. Zvezdin and V. A. Kotov, *Modern Magnetooptics and Magnetooptical Materials*, 1st ed. (Taylor & Francis Group, New York, NY, USA, 1997), pp. 1–109.
13. D. Imaizumi, T. Hayakawa, and M. Nogami, "Faraday rotation effects of Mn2+-modified Tb2O3-B2O3 glass in pulsed magnetic field," *J. Lightwave Technol.* 20(4), 740–744 (2002).
14. P. S. Pershan, "Magneto-optical effects," *J. Appl. Phys.* 38, 1482–1490 (1967), doi: 10.1063/1.1709678
15. M. Freiser, "A survey of magnetooptic effects," *IEEE Trans. Magn.* 4(2), 152–161 (1968).
16. M. N. Deeter, A. H. Rose, and G. W. Day, "Sensitivity limits to ferrimagnetic Faraday effect magnetic field sensors," *J. Appl. Phys.* 70, 6407 (1991), doi: 10.1063/1.349910
17. D.-X. Chen, J. A. Brug, and R. B. Goldfarb, "Demagnetization factors for cylinders," *IEEE Trans. Magn.* 27(4), 3601–3619 (1991).
18. R. C. Booth and E. A. D. White, "Magneto-optic properties of rare earth iron Garnet crystals in the wavelength range 1.1–1.7 μm and their use in device fabrication," *J. Phys. D Appl. Phys.* 17, 579–587, 1984. doi: 10.1088/0022-3727/17/3/015

19. C. Z. Tan and J. Arndt, "The Faraday effect in silica glasses," *Physica B Condens. Matter* 233(1), 1–7 (1997).
20. C. Z. Tan and J. Arndt, "Wavelength dependence of the Faraday effect in glassy SiO2," *J. Phys. Chem. Solids* 60, 1689–1692 (1999).
21. R. Yasuhara, S. Tokita, J. Kawanaka, et al. "Cryogenic temperature characteristics of Verdet constant on terbium gallium garnet ceramics," *Opt. Express* 15(18), 11255–11261 (2007).
22. J. L. Flores and J. A. Ferrari, "Verdet constant dispersion measurement using polarization-stepping techniques," *Appl. Opt.* 47(24), 4396–4399 (2008).
23. K. Turvey, "Determination of Verdet constant from combined ac and dc measurements," *Rev. Scient. Instrum.* 64(6), 1561–1568 (1993),. doi: doi.org/10.1063/1.1144027
24. R. B. Wagreich and C. C. Davis, "Accurate magneto-optic sensitivity measurements of some diamagnetic glasses and ferrimagnetic bulk crystals using small applied AC magnetic fields," *IEEE Trans. Magn.* 33(3), 2356–2361 (1997), doi: 10.1109/20.573856
25. M. Sofronie, M. Elisa, B. A. Sava, L. Boroica, M. Valeanu, and V. Kuncser, "Rapid determination of Faraday rotation in optical glasses by means of secondary Faraday modulator," *Rev. Scient. Instrum.* 86(5), 053905 (2015).
26. A. Shaheen, H. Majeed, and M. S. Anwar, "Ultralarge magneto-optic rotations and rotary dispersion in terbium gallium garnet single crystal," *Appl. Opt.* 54, 5549–5554 (2015).
27. A. Kruk and M. Mrozek, "The measurement of Faraday effect of translucent material in the entire visible spectrum," *Measurement* 162, 107912 (2020), doi: 10.1016/j.measurement.2020.107912
28. P. A. Williams, A. H. Rose, G. W. Day, T. E. Milner, and M. N. Deeter, "Temperature dependence of the Verdet constant in several diamagnetic glasses," *Appl. Opt.* 30(10), 1176–1178 (1991).
29. S. Kumari and S. Chakraborty, "Study of different magneto-optic materials for current sensing applications," *J. Sens. Sens. Syst.* 7(1), 421–431 (2018), doi: 10.5194/jsss-7-421-2018
30. J. L. Cruz, M. V. Andres, and M. A. Hernandez, "Faraday effect in standard optical fibers: dispersion of the effective Verdet constant," *Appl. Opt.* 35, 922–927 (1996).
31. K. Kurosawa, K. Yamashita, T. Sowa, and Y. Yamada, "Flexible fiber Faraday effect current sensor using flint glass fiber and reflection scheme," *IEICE Trans. Electron.* E83-C(3), 326–330 (2000).
32. K. Kurosawa, "Optical transducer using flint glass fiber as the Faraday sensor element," *Opt. Rev.* 4(1A), 38–44 (1997).
33. H. Katsukawa, H. Ishikawa, H. Okajima, and T. W. Cease, "Development of an optical current transducer with a bulk type Faraday sensor for metering," *IEEE Trans. Power Deliv.* 11(2), 702–707 (1996). doi: 10.1109/61.489326
34. S. Petricevic and P. Mihailovic, "Compensation of Verdet constant temperature dependence by crystal core temperature measurement," *Sensors* 16, 1627 (2016).
35. K. Kyuma, S. Tai, M. Nunoshita, T. Takioka, and Y. Ida, "Fiber optic measuring system for electric using a magnetooptic sensor," *IEEE J. Quant. Electron.* 18(10), 1619–1623 (1982).
36. C. Li and T. Yoshino, "Simultaneous measurement of current and voltage by use of one bismuth germanate crystal," *Appl. Opt.* 41(25), 5391–5397 (2002).
37. A. B. Villaverde and E. C. C. Vasconcellos, "Magnetooptical dispersion of Hoya glasses: AOT-5, AOT-44B, and FR-5," *Appl. Opt.* 21, 1347–1348 (1982).
38. D. Huang, S. Srinivasan, and J. E. Bowers, "Compact Tb doped fiber optic current sensor with high sensitivity," *Opt. Express* 23, 29993–29999 (2015).
39. Q. Chen, Q. Ma, H. Wang, and Q. Chen, "Diamagnetic tellurite glass and fiber based magneto-optical current transducer," *Appl. Opt.* 54, 8664–8669 (2015).
40. N. P. Barnes and L. B. Petway, "Variation of the Verdet constant with temperature of terbium gallium garnet," *J. Opt. Soc. Am. B* 9, 1912–1915 (1992).
41. O. Slezák, R. Yasuhara, A. Lucianetti, and T. Mocek, "Temperature-wavelength dependence of terbium gallium garnet ceramics Verdet constant," *Opt. Mater. Express* 6(11), 3683–3691 (2016).

42. A. H. Rose, M. N. Deeter, and G. W. Day, "Submicroampere-per-root-hertz current sensor based on the Faraday effect in Ga:YIG," *Opt. Lett.* 18, 1471–1473 (1993).
43. G. V. Golodolinski, "Application of the Faraday effect for current measurement," *Electrichestvo* 8 (1956).
44. S. Saito, Y. Fujii, K. Yokoyama, J. Hamasaki, and Y. Ohno, "C1 – The laser current transformer for EHV power transmission lines," *IEEE J. Quantum Electron.* 2(8), 255–259 (1966). doi: 10.1109/JQE.1966.1074032
45. S. Saito, J. Hamasaki, Y. Fujii, and Y. Ohno, "12.4 – Development of the laser current transformer for the extra-high-voltage power transmission lines," *IEEE J. Quantum Electron.* 3(11), 589–597 (1967). doi: 10.1109/JQE.1967.1074399
46. S. Saito, "Current transformer…," U.S. Patent 3597683, Nov. 17, 1965.
47. Y. Pelenc, et al., "Apparatus for current measurement by means of the Faraday effect," U.S. Patent 3419802, Apr. 10, 1965.
48. Y. Pelenc and G. Bernard, "Prototype industriel de transformateur de courant a effet magneto-optique," *Revue Générale de l'Electricité (France)* 76(7–8), 1055–1064 (1967).
49. S. Takeshita and T. Sasano, "Measurement of impulse current by laser current transformer," *Proc. IEEE* 56(8) 1404–1405 (1968). doi: 10.1109/PROC.1968.6622
50. K. Mollenbeck, "Device for measuring currents in a high voltage conductor….," U.S. Patent 3679974, Aug. 1, 1969.
51. G. Lesueur, "Current reducers using the Faraday effect", U.S. Patent 3743929, Nov. 6, 1969.
52. G. Lesueur, "Optical current transformer", U.S. Patent 3708747, Oct. 23, 1970.
53. A. Jaecklin and M. Lietz, "Elimination of disturbing birefringence effects on Faraday rotation," *Appl. Opt.* 11(3), 617–621 (1972).
54. A. Jaecklin, U.S. Patent 3693082, Dec. 23, 1969.
55. H. Harms and E. Feldtkeller, "Automatic recording of Faraday rotation and circular dichroism," *Rev. Sci. Instrum.* 44, 742 (1973).
56. A. J. Rogers, "Optical technique for measurement of current at high voltage," *Proc. IEEE* 120(2), 261–267 (1973).
57. A. J. Rogers, "Method for simultaneous measurement of current and voltage on high voltage lines using optical techniques," *Proc. IEE* 123, 957–960, 1976.
58. A. J. Rogers, "Optical methods for measurement of voltage and current on power systems," *Opt. Laser Technol.* 9, 273 (1977).
59. G. A. Massey, D. C. Erickson, and R. A. Kadlec, "Electromagnetic field components: their measurement using linear electrooptic and magnetooptic effects," *Appl. Opt.* 14(11), 2712–2719 (1975).
60. K. Kyuma, S. Tai, and M. Nunoshita, "Development of fibre optic sensing systems – a review," *Opt. Lasers Eng.* 3(3), 155–182, (1982).
61. G. Li, M. G. Kong, G. R. Jones, and J. W. Spencer, "Sensitivity improvement of an optical current sensor with enhanced Faraday rotation," *J. Lightwave Technol.* 15(12), 2246–2252 (1997), doi: 10.1109/50.643549
62. G. D. Li, R. A. Aspey, M. G. Kong, J. R. Gibson, and G. R. Jones, "Elliptical polarization effects in a chromatically addressed Faraday current sensor," *Meas. Sci. Technol.* 10(1), 25–30 (1999).
63. B. Yi, C. B. Chu, K. S. Chiang, and H. S. H. Chung, "New design of optical electric-current sensor for sensitivity improvement," *IEEE Trans. Instrum. Meas.* 49(2), 418–423 (2000), doi: 10.1109/19.843089
64. W. C. Michie, A. Cruden, P. Niewczas, et al., "Harmonic analysis of current waveforms using optical current sensor," *IEEE Trans. Instrum. Meas.* 51(5), 1023–1026 (2002).
65. N. Inoue and K. Yamasawa, "Stabilizing the temperature dependencies of the Verdet constant of a Bi doped garnet and development of a highly sensitive optical magnetic field sensor," *Trans. IEE Jpn.* 115-A, 1114–1120 (1995).
66. K. Kyuma, Shuichi Tai, M. Nunoshita, N. Mikami, and Y. Ida, "Fiber-optic current and voltage sensors using a Bi12GeO20 single crystal," *J. Lightwave Technol.* 1(1), 93–97 (1983). doi: 10.1109/JLT.1983.1072069

67. Y. Hasegawa, Y. Ichikawa, H. Katsukawa, N. Tanaka, and Y. Sakurai, "Development of a new type of optical transducer for measuring fault current," *IEEE Trans. Power Deliv.* 9(3), 1245–1252 (1994). doi: 10.1109/61.311150
68. C. Li and X. Cui, "An optical voltage and current sensor with electrically switchable quarter waveplate," *Sens. Actuator. A Phys.* 126(1), 62–67 (2006).
69. C. Li and T. Yoshino, "Single-crystal magneto-optic sensor with electrically adjustable sensitivity," *Appl. Opt.* 51(21), 5119–5125 (2012).
70. L. Wang, Y. Huang, C. Deng, C. Hu, and T. Wang, "A compact polarimetric fiber-optic sensor based on Bi4Ge3O12 crystal for ultra-high surge current sensing," in *26th Int. Conf. on Optical Fiber Sensors*, OSA Technical Digest (Optical Society of America, 2018), paper WF96.
71. P. M. Mihailovic, S. J. Petricevic, and J. B. Radunovic, "Compensation for temperature-dependence of the Faraday effect by optical activity temperature shift," *IEEE Sens. J.* 13(2), 832–837 (2013), doi: 10.1109/JSEN.2012.2230322
72. A. C. S. Brigida, I. M. Nascimento, S. Mendonça, et al., "Experimental and theoretical analysis of an optical current sensor for high power systems," *Photonic Sens.* 3, 26–34 (2013), doi: 10.1007/s13320-012-0092-1
73. L. N. Bjorn, "Faraday effect current sensor," U.S. Patent application US2010/0711004A1, Nov. 30, 2007.
74. C. D. Perciante and J. A. Ferrari, "Magnetic crosstalk minimization in optical current Sensors," *IEEE Trans. Instrum. Meas.* 57(10), 2304–2308 (2008). doi: 10.1109/TIM.2008.919913
75. L. Di Rienzo, R. Bazzocchi, and A. Manara, "Circular arrays of magnetic sensors for current measurement," *IEEE Trans. Instrum. Meas.* 50(5), 1093–1096 (2001). doi: 10.1109/19.963165
76. X. Ma, Y. Guo, X. Chen, Y. Xiang, and K. Chen, "Impact of coreless current transformer position on current measurement," *IEEE Trans. Instrum. Meas.* 68(10), 3801–3809 (2019). doi: 10.1109/TIM.2019.2927544
77. P. Niewczas, W. I. Madden, W. C. Michie, A. Cruden, and J. R. McDonald, "Magnetic crosstalk compensation for an optical current transducer," *IEEE Trans. Instrum. Meas.* 50(5), 1071–1075, (2001), doi: 10.1109/19.963160
78. Y. Li, W. Zhang, M. Wang, and J. Liu, "Optical current sensing mechanism under a non-uniform magnetic field," *Appl. Opt.* 58(20), 5472–5478 (2019).
79. C. D. Perciante, "Evolution of the polarization state in material media with uniform linear birefringence and an applied nonhomogeneous external magnetic field: application to bulk-type Faraday current sensors," *J. Opt. Soc. Am. A* 25(8), 1994–2000 (2008).
80. D. C. Erickson, "The use of fiber optics for communications, measurement and control within high voltage substations," *IEEE Trans. Power Appar. Syst.1* PAS-99(3), 1057–1065 (1980).
81. T. Sawa, K. Kurosawa, T. Kaminishi, and T. Yokota, "Development of optical instrument transformers," *IEEE Trans. Power Deliv.* 5(2), 884–891 (1990), doi: 10.1109/61.53098
82. A. Cruden, Z. J. Richardson, J. R. NcDonald, and I. Andonovic, "Optical crystal based devices for current and voltage measurement," *IEEE Trans. Power Deliv.* 10(3), 1217–1223 (1995), doi: 10.1109/61.400899
83. A. Cruden, Z. J. Richardson, J. R. McDonald, I. Andonovic, W. Laycock, and A. Bennett, "Compact 132 kV combined optical voltage and current measurement system," *IEEE Trans. Instrum. Meas.* 47(1), 219–223 (1998), doi: 10.1109/19.728822
84. G. W. Day, P. D. Hale, M. Deeter, T. E. Milner, D. Conrad, and S. M. Etzel, "Optical power line voltage and current measurement systems. Volume 1: limits to precision of electro-optic and magneto-optic sensors," Final Report National Bureau of Standards, Washington, DC, Sept. 1987.
85. G. Katsuta, K. Muraoka, N. Inoue, S. Sakai, T. Tsunekage, and K. Ando, "Fault section detection system for 66 kV underground branch transmission lines using optical magnetic field sensors," *IEEE Trans. Power Deliv.* 7(1), 1–9, (1992). doi: 10.1109/61.108882
86. Y. Yoshida, S. Kawazoe, K. Ibuki, K. Yamada, and N. Ochi, "New fault locating system for air-insulated substations using optical current detector," *IEEE Trans. Power Deliv.* 7(4), 1805–1812 (1992). doi: 10.1109/61.156982

87. N. Inoue, T. Tsunekage, and S. Sakai, "Fault section detection system for 275-kV XLPE-insulated cables with optical sensing technique," *IEEE Trans. Power Deliv.* 10(3), 1148–1155, (1995). doi: 10.1109/61.400891
88. F. Sorensen, EP Patent application EP 2128631A1, May 30, 2008.
89. I. M. Nascimento, A. C. S. Brígida, J. M. Baptista, et al., "Novel optical current sensor for metering and protection in high power applications," *Instrum. Sci. Technol.* 44(2), 148–162, 2016.
90. J. Zubia, L. Casado, G. Aldabaldetreku, et al., "Design and development of a low-cost optical current sensor," *Sensors* 13, 13584–13595 (2013).
91. S. H. Zaidi and R. P. Tatam, "Faraday-effect magnetometry: compensation for the temperature-dependent Verdet constant," *Meas. Sci. Technol.* 5, 1471 (1994).
92. C. D. Perciante and J. A. Ferrari, "Faraday current sensor with temperature monitoring," *Appl. Opt.* 44(32), 6910–6912 (2005).
93. H. Wang, J. Jiang, H. Hu, et al. "Temperature compensation of optical alternating magnetic field sensor via a novel method for on-line measuring," *Opt. Express* 28(9), 13682–13693 (2020).
94. J. Lenz and A. S. Edelstein, "Magnetic sensors and their applications," *IEEE Sens. J.* 6(3), 631–649 (2006).
95. P. Ripka and M. Janosek, "Advances in magnetic field sensors," *IEEE Sens. J.* 10(6), 1108–1116 (2010), doi: 10.1109/JSEN.2010.2043429
96. C. P. Gooneratne, B. Li, and T. E. Moellendick, "Downhole applications of magnetic sensors." *Sensors* 17(10), 2384 (2017), doi:10.3390/s17102384
97. M. Iwanami, M. Nakada, H. Tsuda, and K. Ohashi, "Ultra small magneto-optic field probe fabricated by aerosol deposition," *IEICE Electron. Express* 4(17), 542–548 (2007), doi: 10.1587/elex.4.542
98. S. J. Lee, S. H. Song, D. C. Jiles, and H. Hauser, "Magnetooptic sensor for remote evaluation of surfaces," *IEEE Trans. Magn.* 41(7), 2257–2259 (2005), doi: 10.1109/TMAG.2005.851858
99. D. Gerber and J. Biela, "High-dynamic and high-precise optical current measurement system based on the Faraday effect," *IEEE Trans. Plasma Sci.* 43(10), 3550–3554 (2015), doi: 10.1109/TPS.2015.2437395
100. S. Rietman and J. Biela, "Sensor design for a current measurement system with high bandwidth and high accuracy based on the Faraday effect," in *2019 21st European Conference on Power Electronics and Applications (EPE '19 ECCE Europe)*, Genova, Italy, 2019, pp. P.1–P.10, doi: 10.23919/EPE.2019.8915122
101. M. N. Deeter, A. H. Rose, and G. W. Day, "Fast, sensitive magnetic-field sensors based on the Faraday effect in YIG," *J. Lightwave Technol.* 8(12), 1838–1842 (1990). doi: 10.1109/50.62880
102. M. N. Deeter, "High sensitivity fiber-optic magnetic field sensors based on iron garnets," *IEEE Trans. Instrum. Meas.* 44(2), 464–467 (1995). doi: 10.1109/19.377881
103. M. N. Deeter, "Fiber-optic Faraday-effect magnetic-field sensor based on flux concentrators," *Appl. Opt.* 35(1), 154–157 (1996).
104. N. Itoh, H. Minemoto, D. Ishiko, and S. Ishizuka, "Small optical magnetic field sensor probe using rare-earth iron garnet crystals," *J. Magn. Soc. Jpn.* 21(4), 637–640 (1997).
105. O. Kamada, T. Nakaya, and S. Higuchi, "Magnetic field optical sensors using Ce:YIG single crystals as a Faraday element," *Sens. Actuator.* A119, 345–348 (2005).
106. E. J. Casey and C. H. Titus, "Magneto-optical electric current sensing arrangement," U.S. Patent 3324393, July 26, 1963.
107. E. Feldtkeller, "Magneto-optic measuring transducer for very high currents/voltages," U.S. Patent 3980949, Sept. 17, 1973.
108. M. Hercher, "Ultra-high resolution interferometric sensor," *OSA Optics & Photonics News*, Nov. 1991, pp. 24–29.
109. A. E. Petersen, "Portable optical AC- and proposed DC-current sensor for high voltage applications," *IEEE Trans. Power Deliv.* 10(2), 595–599 (1995), doi: 10.1109/61.400873

110. N. Itoh, H. Minemoto, D. Ishiko, and S. Ishizuka, "Small optical magnetic-field sensor that uses rare-earth iron garnet films based on the Faraday effect," *Appl. Opt.* 38(10), 2047–2052 (1999).
111. T. Yoshino, K. Minegishi, and M. Nitta, "A very sensitive Faraday effect current sensor using a YIG/ring-core transformer in a transverse configuration," *Meas. Sci. Technol.* 12(7), 850–853 (2001).
112. B. Yi, B. C. B. Chu, and K. S. Chiang, "Magneto-optical electric-current sensor with enhanced sensitivity," *Meas. Sci. Technol.* 13, N61–N64 (2002).
113. T. Yoshino, S. Torihata, M. Yokota, and N. Tsukada, "Faraday-effect optical current sensor with a garnet film/ring core in a transverse configuration," *Appl. Opt.* 42(10), 1769–1772 (2003).
114. P. Mihailovic, S. Petricevic, Z. Stojkovic, and J. B. Radunovic, "Development of a portable fiber-optic current sensor for power systems monitoring," *IEEE Trans. Instrum. Meas.* 53(1), 24–30 (2004), doi: 10.1109/TIM.2003.821500
115. S. J. Petricevic, Z. Stojkovic, and J. B. Radunovic, "Practical application of fiber-optic current sensor in power system harmonic measurement," *IEEE Trans. Instrum. Meas.* 55(3), 923–930 (2006), doi: 10.1109/TIM.2006.873793
116. X. Wan, J. Zhao, and M. Wang, "A novel design of the sensor head for avoiding the influence of the reflection phase shift in optical current transducer," *Opt. Lasers Eng.* 48(3), 325–328 (2010).
117. M. H. Samimi, S. Bahrami, A. A. S. Akmal, and H. Mohseni, "Effect of nonideal linear polarizers, stray magnetic field, and vibration on the accuracy of open-core optical current transducers," *IEEE Sens. J.* 14(10), 3508–3515 (2014), doi: 10.1109/JSEN.2014.2328873
118. M. H. Samimi, A. A. S. Akmal, H. Mohseni, and J. Jadidian, "Open-core optical current transducer: modeling and experiment," *IEEE Trans. Power Deliv.* 31(5), 2028–2035 (2016), doi: 10.1109/TPWRD.2015.2490720
119. P. B. Davis et al., "Electro-optic current sensor with high dynamic range and accuracy," U.S. patent 10006944B2, Nov. 6, 2017.
120. R. Renz, "Apparatus for measuring very high currents particularly direct currents," U.S. Patent 3746983, July 20, 1970.
121. T. Sato, G. Takahashi, and Y. Inui, "Method and apparatus for optically measuring a current," U.S. Patent 4564754, priority Mar. 8, 1982.
122. M. Kanoi, G. Takahashi, T. Sato, M. Higaki, E. Mori, and K. Okumura, "Optical voltage and current measuring system for electric power systems," *IEEE Trans. Power Deliv.* 1(1), 91–97 (1986), doi: 10.1109/TPWRD.1986.4307893
123. E. A. Ulmer and T. J. Meyer, "Ring optical current transducer," U.S. Patent 5124634, April 25, 1990.
124. E. A. Ulmer, "A high accuracy optical current transducer for electric power systems," *IEEE Trans. Power Deliv.* 5(2), 892–898 (1990).
125. T. W. Cease and P. Johnston, "A magneto-optic current transducer," *IEEE Trans. Power Deliv.* 5(2), 548–555 (1990), doi: 10.1109/61.53056
126. T. W. Cease, J. G. Driggans and S. J. Weikel, "Optical voltage and current sensors used in a revenue metering system," *IEEE Trans. Power Deliv.* 6(4), 1374–1379 (1991).
127. T. D. Maffetone and T. M. McClelland, "345 kV substation optical current measurement system for revenue metering and protective relaying," *IEEE Trans. Power Deliv.* 6(4), 1430–1437 (1991), doi: 10.1109/61.97673
128. K. Bohnert, P. Gabus, H. Brändle, and A. Khan, "Fiber-optic current and voltage sensors for high-voltage substations," in *Proc. 16th Int. Conference on Optical Fiber Sensors*, 2003, pp. 752–755.
129. S. P. Bush and D. A. Jackson, "Dual-channel Faraday-effect current sensor capable of simultaneous measurement of two independent currents," *Opt. Lett.* 16(12), 955–957 (1991).
130. Y. N. Ning, B. C. B. Chu, and D. A. Jackson, "Miniature Faraday current sensor based on multiple critical angle reflections in a bulk-optic ring," *Opt. Lett.* 16(24), 1996–1998 (1991).
131. S. P. Bush and D. A. Jackson, "Numerical investigation of the effects of birefringence and total internal reflection on Faraday effect current sensors," *Appl. Opt.* 31(25), 5366–5374 (1992).

132. B. C. B. Chu, Y. N. Ning, and D. A. Jackson, "Faraday current sensor that uses a triangular-shaped bulk-optic sensing element," *Opt. Lett.* 17(16), 1167–1169 (1992).
133. Y. N. Ning and D. A. Jackson, "Faraday effect optical current clamp using a bulk-glass sensing element," *Opt. Lett.* 18(10), 835–837 (1993).
134. Y. N. Ning, Z. P. Wang, A. W. Palmer, K. T. V. Grattan, and D. A. Jackson, "Recent progress in optical current sensing techniques," *Rev. Sci. Instrum.* 66(5), 3097–3111, (1995).
135. A. G. Podoleanu, N. E. Fisher, D. A. Jackson, "A single fibre-optic down-lead Faraday current sensor," *Sens. Actuator. A Phys.* 58(3), 225–228 (1997).
136. A. J. Rogers, J. Xu, and J. Yao, "Vibration immunity for optical-fibre current measurement," in *Proc. SPIE 2360, Tenth Int. Conf. Optical Fibre Sensors*, 1994, doi: 10.1117/12.184958
137. J. Song, P. G. McLaren, D. J. Thomson, and R. L. Middleton, "A prototype clamp-on magneto-optical current transducer for power system metering and relaying," *IEEE Trans. Power Deliv.* 10(4), 1764–1770 (1995), doi: 10.1109/61.473382
138. N. E. Fisher and D. A. Jackson, "Vibration immunity and Ampere's circuital law for a near perfect triangular Faraday current sensor," *Meas. Sci. Technol.* 7(8), 1099–1102 (1996).
139. G. A. Woolsey, N. E. Fisher, and D. A. Jackson, "Tuning a bulk-glass optical current sensor by controlling conditions external to its reflecting surfaces," *Sens. Actuator. A Phys.* 63(1), 27–32 (1997).
140. N. E. Fisher, D. A. Jackson, and G. A. Woolsey, "Faraday current sensors and the significance of subtended angles," *Sens. Actuator. A Phys.* 63(2), 5 (1997).
141. Z. P. Wang, W. M. Sun, Z. J. Huang, C. Kang, S. L. Ruan, Y. H. Luo, A. W. Palmer, and K. T. V. Grattan, "Effects of reflection-induced retardance on the immunity of bulk optic-material current sensors," *Appl. Opt.* 37(31), 7293–7297 (1998).
142. I. W. Madden, W. C. Michie, A. Cruden, P. Niewczas, and J. R. McDonald, "Temperature compensation for optical current sensors," *Opt. Eng.* 38(10), 1699–1707 (1999), doi: 10.1117/1.602222
143. P. Niewczas, A. C. W. Craig Michie, I. W. Madden, and J. R. McDonald, "Vibration compensation technique for an optical current transducer," *Opt. Eng.* 38(10), 1708–1714 (1999), doi: 10.1117/1.602296
144. T. Bosselmann and P. Menke, "Intrinsic temperature compensation of magnetooptic AC current transformers with glass ring sensor head," in *Proc. SPIE 2360, pp. 20-23, 10th Int. Conf. on Optical Fibre Sensors*, 1994, doi: 10.1117/12.185043
145. P. Menke and T. Bosselmann, "Temperature compensation in magnetooptic AC current sensors using an intelligent AC-DC signal evaluation," *J. Lightwave Technol.* 13(7), 362–1370 (1995), doi: 10.1109/50.400675
146. M. Willsch and T. Bosselmann, "Vibration compensation for a glass ring type magnetooptic current sensor," in *Optical Fiber Sensors* (Optical Society of America, 1996), paper We14.
147. T. Sato and I. Sone, "Development of bulk-optic current sensor using glass ring type faraday cells," *Opt. Rev.* 4, A35 (1997), doi: 10.1007/BF02935986
148. J. L. Arce-Diego, R. López-Ruisánchez, J. M. López-Higuera, and M. A. Muriel, "Model of an openable Faraday-effect hybrid-current optical transducer based on a square-shaped structure with internal mirror," *Appl. Opt.* 36, 6242–6245 (1997).
149. L. Erning, J. M. K. MacAlpine, L. Yanbing, and Y. Benshun, "A novel optical current transducer for power systems," *Electr. Power Syst. Res.* 46(1), 21–25 (1998).
150. T. Fujimoto, M. Shimizu, H. Nakagawa, I. Sone, K. Kawashima, and E. Mori, "Development of an optical current transformer for adjustable speed pumped storage systems," *IEEE Trans. Power Deliv.* 12(1), 45–50 (1997), doi: 10.1109/61.568224
151. T. Yoshino, M. Gojyuki, Y. Takahashi, and T. Shimoyama, "Single glass block faraday effect current sensor with homogeneous isotropic closed optical circuit," *Appl. Opt.* 36, 5566–5573 (1997).
152. T. Yoshino, Y. Takahashi, and M. Gojyuki. "Imperfection errors in glass block Faraday effect current sensor," *Opt. Rev.* 4(1), A108–A110, (1997).

153. T. Yoshino, M. Yokota, K. Aoki, K. Yamamoto, S. Itoi, and M. Ohtaka, "Fiber-linked Faraday effect current sensor by use of a flint glass cell with dielectric-coated retardation-compensated total reflection surfaces," *Appl. Opt.* 41, 5963–5968 (2002).
154. X. Ma and C. Luo, "A method to eliminate birefringence of a magneto-optic AC current transducer with glass ring sensor head," *IEEE Trans. Power Deliv.* 13(4), 1015–1019 (1998), doi: 10.1109/61.714435
155. Z. P. Wang, Z. J. Huang, C Kang, et al., "Optical current sensing element with single medium layers for high voltage applications," *Opt. Laser Technol.* 31(6), 455–458 (1999).
156. B. Yi, B. Chu, K. S. Chiang, "New design of a detachable bulk-optic Faraday effect current clamp," *Opt. Eng.* 40(6) (2001), doi: 10.1117/1.1365106
157. K. Kong, B. Chu, H. Chung, K. S. Chiang, "Triangular-shaped bulk-optic glass sensor for simultaneous measurement of three ac currents," *Opt. Eng.* 42(2), 421 (2003), doi: 10.1117/1.1533037
158. D. Chatrefou, "Device for optically measuring current intensity", European Patent 0613017B1, Feb. 22, 1994.
159. D. Chatrefou and G. F. Montillet, "A series of implementation of optical sensors in high voltage substations," in *2003 IEEE PES Transmission and Distribution Conference and Exposition* (IEEE Cat. No.03CH37495), Dallas, TX, USA, 2003, vol. 2, pp. 792–797, doi: 10.1109/TDC.2003.1335376
160. Z. P. Wang, X. Wang, X. Liu, C. Ouyang, and Q. Tan, "Effect of the spectral width of optical sources upon the output of an optical current sensor," *Meas. Sci. Technol.* 16(8), 1588–1592 (2005).
161. Z. P. Wang, Q. B. Li, Y. Qi, Z. J. Huang, J. H. Shi, "Wavelength dependence of the sensitivity of a bulk-glass optical current transformer," *Opt. Laser Technol.* 38(2), 87–93 (2006).
162. Z. P. Wang, Q. B. Li, and Q. Wu, "Effects of the temperature features of linear birefringence upon the sensitivity of a bulk glass current sensor," *Opt. Laser Technol.* 39(1), 8–12 (2007).
163. Z. P. Wang, X. Y. Liu, Y. M. Zhang, "Theoretical analysis of the effects of linear birefringence upon optical glass current transformers with different optical designs," *Measurement* 40(9–10), 811–815 (2007).
164. Z. Wang, C. Kang, Z. Huang, and X. Zhang, "Faraday mirror-typed optical current transformers and its theoretical analysis," *Opt. Laser Technol.* 39(2), 368–371 (2007).
165. X. Deng, Z. Li, Q. Peng, J. Liu, and J. Tian, "Research on magneto-optic current sensor for high-current pulses," *Rev. Scient. Instrum.* 79, 083106 (2008).
166. M. Wang, J. Zhao, S. Liu, et al., "Optical current sensor immune to reflection phase shift based on graded-index magneto-optical glass," *Appl. Opt.* 48(32), 6264–6270 (2009).
167. N. Correa, H. Chuaqui, E. Wyndham, et al., "Current measurement by Faraday effect on GEPOPU," *Appl. Opt.* 51(6), 758–762 (2012).
168. K. B. Rochford, A. H. Rose, M. N. Deeter, and G. W. Day, "Faraday effect current sensor with improved sensitivity–bandwidth product," *Opt. Lett.* 19, 1903–1905 (1994).
169. Y. Li, J. Wan, W. Zhang, and Jun Liu, "Analysis of antimagnetic interference properties and structural optimization of magneto-optical glass current sensors," *Appl. Opt.* 59, 1123–1129 (2020).
170. E. Hecht and A. Zajac, *Optics* (Addison-Wesley Publishing Company, 1974).
171. A. M. Smith, "Polarization and magnetooptic properties of single-mode optical fiber," *Appl. Opt.* 17(1), 52, 1978.
172. A. M. Smith, "Birefringence induced by bends and twists in single-mode optical fiber," *Appl. Opt.* 19, 2606–2611 (1980).
173. R. Ulrich, S. C. Rashleigh, and W. Eickhoff, "Bending-induced birefringence in single-mode fibers," *Opt. Lett.* 5, 273–275 (1980).
174. S. C. Rashleigh and R. Ulrich, "Magneto-optic current sensing with birefringent fibers," *Appl. Phys. Lett.* 34, 768–770 (1979).

175. S. C. Rashleigh and R. Ulrich, "High birefringence in tension-coiled single-mode fibers," *Opt. Lett.* 5, 354–356 (1980).
176. S. C. Rashleigh, "Origins and control of polarization effects in single-mode fibers," *J. Lightwave Technol.* 1(2), 312–331 (1983).
177. N. Imoto, N. Yoshizawa, J. Sakai, and H. Tsuchiya, "Birefringence in single-mode optical fiber due to elliptical core deformation and stress anisotropy," *IEEE J. Quantum Electron.* 16(11), 1267–1271 (1980), doi: 10.1109/JQE.1980.1070382
178. J. Sakai and T. Kimura, "Birefringence and polarization characteristics of single-mode optical fibers under elastic deformations," *IEEE J. Quantum Electron.* 17(6), 1041–1051 (1981), doi: 10.1109/JQE.1981.1071213
179. J. Sakai and T. Kimura, "Polarization behavior in multiply perturbed single-mode fibers," *IEEE J. Quantum Electron.* 18(1), 59–65 (1982), doi: 10.1109/JQE.1982.1071368
180. Z. B. Ren, P. Robert, and P.-A. Paratte, "Temperature dependence of bend- and twist-induced birefringence in a low-birefringence fiber," *Opt. Lett.* 13, 62–64 (1988).
181. A. J. Rogers, "A vibration-insensitive optical fiber current sensor," *J. Opt. Sens.* 1, 361–370 (1986).
182. T. W. MacDougall and T. F. Hutchinson, "Stray magnetic-field response of linear birefringent optical current sensors," *Appl. Opt.* 34(21), 4373–4379 (1995).
183. S. Cheng, Z.-Z. Guo, G.-Q. Zhang, W.-B. Yu, and Y. Shen, "Distributed parameter model for characterizing magnetic crosstalk in a fiber optic current sensor," *Appl. Opt.* 54, 10009–10017 (2015).
184. D. Marcuse, *Theory of Dielectric Optical Waveguides* (Academic Press, New York, 1974).
185. R. Ulrich, "Representation of codirectional coupled waves," *Opt. Lett.* 1(3), 109–111 (1977).
186. R. Ulrich and A. Simon, "Polarization optics of twisted single mode fibers," *Appl. Opt.* 18(13), 2241–2251 (1979).
187. A. Yariv, "Coupled-mode theory for guided-wave optics," *IEEE J. Quantum Electron.* QE-9, 919–933 (1973).
188. M. Monerie and L. Jeunhomme, "Polarization mode coupling in long single-mode fibres," *Opt. Quantum Electron.* 12, 449–461 (1980).
189. R. Dändliker, "Rotational effects of polarization in optical fibers," in *Optical Wave Sciences and Technology, Anisotropic and Nonlinear Optical Waveguides*, C. G. Someda and G. Stegeman, Eds. (Elsevier, 1992), pp. 39–76. ISSN 09275479, ISBN9780444884893, doi: 10.1016/B978-0-444-88489-3.50006-1
190. C. R. Menyuk and P. K. A. Wai, "Polarization evolution and dispersion in fibers with spatially varying birefringence," *J. Opt. Soc. Am. B* 11, 1288–1296 (1994).
191. J. N. Nye, *Physical Properties Of Crystals* (Oxford University Press, London, 1967).
192. Z. Fang, F. Yang, H. Cai, and R. Qu, "Coupled-mode equation of polarization modes of twisted birefringent fibers in a unified coordinate," *Appl. Opt.* 52(3), 530–537 (2013).
193. D. Tang, A. H. Rose, G. W. Day, and S. M. Etzel, "Annealing of linear birefringence in single-mode fiber coils: applications to optical fiber current sensors," *J. Lightwave Technol.* 9(8), 1031–1037 (1991).
194. Y. Namihira, "Opto-elastic constant in single-mode optical fibers," *J. Lightwave Technol.* 3, 1078–1083 (1983).
195. W. J. Tabor and F. S. Chen, "Electromagnetic propagation through materials possessing both Faraday rotation and birefringence: experiments with Ytterbium Orthoferrite," *J. Appl. Phys.* 40, 2760–2765 (1969).
196. A. Papp and H. Harms, "Magneto-optical current transformer. 1: Principles," *Appl. Opt.* 19(22), 3729–3734 (1980).
197. Z. B. Ren and Ph. Robert, "Polarization multiplexing applied to a fiber current sensor," *Opt. Lett.* 14, 1228–1230 (1989).

198. E. Feldtkeller, H. Harms, and A. Papp, "Magneto-optic high voltage current measuring transducer," U.S. Patent 4070620, Sept. 15, 1975.
199. H. Harms and E. Feldtkeller, "Magneto-optic high voltage current measuring transducer," U.S. Patent 4070622, Sept. 26, 1975.
200. H. Schneider, H. Harms, A. Papp, and H. Aulich, "Low-birefringence single-mode optical fibers: preparation and polarization characteristics," *Appl. Opt.* 17(19), 3035–3037 (1978).
201. H. Aulich, W. Beck, N. Douklias, H. Harms, A. Papp, and H. Schneider, "Magneto-optical current transformer. 2: Components," *Appl. Opt.* 19(22), 3735–3740 (1980).
202. C. D. Perciante, S. Aparicio, R. Illa, and J. A. Ferrari, "Nonplanar fiber-optic sensing head for the compensation of bending-induced birefringence in Faraday current sensors," *Appl. Opt.* 54, 5708–5714 (2015).
203. M. J. Marrone, R. D. Esman, and A. D. Kersey, "Fiber-optic magnetic field sensor with an orthoconjugating loop mirror," *Opt. Lett.* 18, 1556–1558 (1993).
204. S. Zhou and X. Zhang, "Simulation of linear birefringence reduction in fiber-optical current sensor," *IEEE Photonics Technol. Lett.* 19(19), 1568–1570 (2007), doi: 10.1109/LPT.2007.904344
205. P. Drexler and P. Fiala, "Utilization of Faraday mirror in fiber-optic current sensors," *Radioengineering* 17(4), 101–107 (2008).
206. A. Ben-Kish, M. Tur, and E. Shafir, "Geometrical separation between the birefringence components in Faraday-rotation fiber-optic current sensors," *Opt. Lett.* 16(9), 687–689 (1991).
207. J. W. Dawson and T. W. MacDougall, "Concatenated magneto-optic field sensors," U.S. patent 5736737, filed Nov. 22, 1995.
208. P. R. Forman and F. C. Jahoda, "Linear birefringence effects on fiber-optic current sensors," *Appl. Opt.* 27(15), 3088–3096 (1988).
209. P. Menke and T. Bosselmann, "Temperature compensation in magnetooptic AC current sensors using an intelligent AC-DC signal evaluation," *J. Lightwave Technol.* 13(7), 1362–1370 (1995), doi: 10.1109/50.400675
210. S. Mohr and T. Bosselmann, "A high dynamic magnetooptic current transformer with advanced signal processing," *IEEE Sens. J.* 3(1), 87–94 (2003), doi: 10.1109/JSEN.2003.810104
211. Y. Liu, L. Ma, and Z. He, "Birefringence variation independent fiber-optic current sensor using real-time SOP measurement," *IEEE Photonics J.* 9(5), 1–9 (2017), art no. 7105409, doi: 10.1109/JPHOT.2017.2746764
212. J. L. Flores-Nuñez, J. A. Ferrari, and C. D. Perciante, "Faraday current sensor using space-variant," *Opt. Eng.* 47(12) 123603 (2008), doi: 10.1117/1.3041772
213. N. E. Fisher and D. A. Jackson, "Improving the sensitivity of a Faraday current sensor by varying its operating point," *Meas. Sci. Technol.* 6(10) 1508, (1995).
214. G. W. Day, D. N. Payne, A. J. Barlow, and J. J. Ramskov-Hanson, "Faraday rotation in coiled mono-mode optical fibers: isolators, filters, and magnetic sensors," *Opt. Lett.* 7(5), 238–240 (1982).
215. R. H. Stolen and E. H. Turner, "Faraday rotation in highly birefringent fibers," *Appl. Opt.* 19(6), 842–845 (1980).
216. C. D. Perciante and J. A. Ferrari, "Cancellation of bending-induced birefringence in single-mode fibers: application to Faraday sensors," *Appl. Opt.* 45, 1951–1956 (2006).
217. Y. Huang, L. Xia, F. Pang, Y. Yuan, and J. Ji, "Self-compensative fiber optic current sensor," *J. Lightwave Technol.* 39(7), 2187–2193 (2021), doi: 10.1109/JLT.2020.3044935
218. P. McIntyre and A. W. Snyder, "Light propagation in twisted anisotropic media: Application to photoreceptors," *J. Opt. Soc. Am.* 68, 149–157 (1978).
219. L. Jeunhomme and M. Monerie, "Polarisation-maintaining single-mode fibre cable design," *Electron. Lett.* 16(24), 921–922 (1980), doi: 10.1049/el:19800656
220. M. Monerie and P. Lamouler, "Birefringence measurement in twisted single-mode fibres," *Electron. Lett.* 17(7), 252–253 (1981), doi: 10.1049/el:19810178

221. A. J. Barlow and D. N. Payne, "Polarisation maintenance in circularly birefringent fibres," *Electron. Lett.* 17(11), 388–389 (1981), doi: 10.1049/el:19810272
222. A. J. Barlow, J. J. Ramskov-Hansen, and D. N. Payne, "Birefringence and polarization mode-dispersion in spun single-mode fibers," *Appl. Opt.* 20, 2962–2968 (1981).
223. J. Lizet, S. Valette, and D. Langeac, "Reduction of temperature and vibration sensitivity of a polarimetric current sensor," *Electron. Lett.* 19(15), 578–579 (1983), doi: 10.1049/el:19830394
224. G. I. Chandler and F. C. Jahoda, "Current measurements by Faraday rotation in single-mode optical fibers," *Rev. Scient. Instr.* 56, 852 (1985), doi: 10.1063/1.1138070
225. G. I. Chandler, P. R. Forman, F. C. Jahoda, and K. A. Klare, "Fiber-optic heterodyne phase-shift measurement of plasma current," *Appl. Opt.* 25(11), 1770–1774 (1986).
226. X. P. Dong, B. C. B. Chu, and K. S. Chiang, "An electric-current sensor employing twisted fibre with compensation for temperature and polarization fluctuations," *Meas. Sci. Technol.* 8(6), 606–610 (1997).
227. M. Takahashi, H. Noda, and K. Terai, et al., "Optical current transformer for gas insulated switchgear using silica optical fiber," *IEEE Trans. Power Deliv.* 12(4), 1422–1427 (1997), doi: 10.1109/61.634155
228. H. S. Kang, J. H. Lee, and K. S. Lee, "A stabilization method of the Sagnac optical fiber current sensor with twist control," *IEEE Phot. Technol. Lett.* 10(10), 1464–1466 (1998). doi: 10.1109/68.720295
229. X. Dong, B. C. B. Chu, K. H. Kong, and K. S. Chiang, "Phase drift compensation for electric current sensor employing a twisted fiber or a spun highly birefringent fiber," *IEEE J. Sel. Top. Quantum Electron.* 6(5), 803–809 (2000), doi: 10.1109/2944.892621
230. T. Chartier, A. Hideur, C. Özkul, F. Sanchez, and G. M. Stéphan, "Measurement of the elliptical birefringence of single-mode optical fibers," *Appl. Opt.* 40, 5343–5353 (2001).
231. D. Tentori and A. Garcia-Weidner, "Jones birefringence in twisted single-mode optical fibers," *Opt. Express* 21, 31725–31739 (2013).
232. P. Dey, R. Shukla, and A. Sharma, "Simplified high-current measurement probe based on a single-mode optical fiber," *Rev. Scient. Instrum.* 89, 103110 (2018), doi: 10.1063/1.5022228
233. A. H. Rose, Z. B. Ren and G. W. Day, "Twisting and annealing optical fiber for current sensors," *J. Lightwave Technol.* 14(11), 2492–2498 (1996), doi: 10.1109/50.548146
234. Y. V. Przhiyalkovsky, S. A. Vasiliev, O. I. Medvedkov, S. K. Morshnev, and E. M. Dianov, "Polarization state evolution in spun birefringent optical fibers," *J. Appl. Phys.* 122(12), 123104 (2017).
235. Z. Ren, Y. Wang, and P. Robert, "Faraday rotation and its temperature dependence measurements in low-birefringence fibers," *J. Lightwave Technol.* 7(8), 1275–1278 (1989), doi: 10.1109/50.32393
236. W.-D. Barkmann and H. Winterhoff, German Patent DE2924804C2, filed June 20, 1979.
237. S. C. Rashleigh and R. Ulrich, German Patent application DE 3116149A, Apr. 23, 1981.
238. H. Winterhoff and W.-D. Barkmann, German Patent DE3010005C2, filed Mar. 15, 1980.
239. L. Li, J. R. Qian, and D. N. Payne, "Current sensors using highly birefringent bow-tie fibres," *Electron. Lett.* 22(21), 1142–1144 (1986).
240. L. Li, J. R. Qian, and D. N. Payne, "Miniature multi-turn fibre current sensors," *Int. J. Opt. Sens.* 2(1), 25–31 (1987).
241. R. I. Laming and D. N. Payne, "Electric current sensors employing spun highly birefringent optical fibers," *J. Lightwave Technol.* 7(12), 2084–2094 (1989), doi: 10.1109/50.41634
242. I. M. Bassett, "Design principle for a circularly birefringent optical fiber," *Opt. Lett.* 13(10), 844–846 (1988), doi: 10.1364/OL.13.000844
243. S. B. Poole, J. E. Townsend, D. N. Payne, et al., "Characterization of special fibers and fiber devices," *J. Lightwave Technol.* 7(8), 1242–1255, (1989), doi: 10.1109/50.32389
244. I. G. Clarke, "Temperature-stable spun elliptical-core optical-fiber current transducer," *Opt. Lett.* 18(2), 158–160 (1993), doi: 10.1364/ol.18.000158

245. J. R. Qian, Q. Guo, and L. Li, "Spun linear birefringence fibres and their sensing mechanism in current sensors with temperature compensation," *IEE Proc. Optoelectron.* 141(6), 373–380 (1994).
246. J. R. Qian, "A note on the beat length of spun linear birefringence fiber," *Microw. Opt. Technol. Lett.* 16(4), 225–227 (1997).
247. H.-C. Huang, "Practical circular-polarization-maintaining optical fiber," *Appl. Opt.* 36(27), 6968–6975 (1997), doi: 10.1364/AO.36.006968
248. X. Dong, B. C. B. Chu, K. H. Kong, and K. S. Chiang, "Phase drift compensation for electric current sensor employing a twisted fiber or a spun highly birefringent fiber," *IEEE J. Sel. Top. Quantum Electron.* 6(5), 803–809 (2000), doi: 10.1109/2944.892621
249. P. Polynkin and J. Blake, "Polarization evolution in bent spun fiber," *J. Lightwave Technol.* 23(11), 3815–3820 (2005).
250. Y. Wang, C.-Q. Xu, and V. Izraelian, "Characterization of spun fibers with millimeter spin periods," *Opt. Express* 13(10), 3841–3851 (2005).
251. V. P. Gubin, V. A. Isaev, S. K. Morshnev, et al., "Use of spun optical fibres in current sensors," *Quantum Electron.* 36(3), 287–291 (2006).
252. A. Michie, J. Canning, I. Bassett, J. Haywood, et al., "Spun elliptical birefringent photonic crystal fibre," *Opt. Express* 15(4), 1811–1816 (2007).
253. I. L. Lovchii, "Numerical modelling and investigation of a polarimetric current transducer with a spun-type light-guide," *J. Opt. Technol.* 77(12), 755–761 (2010).
254. N. Peng Y. Huang, S. Wang, et al., "Fiber optic current sensor based on special spun highly birefringent fiber," *IEEE Photonics Technol. Lett.* 25(17), 1668–1671 (2013), doi: 10.1109/LPT.2013.2272974
255. Y. V. Przhiyalkovsky, S. K. Morshnev, N. I. Starostin, and V. P. Gubin, "Propagation of broadband optical radiation in a spun high-birefringent fibre," *Quantum Electron.* 43(2), 167–173 (2013).
256. S. K. Morshnev, V. P. Gubin, Y. V. Przhiyalkovsky, N. I. Starostin, "Temperature dependences of phase and group birefringence in spun fibres," *Quantum Electron.* 43(12), 1143–1148 (2013).
257. Y. V. Przhiyalkovsky, Sergey K. Morshnev, Nikolay I. Starostin, and V. P. Gubin, "Modified sensing element of a fibre-optic current sensor based on a low-eigen ellipticity spun fibre," *Quantum Electron.* 44(10), 957–964 (2014).
258. Y. V. Przhiyalkovsky, S. K. Morshnev, N. I. Starostin, and V. P. Gubin, "Propagation of polarized light in bent hi-bi spun fibers," *Quantum Electron.* 45(11), 1075 (2015).
259. S. K. Morshnev, V. P. Gubin, N. I. Starostin, Y. V. Przhiyalkovsky, and A. I. Sazonov, "Effect of protective coating on random birefringence variations in anisotropic optical fibres in response to temperature changes," *Quantum Electron.* 46(10), 911–918 (2016).
260. S. A. Vasiliev, Y. V. Przhiyalkovsky, P. I. Gnusin, O. I. Medvedkov, and E. M. Dianov, "Measurement of high-birefringent spun fiber parameters using short-length fiber Bragg gratings," *Opt. Express* 24(11), 11290–11298 (2016), doi: 10.1364/OE.24.011290
261. Y. V. Przhiyalkovsky, V. P. Gubin, N. I. Starostin, and S. K. Morshnev, "Detection of electric current pulses by a fibre-optic sensor using spun fibre," *Quantum Electron.* 48(1), 62–69 (2018).
262. S. K. Morshnev, V. P. Gubin, N. I. Starostin, Y. V. Przhiyalkovsky, and A. I. Sazonov, "Temperature evolution of interferometer visibility in a small radius fibre coil current sensor," *Quantum Electron.* 48(3), 275–282 (2018).
263. Y. V. Przhiyalkovskiy, N. I. Starostin, S. K. Morshnev, and A. I. Sazonov, "Polarization dynamics of light propagating in bent spun birefringent fiber," *J. Lightwave Technol.* 38(24), 6879–6885 (2020), doi: 10.1109/JLT.2020.3017795
264. J. Wen, W. Liu, Y. Huang, et al., "Spun-related effects on optical properties of spun silica optical fibers," *J. Lightw. Technol.* 33(12), 2674–2678 (2015).
265. G. M. Müller, X. Gu, L. Yang, A. Frank, and K. Bohnert, "Inherent temperature compensation of fiber-optic current sensors employing spun highly-birefringent fiber," *Opt. Express* 24(10), 11164–11173 (2016).

266. G. M. Müller, A. Frank, L. Yang, X. Gu, and K. Bohnert, "Temperature compensation of interferometric and polarimetric fiber-optic current sensors with spun highly birefringent fiber," *J. Lightwave Technol.* 37(18), 4507–4513 (2019), doi: 10.1109/JLT.2019.2907803
267. H.-C. Chen, J.-X. Wen, Y. Huang, et al., "Influence of linear birefringence on Faraday effect measurement for optical fibers," *Optoel. Lett.* 13, 147–150 (2017).
268. Z. Xu, X. S. Yao, Z. Ding, et al., "Accurate measurements of circular and residual linear birefringences of spun fibers using binary polarization rotators," *Opt. Express* 25, 30780–30792 (2017), doi: 10.1364/OE.25.030780
269. D. Kowal, G. Statkiewicz-Barabach, M. Napiorkowski, et al., "Measurement of birefringence and ellipticity of polarization eigenmodes in spun highly birefringent fibers using spectral interferometry and lateral point-force method," *Opt. Express* 26, 34185–34199 (2018).
270. G. B. Malykin and V. I. Pozdnyakova, "Linear transformation of the polarization modes in coiled optical spun-fibers with strong unperturbed linear birefringence. I. Nonresonant transformation," *Opt. Spectrosc.* 124, 360–372 (2018), doi: 10.1134/S0030400X18030177
271. G. B. Malykin and V. I. Pozdnyakova, "Linear transformation of the polarization modes in coiled optical spun fibers. II. Resonant transformation," *Opt. Spectrosc.* 125(4), 543–550 (2018).
272. H. Hu, J. Huang, Y. Huang, L. Xia, and J. Yu, "Modeling of the birefringence in spun fiber," *Opt. Comm.* 473, 125919 (2020).
273. A. H. Rose, N. Feat, and S. M. Etzel, "Wavelength and temperature performance of polarization-transforming fiber," *Appl. Opt.* 42, 6897–6904 (2003).
274. X. Zhu and R. Jain, "Detailed analysis of evolution of the state of polarization in all-fiber polarization transformers," *Opt. Express* 14, 10261–10277 (2006).
275. Z. Shi, S. Bi, M. Ji, and L. Wang, "Performance analysis for all-fiber polarization transformer with specific spun rate profile," *Opt. Comm.* 392, 20–24 (2017).
276. M. H. Kang, Y. L. Wang, L. Y. Ren, et al., "Design of vibration-insensitive Sagnac fiber-optic current sensors using spun high-birefringence fibers," *J. Mod. Opt.* 61(14), 1120–1126 (2014).
277. A. Kumar and R. K. Varshney, "Propagation characteristics of highly elliptical core waveguides: a perturbation approach," *Opt. Quantum Electron.* 16, 349–354 (1984).
278. J. N. Blake, "Fiber-optic current sensor with improved isolation," U.S. patent 6,356,351B1, Mar. 12, 2002.
279. Y. Fujii and C. D. Hussey, "Design considerations for circularly form-birefringent optical fibers," *IEE Proc. J.* 133, 249–255 (1986), doi: 10.1049/ip-j.1986.0040
280. A. Michie, K. Digweed, J. Canning et al., "Spun highly birefringent photonic crystal fibre for current sensing," in *Optical Fiber Sensors*, OSA Technical Digest (CD) (Optical Society of America, 2006), paper ThE56.
281. A. Michie, J. Canning, I. Bassett, et al., "Spun elliptically birefringent photonic crystal fibre for current sensing," *Meas. Sci. Technol.* 18(10), 3070–3074 (2007).
282. A. Michie, J. Canning, I. Bassett, et al., "Spun elliptical birefringent photonic crystal fibre," *Opt. Express* 15(4), 1811–1816 (2007).
283. Y. K. Chamorovskiy, N. I. Starostin, M. V. Ryabko et al., "Miniature microstructured fiber coil with high magneto-optical sensitivity," *Opt. Comm.* 282(23), 4618–4621 (2009).
284. M. A. Schmidt, L. Wondraczek, H. W. Lee, N. Granzow, N. Da, and P. S. J. Russell, "Complex Faraday rotation in microstructured magneto-optical fiber waveguides," *Adv. Mater.* 23(22–23), 2681–2688 (2011), doi: 10.1002/adma.201100364
285. A. Argyros, J. Pla, F. Ladouceur, and L. Poladian, "Circular and elliptical birefringence in spun microstructured optical fibres," *Opt. Express* 17, 15983–15990 (2009).
286. V. P. Gubin, S. K. Morshnev, N. I. Starostin, et al., "Efficient direct magneto-optical phase modulation of light waves in spun microstructured fibres," *Quantum Electron.* 41(9), 815–820 (2011).
287. R. Beravat, G. K. L. Wong, X. M. Xi, M. H. Frosz, and P. St. J. Russell, "Current sensing using circularly birefringent twisted solid-core photonic crystal fiber," *Opt. Lett.* 41, 1672–1675 (2016).

288. P. St. J. Russell, R. Beravat, and G. K. L. Wong, "Helically twisted photonic crystal fibres," *Phil. Trans. R. Soc. A*. 375: 20150440 (2016), doi: 10.1098/rsta.2015.0440
289. H. Gao, G. Wang, W. Gao, et al. "A chiral photonic crystal fiber sensing coil for decreasing the polarization error in a fiber optic current sensor." *Opt. Commun*. 469, 125755 (2020).
290. M. V. Berry, "Quantal phase factors accompanying adiabatic changes," *Proc. R. Soc. Lond. Ser. A* 392, 45–57 (1984).
291. J. N. Ross, "The rotation of the polarization in low birefringence monomode optical fibers due to geometric effects," *Opt. Quantum Electron*. 16, 455–461 (1984), doi: 10.1007/BF00619638
292. M. P. Varnham, R. D. Birch, and D. N. Payne, "Helical-core circularly-birefringent fibers," in *Proceedings of the International Conference on Integrated Optics and Optical Fibre Communications–European Conference on Optical Communications* (Instituto Internazionale delle Communicazione, Genova, Italy, 1985), pp. 135–138.
293. A. Tomita and R. Y. Chiao, "Observation of Berry's topological phase by use of an optical fibre," *Phys. Rev. Lett*. 57, 937–940 (1986).
294. X.-S. Fang and Z.-Q. Lin "Field in single-mode helically wound optical fibers," *IEEE Trans. Microw. Theory Tech*. MTT-33, 1150–1154 (1985).
295. X.-S. Fang and Z.-Q. Lin, "A coupled-mode approach to the analysis of fields in space-curved and twisted waveguides," *IEEE Trans. Microw. Theory Tech*. 35(11), 978–983 (1987).
296. J. Qian, "Coupled-mode theory for helical fibres," *IEE Proc. J*. 135, 178–182 (1988).
297. T. Eftimov and T. Kortenski, "State of polarization in open- and closed-loop helices of single-mode fibres," *J. Mod. Opt*. 36, 287 (1989), doi: 10.1080/09500348914550331
298. R. Castelli, F. Irrera, and C. G. Someda, "Circularly birefringent optical fibres: new proposals – Part I. Field analysis," *Opt. Quantum Electron*. 21, 35–46 (1989).
299. C. G. Someda, "Circularly birefringent optical fibres: new proposals – Part II. Birefringence and coupling loss," *Opt. Quantum Electron*. 23, 713–725 (1991).
300. G. Chen and Q. Wang, "Local fields in single-mode helical fibres," *Opt. Quantum Electron*. 27, 1069–1074 (1995).
301. E. M. Frins and W. Dultz, "Rotation of the polarization plane in optical fibers," *J. Lightwave Technol*. 15, 144–147 (1997).
302. F. Wassmann and A. Ankiewicz, "Berry's phase analysis of polarization rotation in helicoidal fibers," *Appl. Opt*. 37, 3902–3911 (1998).
303. D. Tentori, C. Ayala-Díaz, F. Treviño-Martínez, F. J. Mendieta-Jiménez, and H. Soto-Ortiz, "Birefringence evaluation of helically wound optical fibres," *J. Mod. Opt*., 48(11), 1767–1780, (2001), doi: 10.1080/09500340108231432
304. C. H. Tang, "An orthogonal coordinate system for curved pipes (Correspondence)," *IEEE Trans. Microw. Theory Tech*. 18(1), 69–69 (1970), doi: 10.1109/TMTT.1970.1127150
305. F. Maystre and A. Bertholds, "Magneto-optic current sensor using a helical-fiber Fabry–Perot resonator," *Opt. Lett*. 14, 587–589 (1989).
306. A. Papp and H. Harms, "Polarization optics of liquid-core optical fibers," *Appl. Opt*. 16, 1315 (1977).
307. S. Xu, W. Li, Y. Wang, and F. Xing, "Stray current sensor with cylindrical twisted fiber," *Appl. Opt*. 53(24), 5486–5492 (2014).
308. S. Xu, W. Li, F. Xing, et al., "An elimination method of temperature-induced linear birefringence in a stray current sensor," *Sensors* 17, 551 (2017).
309. S. X. Short, J. U. de Arruda, A. A. Tselikov, and J. N. Blake, "Elimination of birefringence induced scale factor errors in the in-line Sagnac interferometer current sensor," *J. Lightwave Technol*. 16, 1844–1850 (1998).
310. C. Zhang, C. Li, X. Wang, et al., "Design principle for sensing coil of fiber-optic current sensor based on geometric rotation effect," *Appl. Opt*. 51(18), 3977–3988 (2012).
311. H. Winterhoff and W.-D. Barkmann, German Patent DE2855337C2, filed Dec. 21, 1978.
312. C. D. Hussey, R. D. Birch, and Y. Fujii, "Circularly birefringent single-mode optical fibres," *Electron. Lett*. 22(3), 129–130 (1986), doi: 10.1049/el:19860090

313. J. Qian and C. D. Hussey, "Circular birefringence in helical-core fibre," *Electron. Lett.* 22(10), 515–517 (1986).
314. Y. Fujii and C. D. Hussey, "Design considerations for circularly form-birefringent optical fibres," *IEE Proc. J.* 133, 249 (1986).
315. R. D. Birch, "Fabrication and characterisation of circularly birefringent helical fibres," *Electron. Lett.* 23(1), 50–52 (1987), doi: 10.1049/el:19870037
316. M. Napiorkowski and W. Urbanczyk, "Rigorous simulations of a helical core fiber by the use of transformation optics formalism," *Opt. Express* 22, 23108–23120 (2014), doi: 10.1364/OE.22.023108
317. P. Wang, L. J. Cooper, J. K. Sahu, and W. A. Clarkson, "Efficient single-mode operation of a cladding-pumped ytterbium-doped helical-core fiber laser," *Opt. Lett.* 31(2), 226–228 (2006).
318. Y. Chen and P. St. J. Russell, "Frenet–Serret analysis of helical Bloch modes in N-fold rotationally symmetric rings of coupled spiraling optical waveguides," *J. Opt. Soc. Am. B* 38, 1173–1183 (2021).
319. S. R. Norman, D. N. Payne, M. J. Adams, and A. M. Smith, "Fabrication of single-mode fibres exhibiting extremely low polarisation birefringence," *Electron. Lett.* 15(11), 309–311, (1979), doi: 10.1049/el:19790219
320. A. M. Smith, "Optical fibres for current measurement applications," *Opt. Laser Technol.* 12(1), 25–29 (1980).
321. A. J. Barlow, D. N. Payne, M. R. Hadley, and R. J. Mansfield, "Production of single-mode fibres with negligible intrinsic birefringence and polarisation mode dispersion," *Electron. Lett.* 17(20), 725–726 (1981).
322. D. N. Payne, A. J. Barlow, J. J. Ramskov-Hansen, M. R. Hadley, and R. J. Mansfield, "Fabrication and properties of low birefringence spun fibers," in *Fiber-Optic Rotation Sensors and Related Technologies*, S. Ezekiel, H. J. Arditty, Eds. (Springer Series in Optical Sciences, Berlin, Heidelberg, Germany, 1982), vol. 32, pp. 185–195, doi: 10.1007/978-3-540-39490-7_20
323. D. N. Payne, A. J. Barlow and J. J. Ramskov-Hansen, "Development of low- and high-birefringence optical fibers," *IEEE J. Quantum Electron.* QE-18, 477–487 (1982).
324. C. D. Poole, "Statistical treatment of polarization dispersion in single-mode fiber," *Opt. Lett.* 13(8), 687–689 (1988).
325. L. Palmieri, "Polarization properties of spun single-mode fibers," *J. Lightwave Technol.* 24, 4075–4088 (2006).
326. K. Bohnert, P. Gabus, J. Nehring, H. Brändle, and M. G. Brunzel, "Fiber-optic current sensor for electrowinning of metals," *J. Lightwave Technol.* 25(11), 3602–3609 (2007), doi: 10.1109/JLT.2007.906795
327. K. Bohnert, A. Frank, G. M. Müller, L. Yang, M. Lenner, P. Gabus, X. Gu, and S. V. Marchese, "Fiber optic current and voltage sensors for electric power transmission systems," in *Fiber Optic Sensors and Applications XV*, A. Mendez, C. S. Baldwin, and Henry H. Du, Eds., *Proc. of SPIE*, vol. 10654, 1065402, 2018, doi: 10.1117/12.2303945
328. K. Bohnert, A. Frank, L. Yang, X. Gu, and G. M. Müller, "Polarimetric fiber-optic current sensor with integrated-optic polarization splitter," *J. Lightwave Technol.* 37(14), 3672–3678 (2019), doi: 10.1109/JLT.2019.2919387
329. G. W. Day and S. M. Etzel, "Annealing of bend-induced birefringence in fiber current sensors," in *Tech. Dig. IOOCIECOC '85*, Venice, Italy, pp. 871–874 (1985).
330. J. Stone, "Stress-optic effects, birefringence, and reduction of birefringence by annealing in fiber Fabry-Perot interferometers," *J. Lightwave Technol.* 6(7), 1245–1248 (1988), doi: 10.1109/50.4122
331. A. H. Rose, "Devitrification in annealed optical fiber," *J. Lightwave Technol.* 15(5), 808–814 (1997).
332. T. W. MacDougall, D. R. Lutz, and R. A. Wandmacher, "Development of a fiber optic current sensor for power systems," *IEEE Trans. Power Deliv.* 7(2), 848–852 (1992), doi: 10.1109/61.127089

333. J. W. Dawson, T. W. MacDougall, and E. Hernandez, "Verdet constant limited temperature response of a fiber-optic current sensor," *IEEE Photonics Technol. Lett.* 7(12), 1468–1470 (1995), doi: 10.1109/68.477285
334. D. R. Lutz, T. W. MacDougall, et al., "Faraday effect sensing coil with stable birefringence," U.S. patent 5463312, filed Mar. 3, 1994.
335. L. Yang, A. Frank, R. Wüest, et al., "A study on different types of fiber coils for fiber optic current sensors," in *Key Engineering Materials* (Trans Tech Publications, Ltd., 2014), vol. 605, pp. 283–286, doi: 10.4028/www.scientific.net/kem.605.283
336. K. Bohnert, P. Gabus, J. Nehring, and H. Brändle, "Temperature and vibration insensitive fiber-optic current sensor," *J. Lightwave Technol.* 20(2), 267–276 (2002).
337. M. Lenner, R. Wüest, A. Frank, and K. Bohnert, "Effects of thermal fiber annealing on the temperature compensation of interferometric fiber-optic current sensors," in *SENSORS 2012 IEEE*, Taipei, 2012, pp. 1–4, doi: 10.1109/ICSENS.2012.6411054
338. K. Kurosawa, I. Masuda, and T. Yamashita, "Faraday effect current sensor using flint glass fiber for the sensing element," in *Tech. Dig. 9th Optical Fiber Sensors Conf.*, 1993, pp. 415–418.
339. K. Kurosawa, S. Yoshida, and K. Sakamoto, "Polarization properties of the flint glass fiber," *J. Lightwave Technol.* 13(7), 1378–1384 (1995).
340. S. Yoshida, K. Kurosawa, and O. Sano, "Development of an optical current transducer using a flint glass fiber for a gas circuit breaker," in *Tech. Dig. 11th Optical Fiber Sensors Conf.*, 1996, pp. 172–175.
341. K. Kurosawa, "Development of fiber-optic current sensing technique and its applications in electric power systems," *Photonic Sens.* 4(1), 12–20 (2014), doi: 10.1007/s13320-013-0138-z
342. L. Sun, S. Jiang, J. D. Zuegel, and J. R. Marciante, "Effective Verdet constant in a terbium-doped-core phosphate fiber," *Opt. Lett.* 34, 1699–1701 (2009).
343. L. Sun, S. Jiang, and J. R. Marciante, "All-fiber optical magnetic-field sensor based on Faraday rotation in highly terbium-doped fiber," *Opt. Express* 18, 5407–5412 (2010).
344. Y. Huang, H. Chen, W. Dong, et al., "Fabrication of europium-doped silica optical fiber with high Verdet constant," *Opt. Express* 24, 18709–18717 (2016).
345. H. C. Y. Yu, M. A. van Eijkelenborg, Sergio G. Leon-Saval, et al., "Enhanced magneto-optical effect in cobalt nanoparticle-doped optical fiber," *Appl. Opt.* 47, 6497–6501 (2008).
346. P. R. Watekar, H. Yang, S. Ju, and W.-T. Han, "Enhanced current sensitivity in the optical fiber doped with CdSe quantum dots," *Opt. Express* 17, 3157–3164 (2009).
347. P. R. Watekar, S. Ju, S.-A. Kim, et al., "Development of a highly sensitive compact sized optical fiber current sensor," *Opt. Express* 18, 17096–17105 (2010).
348. S. Ju, J. Kim, K. Linganna, P. R. Watekar, et al., "Temperature and vibration dependence of the Faraday effect of Gd2O3 NPs-doped alumino-silicate glass optical fiber," *Sensors* 2018, 18, 988, doi: 10.3390/s18040988
349. H. O. Edwards, K. P. Jedrzejewski, R.I. Laming, and D. N. Payne, "Optimal design of optical fibers for electric current measurement," *Appl. Opt.* 28, 1977–1979 (1989).
350. Y. Shiyu, J. Lousteau, M. Olivero, et al., "Analysis of Faraday effect in multimode tellurite glass optical fiber for magneto-optical sensing and monitoring applications," *Appl. Opt.* 51, 4542–4546 (2012).
351. Q. Chen, H. Wang, and Q. P. Chen, "Elliptical measurement for Faraday rotation in multimode TZN fiber for magneto-optical current sensor application," *Adv. Mater. Res.* 785, 1367–1373 (2013).
352. H. Wang, Q. Wang, and Q. P. Chen, "Faraday rotation influence factors in tellurite-based glass and fibers," *Appl. Phys. A* 120, 1001–1010 (2015).
353. A. Papp and H. Harms, "Polarization optics of index-gradient optical waveguide fibers," *Appl. Opt.* 14, 2406–2411 (1975).
354. H. Harms, A. Papp, and K. Kempter, "Magnetooptical properties of index-gradient optical fibers," *Appl. Opt.* 15(3), 799–801 (1976).

355. F. J. Wessel, N. C. Wild, H. U. Rahman, A. Ron, and F. S. Felber, "Faraday rotation in a multimode optical fiber in a fast rise-time high magnetic field," *Rev. Sci. Instrum.* 57(9), 2246–2249 (1986).
356. W. A. Gambling, D. N. Payne, and H. Matsumura, "Mode excitation in a multimode optical-fibre waveguide," *Electron. Lett.* 9(18), 412–414 (1973), doi:10.1049/el:19730303
357. S. A. Planas, E. Bochove, and R. Srivastava, "Geometrical characterization of liquid core fibers by measurement of thermally induced mode cutoffs and interference," *Appl. Opt.* 21, 2708-2715 (1982).
358. R. B. Frenkel, "Optical sensing of high electric currents using the Faraday effect in a rigid single turn of liquid," *Meas. Sci. Technol.* 4(9), 976–981 (1993).
359. L. M. Tong, J. Y. Lou, and E. Mazur, "Single-mode guiding properties of subwavelength-diameter silica and silicon wire waveguides," *Opt. Express* 12, 1025–1035 (2004).
360. G. Y. Chen, M. Ding, T. P. Newson, and G. Brambilla, "A review of microfiber and nanofiber based optical sensors," *Open Opt. J.* 7, 32–57 (2013).
361. G. Y. Chen, D. G. Lancaster, and T. M. Monro, "Optical microfiber technology for current, temperature, acceleration, acoustic, humidity and ultraviolet light sensing," *Sensors* 2018, 18, 72.
362. G. Y. Chen, T. Lee, R. Ismaeel, G. Brambilla, and T. P. Newson, "Resonantly enhanced Faraday rotation in an microcoil current sensor," *IEEE Photonics. Technol. Lett.* 24, 860–862 (2012).
363. G. Y. Chen, G. Brambilla, amd T. P. Newson, "Spun optical microfiber," *IEEE Photonics Technol. Lett.* 24, 1663–1666 (2012).
364. G. Y. Chen, T. P. Newson, and G. Brambilla, "Birefringence treatment of non-ideal optical microfibre coils for continuous Faraday rotation," *Electron. Lett.* 49, 714–715 (2013).
365. G. Y. Chen, T. P. Newson, and G. Brambilla, "Optical microfibers for fast current sensing," *Opt. Fiber Technol.* 19(6), Part B, 802–807 (2013).
366. M. Belal, Z.-Q. Song, Y. Jung, G. Brambilla, and T. Newson, "An interferometric current sensor based on optical fiber micro wires," *Opt. Express* 18, 19951–19956 (2010).
367. M. Belal, Z. Song, Y. Jung, G. Brambilla, and T. P. Newson, "Optical fiber microwire current sensor," *Opt. Lett.* 35, 3045–3047 (2010).
368. K. S. Lim, S. W. Harun, S. S. A. Damanhuri, et al., "Current sensor based on microfiber knot resonator," *Sens. Actuator. A Phys.* 167, 60–62 (2011).
369. X. Xie, J. Li, L.-P. Sun, et al., "A high-sensitivity current sensor utilizing CrNi wire and microfiber coils," *Sensors* 14(5), 8423–8429 (2014).
370. S. Yan, B. Zheng, J. Chen, F. Xu, and Y. Lu, "Optical electrical current sensor utilizing a graphene-microfiber-integrated coil resonator," *Appl. Phys. Lett.* 107, 053502 (2015).
371. S. Yan, Y. Chen, C. Li, F. Xu, and Y. Lu, "Differential twin receiving fiber-optic magnetic field and electric current sensor utilizing a microfiber coupler," *Opt. Express* 23, 9407–9414 (2015).
372. A. A. Jasim, J. Faruki, M. F. Ismail, and H. Ahmad, "Fabrication and characterization of microbent inline microfiber interferometer for compact temperature and current sensing applications," *J. Lightwave Technol.* 35, 2150–2155 (2017).
373. S. Yoshikawa and A. Ueki, Japanese patent application 83680/68, priority Nov. 16, 1968, US patent 3605013, filed, Nov. 7, 1969.
374. H. Harms and A. Papp, "Magnetooptical current transformer. 3: Measurements," *Appl. Opt.* 19, 3741–3745 (1980).
375. Z. Wang, Y. Liao, S. Lai, H. Zhao, and X. Chen, "A novel method for simultaneous measurement of current and voltage using one low-birefringence fiber," *Opt. Laser Technol.* 30(5), 257–262 (1998), doi: 10.1016/S0030-3992(98)00018-8
376. Y. Li, X. Liu, W. Zhang, and J. Liu, "Error characteristic analysis and experimental research on a fiber optic current transformer," *Appl. Opt.* 57, 8359–8365 (2018).
377. A. J. Rogers, J. Xu, and J. Yao, "Vibration immunity for optical-fiber current measurement," *J. Lightwave Technol.* 13, 1371–1377 (1995).

378. X. Fang, A. Wang, R. G. May and R. O. Claus, "A reciprocal-compensated fiber-optic electric current sensor," *J. Lightwave Technol.* 12(10), 1882–1890 (1994), doi: 10.1109/50.337503
379. X. Fang and R.O. Claus, "Optimal design of IRIS-based polarimetric intrinsic fiber optic current sensors," *J. Lightwave Technol.* 14(7), 1664–1673 (1996).
380. M. Willsch, H. Hertsch, U. Augustad, P. Kraemmer, and T. Bosselmann, "Fiber optical current sensor design for accurate measurement of DC currents," in *Technical Digest, 16th Int. Conference on Optical Fiber Sensors*, Nara, Japan 2003, paper WEP-2.
381. M. Berwick, J. D. C. Jones, and D. A. Jackson, "Alternating-current measurement and noninvasive data ring utilizing the Faraday effect in a closed-loop fiber magnetometer," *Opt. Lett.* 12, 293–295 (1987).
382. A. Kersey and D. Jackson, "Current sensing utilizing heterodyne detection of the Faraday effect in single-mode optical fiber," *J. Lightwave Technol.* 4(6), 640–644 (1986), doi: 10.1109/JLT.1986.1074778
383. J. A. Ferrari, A. Dubra, A. Arnaud, and D. Perciante, "Current sensor using heterodyne detection," *Appl. Opt.* 38, 2808–2811 (1999).
384. N. C. Pistoni and M. Martinelli, "Vibration-insensitive fiber-optic current sensor," *Opt. Lett.* 18, 314–316 (1993).
385. D. Alasia and L. Thévenaz, "A novel all-fibre configuration for a flexible polarimetric current sensor," *Meas. Sci. Technol.* 15(8), 1525–1530 (2004).
386. H. Zhang et al., "Temperature and vibration robustness of reflecting all-fiber current sensor using common single-mode fiber," *J. Lightwave Technol.* 32(22), 4311–4317 (2014), doi: 10.1109/JLT.2014.2357687
387. S. Xu, W. Li, Y. Wang, and F. Xing, "Effect and elimination of alignment error in an optical fiber current sensor," *Opt. Lett.* 39(16), 4751–4754 (2014).
388. S. Xu, W. Li, F. Xing, and Y. Wang, "Polarimetric current sensor based on polarization division multiplexing detection," *Opt. Express* 22(10), 11985–11994 (2014).
389. H. Zhang, Y. Qiu, Z. Huang, J. Jiang, G. Li, H. Chen, and H. Li, "Temperature and vibration robustness of reflecting all-fiber current sensor using common single-mode fiber," *J. Lightwave Technol.* 32(22), 3709–3715 (2014).
390. M. Aerssens, F. Descamps, A. Gusarov, et al., "Influence of the optical fiber type on the performances of fiber-optics current sensor dedicated to plasma current measurement in ITER," *Appl. Opt.* 54, 5983–5991 (2015).
391. F. Brifford, L. Thevenaz, P.-A. Nicati, A. Küng, and P. A. Robert, "Polarimetric current sensor using an in-line Faraday rotator," *IEICE Trans. Electron.* E83-C(3), 331–334 (2000).
392. K. Kurosawa et al., "Current measuring device," U.S. patent 7,176,671B2, priority date: Mar. 1, 2002.
393. R. Kondo and K. Kurosawa Kiyoshi, "A method for improving temperature dependence of an optical fiber current sensor," *IEEJ Trans. Power Energy* 130(4), 414–420 (2010), doi: 10.1541/ieejpes.130.414
394. G. Müller, X. Gu, A. Frank, and K. Bohnert, "Fiber-optic current sensor with passive phase biasing employing highly birefringent spun fiber," *OSA Applied Industrial Optics*, 2013, paper ATuB.4.
395. R. Bergh, H. Lefevre, and H. Shaw, "An overview of fiber-optic gyroscopes," *J. Lightwave Technol.* 2(2), 91–107 (1984), doi: 10.1109/JLT.1984.1073580
396. H. Lefèvre, *The Fiber-Optic Gyroscope* (Artech House, 1993).
397. P. Ferdinand and J. Lesne, "Induced circular birefringence and ellipticity measurement in a Faraday effect fiber ring interferometer," in *Fiber-Optic Rotation Sensors*, S. Ezekiel and H. J. Arditty, Eds. (Springer Verlag, Berlin 1982), pp. 215–221, doi: 10.1007/978-3-540-39490-7_24
398. P. A. Leilabady, A. P. Wayte, M. Berwick, J. D. C. Jones, and D. A. Jackson, "A pseudo reciprocal fiber optic Faraday rotation sensor: current measurement and data communication applications," *Opt. Comm.* 59, 173–176 (1986).

399. A. D. Kersey and A. Dandridge, "Optical fibre Faraday rotation current sensor with closed-loop operation," *Electron. Lett.* 21(11), 464–466 (1985), doi: 10.1049/el:19850329
400. P.-A. Nicati and P. Robert, "Stabilized current sensor using Sagnac interferometer," *J. Phys. E: Sci. Instrum.* 21, 791–796 (1988).
401. P.-A. Nicati and P.-A. Robert, "Numerical analysis of second-order polarization effects in a Sagnac current sensor," *IEEE Trans. Instrum. Meas.* 39(1), 219–224 (1990), doi: 10.1109/19.50448
402. G. Frosio, K. Hug, and R. Dändliker, "All fiber Sagnac interferometer current sensor," in *Opto 92* (ESI Publications, Paris, 1992), pp. 560–564.
403. K. Bohnert, H. Brändle, and G. Frosio, "Field test of interferometric optical fiber high-voltage and current sensors," in *10th International Conference on Optical Fibre Sensors*, B. Culshaw, J. D. C. Jones, Ed., *Proc. SPIE 2360*, 1994, pp. 16–19.
404. J. Blake, P. Tantaswadi, and R. T. de Carvalho, "In-line Sagnac interferometer current sensor," *IEEE Trans. Power Deliv.* 11(1), 116–121 (1996).
405. S. X. Short, P. Tantaswadi, R. T. de Carvalho, B. D. Russell, and J. N. Blake, "An experimental study of acoustic vibration effects in optical fiber current sensors," *IEEE Trans. Power Deliv.* 11, 1702–1706 (1996).
406. S. X. Short, A. A. Tselikov, J. U. de Arruda and J. N. Blake, "Imperfect quarter-waveplate compensation in Sagnac interferometer-type current sensors," *J. Lightwave Technol.* 16(7), 1212–1219 (1998).
407. A. Tselikov, J. U. de Arruda, and J. Blake, "Zero-crossing demodulation for open-loop Sagnac interferometers," *J. Lightwave Technol.* 16(9), 1613–1619 (1998).
408. M. Takahashi, K. Sasaki, K. Terai, Y. Hirta and T. Nakajima, "Optical current transformer for 245 kV integrated air insulated switchgear," in *2003 IEEE PES Transmission and Distribution Conference and Exposition* (IEEE Cat. No.03CH37495), 2003, vol. 1, 389–392, doi: 10.1109/TDC.2003.1335254
409. M. Takahashi, K. Sasaki, A. Ohno, Y. Hirata, and K. Terai, "Sagnac interferometer-type fibre-optic current sensor using single-mode fibre down leads," *Meas. Sci. Technol.* 15(8), 1637–1641 (2004).
410. H. C. Lefevre, "Single-mode fibre fractional wave devices and polarization controllers," *Electron. Lett.* 16(20), 778–780 (1980).
411. G. Frosio, "All-fiber adjustable retardation plate," in *EFOC Proc.*, IGI Europe, Boston, MA, 1989, pp. 350–355.
412. E. Kiesel, "Impact of modulation-induced signal instabilities on fiber gyro performance," in *Fiber Optic and Laser Sensors V, Proc. SPIE 838*, 1987, pp. 129–139.
413. M. Skalský, Z. Havránek, and Jirí Fialka, "Efficient modulation and processing method for closed-loop fiber optic gyroscope with piezoelectric modulator," *Sensors* 19, 1710 (2019), doi:10.3390/s19071710
414. H. C. Lefevre, Ph. Graindorge, H.J. Arditty, S. Vatoux and M. Papuchon, "Double closed-loop hybrid fiber gyroscope using digital phase ramp," in *Proc. of OFS*, 1985, pp. PDS7-1-4.
415. G. Spahlinger, "Fiber optic Sagnac interferometer with digital phase ramp resetting via correlation-free demodulator Control," U.S. Patent 5,123,741, 1992.
416. X. Li, Y. Zhang, and Q. Yu, "Four-state modulation in fiber optic gyro," in *2008 IEEE International Conference on Mechatronics and Automation*, 2008, pp. 189–192, doi: 10.1109/ICMA.2008.4798749
417. A. M. Kurbatov and R. A. Kurbatov, "Methods of improving the accuracy of fiber-optic gyros," *Gyroscopy Navig.* 3(2), 132–143 (2012).
418. J. Chamoun and M. J. F. Digonnet, "Pseudo-random-bit-sequence phase modulation for reduced errors in a fiber optic gyroscope," *Opt. Lett.* 41(24), 5664–5667 (2016).
419. L. H. Wang, M. Cao, et al., "Modeling and experimental verification of polarization errors in Sagnac fiber optic current sensor," *Optik* 126, 2743–2746 (2015).
420. S. K. Sheem, "Fibre-optic gyroscope with 3x3 directional coupler," *Appl. Phys. Lett.* 37(10), 869–871 (1980).

421. R. Priest, "Analysis of fiber interferometer utilizing 3x3 fiber coupler," *IEEE J. Quant. Electronics* 18(10), 1601–1603 (1982), doi: 10.1109/JQE.1982.1071386
422. K. P. Koo, A. B. Tveten, and A. Dandridge, "Passive stabilization scheme for fiber interferometers using (3×3) fiber directional couplers," *Appl. Phys. Lett.* 41, 616–618 (1982).
423. J. Pietsch, "Scattering matrix analysis of 3x3 fiber couplers," *J. Lightwave Technol.* 7(2), 303–307 (1989).
424. G. F. Trommer, H. Poisel, W. Buhler, E. Hartl, and R. Muller, "Passive fiber optic gyroscope," *Appl. Opt.* 29, 5360–5365 (1990).
425. L. Veeser and G. Day, "Faraday effect current sensing using a Sagnac interferometer with a 3x3 coupler," in *Proc., Optical Fibre Sensors Conf.*, Sydney, 1990.
426. K. B. Rochford, G. W. Day, and P. R. Forman, "Polarization dependence of response functions in 3x3 Sagnac optical fiber current sensors," *J. Lightwave Technol.* 12(8), 1504–1509 (1994).
427. A. Yu and A. S. Siddiqui, "Practical Sagnac interferometer based fibre optic current sensor," *IEE Proc. Optoelectron.* 141(4), 249–256 (1994).
428. I. M. Bassett, "Passive 3x3 sagnac interferometer with reciprocity and correction for small departures from ideal coupler properties," in *Proc. SPIE 4185, Fourteenth International Conference on Optical Fiber Sensors*, 41852Q, 2000,https://doi.org/10.1117/12.2302241
429. J. H. Haywood, I. M. Bassett, and M. Matar, "Application of the NIMI technique to the 3x3 Sagnac fibre optic current sensor – Experimental results," in *Technical Digest 15th International Optical Fibre Sensors Conference*, Portland, OR, USA, 2002, pp. 553–556.
430. J. Haywood, I. Bassett, M. Matar "Restoring reciprocity to the 3x3 Sagnac interferometer using the NIMI technique," in *Technical Digest 17th International Optical Fibre Sensors Conference*, Bruges, 2002, pp. 553–556.
431. M. Matar, I. M. Bassett, and J. H. Haywood, "Experimental trial of a modulated depolarised Er-doped SFS with the NIMI 3x3 Sagnac fibre optic current sensor," in *Optical Fiber Sensors*, OSA Technical Digest (CD) (Optical Society of America, 2006), paper ThE82.
432. Y. Huang, "All-fiber current sensor," U.S. patent 7,492,977B2, Feb. 17. 2009.
433. K. Goto, T. Sueta, and T. Makimoto, "Traveling-wave light-intensity modulators using the method of polarization-rotated reflection," *IEEE J. Quantum Electron.* 8(6), 486–493 (1972), doi: 10.1109/JQE.1972.1077092
434. A. Enokihara, M. Izutsu, and T. Sueta, "Optical fiber sensors using the method of polarization-rotated reflection," *J. Lightwave Technol.* 5(11), 1584–1590 (1987). doi: 10.1109/JLT.1987.1075449
435. G. Frosio and R. Dändliker, "Reciprocal reflection interferometer for a fiber-optic Faraday current sensor," *Appl. Opt.* 33(25), 6111–6122 (1994).
436. G. Frosio, "Reciprocal interferometers for fiber-optic current sensors," Ph.D. thesis, University of Neuchatel, Switzerland, June 15, 1992.
437. J. Blake, P. Tantaswadi, and R. T. de Carvalho, "In-line Sagnac interferometer for magnetic field sensing," in *Proc. 10th Optical Fibre Sensors Conf.*, 1994, pp. 419–422.
438. F. Rahmatian, G. Polovick, B. Hughes, and V. Aresteanu, "Field experience with high-voltage combined optical voltage and current transducers," *Cigre Session*, 2004, paper A3-111.
439. K. Bohnert, P. Gabus, J. Kostovic, and H. Brändle, "Optical fiber sensors for the electric power industry," *Opt. Lasers Eng.*, Special Issue "Optics in Switzerland," Editor Erwin Hack 43(3–5), 511–526 (2005).
440. K. Bohnert, H. Brändle, M. G. Brunzel, P. Gabus, and P. Guggenbach, "Highly accurate fiber-optic DC current sensor for the electrowinning industry," *IEEE Trans. Ind. Appl.* 43(1), 180–187 (2007), doi: 10.1109/TIA.2006.887311
441. R. Wüest, A. Frank, S. Wiesendanger, P. Gabus, U. E. Meier, J. Nehring, and K. Bohnert, "Influence of residual fiber birefringence and temperature on the high-current performance of an interferometric fiber-optic current sensor," *Proc. SPIE* 7356, 73560K (2009).

442. V. P. Gubin, V. A. Isaev, S. K. Morshnev, et al. "All-fiber optical sensor of electric current with a spun fiber sensing element," *Proc. SPIE* 6251, 1–9 (2010).
443. M. Takahashi, K. Sasaki, Y. Hirata, et al., "Field test of DC optical current transformer for HVDC link," in *Proc. IEEE PES General Meeting* 2010, doi: 10.1109/PES.2010.5589349
444. W. Wang, X. F. Wang, J. L. Xia, "The nonreciprocal errors in fiber optic current sensors," *Opt. Laser Technol.* 43, 1470–1474 (2011).
445. L. Wang, X. Xu, X. Liu, T. Zhang, and J. Yan, "Modelling and simulation of polarization errors in reflective fiber optic current sensor," *Opt. Eng.* 50(7), 074402 (2011).
446. G. M. Müller, A. Frank, M. Lenner, K. Bohnert. G. Gabus, and B. Guelenaltin, "Temperature compensation of fiber-optic current sensors in different regimes of operation," in *Proceedings IEEE Photonics Conference*, 2012, 745–746.
447. Z. Wang, Y. Wang, S. Sun, "Effect of modulation error on all optical fiber current transformers," *J. Sens. Technol.* 2, 172–176 (2012), doi: 10.4236/jst.2012.24024
448. Y. Hirata et al., "Application of an optical current transformer for cable head station of Hokkaido-Honshu HVDC link," *Proc. PES T&D* 1-6 (2012), doi: 10.1109/TDC.2012.6281570
449. J. Yu, C. Zhang, C. Li, X. Wang, Y. Li, and X. Feng, "Influence of polarization-dependent cross-talk on scale factor in the in-line Sagnac interferometer current sensor," *Opt. Eng.* 52(11), 117101 (2013).
450. R. Zhang, H. Lu, Y. Liu, X. Mao, H. Wei, J. Li, and J. Guo, "Polarization-maintaining photonic crystal fiber-based quarter waveplate for temperature stability improvement of fiber optic current sensor," *J. Mod. Opt.* 60(12), 963–969 (2013).
451. K. Sasaki, M. Takahashi, and H. Hirata, "Temperature insensitive Sagnac-type optical current transformer," *J. Lightwave Technol.* 33(12), 2463–2467 (2015).
452. G. M. Müller, Q. Wei, M. Lenner, L. Yang, A. Frank, and K. Bohnert, "Fiber-optic current sensor with self-compensation of source wavelength changes," *Opt. Lett.* 41(12), 2867–2870 (2016).
453. X. Wang, Z. Zhao, C. Li, J. Yu, and Z. Wang, "Analysis and elimination of bias error in a fiber-optic current sensor," *Appl. Opt.* 56, 8887–8895 (2017).
454. K. Bohnert, C.-P. Hsu, L. Yang, A. Frank, G. M. Müller, and P. Gabus, "Fiber-optic current sensor tolerant to imperfections of polarization-maintaining fiber connectors," *J. Lightwave Technol.* 36(11), 2161–2165 (2018).
455. H. Hu, J. Huang, Li Xia, Z., Yan, S. Peng, "The compensation of long-term temperature induced error in the all fiber current transformer through optimizing initial phase delay in λ/4 wave plate," *Microw. Opt. Technol. Lett.* 61(7), 1769–1773 (2019).
456. V. Temkina, A. Medvedev, A. Mayzel, "Research on the methods and algorithms improving the measurements precision and market competitive advantages of fiber optic current sensors," *Sensors* 20, 5995 (2020).
457. M. Wang, N. Zhang, X. Huang, et al., "High sensitivity demodulation of a reflective interferometer-based optical current sensor using an optoelectronic oscillator," *Opt. Lett.* 45, 4519–4522 (2020).
458. M. Lawrence, "Lithium niobate integrated optics," *Rep. Prog. Phys.* 56(3), 363–429 (1993).
459. E. L. Wooten, et al., "A review of lithium niobate modulators for fiber-optic communications systems," *IEEE J. Sel. Top. Quantum Electron.* 6(1), 69–82 (2000), doi: 10.1109/2944.826874
460. M. Lenner, A. Frank, L. Yang, T. M. Roininen, and K. Bohnert, "Long-term reliability of fiber-optic current sensors," *IEEE Sens. Journal* 20(2), 823–832 (2020), doi: 10.1109/JSEN.2019.2944346
461. K. Forrest, S. J. Pagano, and W. Viehmann, "Channel waveguides in glass via silver – sodium field assisted ion exchange," *J. Lightwave Technol.* 4(2), 140–150 (1986).
462. M.-C. Oh, J.-K. Seo, K.-J. Kim, H. Kim, J.-W. Kim, and W.-S. Chu, "Optical current sensors consisting of polymeric waveguide components," *J. Lightwave Technol.* 28, 1851–1857 (2010).

463. M.-C. Oh, W.-S. Chu, K.-J. Kim, and J.-W. Kim, "Polymer waveguide integrated-optic current transducers," *Opt. Express* 19, 9392–9400 (2011).
464. W.-S. Chu, S.-M. Kim, and M.-C. Oh, "Integrated optic current transducers incorporating photonic crystal fiber for reduced temperature dependence," *Opt. Express* 23, 22816–22825 (2015).
465. S.-M. Kim, W.-S. Chu, S.-G. Kim, and M.-C. Oh, "Integrated-optic current sensors with a multimode interference waveguide device," *Opt. Express* 24, 7426–7435 (2016).
466. S.-M. Kim, T.-H. Park, G. Huang, and M.-C. Oh, "Bias-free optical current sensors based on quadrature interferometric integrated optics," *Opt. Expr.* 26(24), 31599–31606 (2018).
467. H. Lin, W.-W. Lin, M.-H. Chen, and S.-C. Huang, "Fiber-optic current sensor using passive demodulation interferometric scheme," *Fiber Integr. Opt.* 18, 79–92 (1999).
468. H. Lin, W. W. Lin, M.-H. Chen, and S.-C. Huang "Vibration insensitive optical fiber current sensor with a modified reciprocal reflection interferometer," *Opt. Eng.* 38(10), 1722–1729 (1999), doi: 10.1117/1.602224
469. H. Lin, W. W. Lin, and M.-H. Chen, "Modified inline Sagnac interferometer with passive demodulation technique for environmental immunity for a fiber-optic current sensor," *Appl. Opt.* 38(13), 2760–2766 (1999).
470. H. Lin and S. C. Huang, "Fiber-optics multiplexed interferometric current sensors," *Sens. Actuator. A Phys.* 121(2), 333–338 (2005).
471. J. N. Blake, Int. patent application WO2007/033057, priority date: Sep. 12, 2005.
472. F. Ferdous, A. H. Rose, and P. Perkins, "Passively biased inline Sagnac interferometer-optical current sensor: theoretical review," *Opt. Eng.* 60(5), 057102 (8 May 2021), doi: 10.1117/1.OE.60.5.057102
473. K. Bohnert, P. Gabus, J. Nehring, S. Wiesendanger, A. Frank, and H. Brändle, "Nonlinearities in the high current response of interferometric fiber-optic current sensors," *Proc. SPIE* 7004, 70040E (2008).
474. Standard of the International Electrotechnical Commission, "IEC60044-8, Instrument transformers – Part 8: Electronic current transformers;" (new: "IEC61689-2, Instrument transformers – Part 2: Additional requirements for current transformers").
475. R. Rosenberg, C. B. Rubinstein, and D. R. Herriott, "Resonant optical Faraday rotator," *Appl. Opt.* 3(9), 1079–1083 (1964).
476. J. Stone, R. M. Jopson, L. W. Stulz, and S. J. Licht, "Enhancement of Faraday rotation in a fiber Fabry–Perot cavity," *Electron. Lett.* 23, 849–851 (1990).
477. R. M. Jopson, J. Stone, L. W. Stulz, and S. J. Licht, "Nonreciprocal transmission in a fiber Fabry-Perot resonator containing a magnetooptic material," *IEEE Photonics Technol. Lett.* 2(10), 702–704 (1990), doi: 10.1109/68.60765
478. H. Y. Ling, "Theoretical investigation of transmission through a Faraday-active Fabry–Perot étalon," *J. Opt. Soc. Am. A* 11, 754–758 (1994).
479. R. B. Wagreich and C. C. Davis, "Magnetic field detection enhancement in an external cavity fiber Fabry-Perot sensor," *J. Lightwave Technol.* 14(10), 2246–2249 (1996), doi: 10.1109/50.541214
480. H. Zhang, Y. Dong, J. Leeson, L. Chen, and X. Bao, "High sensitivity optical fiber current sensor based on polarization diversity and a Faraday rotation mirror cavity," *Appl. Opt.* 50, 924–929 (2011).
481. F. Liu, Q. Ye, J. Geng, R. Qu, and Z. Fang, "Study of fiber-optic current sensing based on degree of polarization measurement," *Chin. Opt. Lett.* 5, 267–269 (2007).
482. H. Zhang, Y. Qiu, Hui Li, et al., "High-current-sensitivity all-fiber current sensor based on fiber loop architecture," *Opt. Express* 20, 18591–18599 (2012).
483. J. Du, Y. Tao, Y. Liu, et al., "Highly sensitive and reconfigurable fiber optic current sensor by optical recirculating in a fiber loop," *Opt. Express* 24, 17980–17988 (2016).
484. H. Zhang, J. Jiang, Y. Zhang, et al., "A loop all-fiber current sensor based on single-polarization single-mode couplers," *Sensors* 17, 2674 (2017).

485. J. Jiang, H. Zhang, Y. He, and Y. Qiu, "Hybrid structure multichannel all-fiber current sensor," *Sensors* 17(8), 1770 (2017).
486. H. K. Kim, S. K. Kim, H. G. Park, B. Y. Kim, "Polarimetric fiber laser sensors," *Opt. Lett.* 18, 317–319 (1993).
487. H. K. Kim, S. K. Kim, B. Y. Kim, "Polarization control of polarimetric fiber-laser sensors," *Opt. Lett.* 18, 1465–1467 (1993).
488. G. A. Ball, G. Meltz, and W. W. Morey, "Polarimetric heterodyning Bragg-grating fiber-laser sensor," *Opt. Lett.* 18, 1976–1978 (1993).
489. K. Bohnert, A. Frank, E. Rochat, K. Haroud, and H. Brändle, "Polarimetric fiber laser sensor for hydrostatic pressure," *Appl. Opt.* 43, 41–48 (2004).
490. J. T. Kringlebotn, W. H. Loh, and R. I. Laming, "Polarimetric Er3+-doped fiber distributed-feedback laser sensor for differential pressure and force measurements," *Opt. Lett.* 21, 1869–1871 (1996).
491. H. Y. Kim, B. K. Kim, S. H. Yun, and Byoung Yoon Kim, "Response of fiber lasers to an axial magnetic field," *Opt. Lett.* 20, 1713–1715 (1995).
492. J. S. Park, S. H. Yun, S. J. Ahn, and B. Y. Kim, "Polarization- and frequency-stable fiber laser for magnetic-field sensing," *Opt. Lett.* 21, 1029–1031 (1996).
493. M. L. Lee, J. S. Park, W. J. Lee, S. H. Yun, Y. H. Lee, B. Y. Kim, "A polarimetric current sensor using an orthogonally polarized dual-frequency fibre laser," *Meas. Sci. Technol.* 9, 952–959, (1998).
494. Y. Takahashi and T. Yoshino, "Fiber ring laser with flint glass fiber and its sensor applications," *J. Lightwave Technol.* 17, 591–597 (1999).
495. L. Cheng, J. Han, Z. Guo, L. Jin, and B.-O. Guan, "Faraday-rotation-based miniature magnetic field sensor using polarimetric heterodyning fiber grating laser," *Opt. Lett.* 38, 688–690 (2013).
496. L. Cheng, J. Han, L. Jin, Z. Guo, and B.-O. Guan, "Sensitivity enhancement of Faraday effect based heterodyning fiber laser magnetic field sensor by lowering linear birefringence," *Opt. Express* 21, 30156–30162 (2013).
497. R. Stierlin, "Optical current transformer," U.S. patent 5304920, priority date: Sep. 28, 1990.
498. P. Debergh and O. Parriaux, "Current sensing by magneto-optic coupling in a birefringent waveguide," in *Proc. of ECIO'93*, Neuchâtel, Switzerland, 1993, pp. 12–36 .
499. V. Minier, D. Persegol, J. L. Lovato, and A. Kévorkian, "Integrated optical current sensor for high-power systems," in *Optical Fiber Sensors* (Optica Publishing Group, 1996), paper We23.
500. V. Minier, D. Persegui, J. L. Lovato, and A. Kévorkian, "Integrated optical current sensor with low-birefringence optical waveguides," in *12th International Conference on Optical Fiber Sensors*, Vol. 16 of 1997 OSA Technical Digest Series (Optica Publishing Group, 1997), paper OWA4.
501. V. Minier, "Waveguide birefringence temperature sensitivity in multi-turn integrated optics current sensors," in *Proc. SPIE 3746, 13th Int. Conf. on Optical Fiber Sensors* 37462C, 1999, doi: 10.1117/12.2302074
502. R. V. Ramaswamy and R. Srivastava, "Ion-exchanged glass waveguides: a review," *J. Lightwave Technol.* 6(6) 984–1000 (1988), doi: 10.1109/50.4090
503. A. D. Kersey and M. J. Marrone, "Fiber Bragg grating high-magnetic-field probe," in *Tenth Int. Conf. Optical Fibre Sensors, Proc. SPIE 2360*, 1994, pp. 53–56.
504. Y. Su, Y. Zhu, B. Zhang, J. Li, and Y. Li, "Use of the polarization properties of magneto-optic fiber Bragg gratings for magnetic field sensing purposes," *Opt. Fiber Technol.* 17(3), 196–200 (2011).
505. H. Peng, Y. Su, and Y. Li, "Evolution of polarization properties in circular birefringent fiber Bragg gratings and application for magnetic field sensing," *Opt. Fiber Technol.* 18(4), 177–182 (2012).
506. F. Descamps, D. Kinet, S. Bette, and C. Caucheteur, "Magnetic field sensing using standard uniform FBGs," *Opt. Express* 24, 26152–26160 (2016).

507. B. Wu, C. Li, K. Qiu, and L. Cheng, "Characteristics of light polarization in magneto-optic fiber Bragg gratings with linear birefringence," *Chin. Opt. Lett.* 9, 010601 (2011).
508. B. Wu, F. Wen, K. Qiu, et al., "Magnetically-induced circular-polarization-dependent loss of magneto-optic fiber Bragg gratings with linear birefringence," *Opt. Fiber Technol.* 19(3), 219–222 (2013).
509. D. Goldstein, "The mathematics of the Mueller matrix," in *Polarized Light*, 2nd ed. (CRC Press, 2003), Chap 9.
510. P. Orr, P. Niewczas, M. Stevenson, and J. Canning, "Compound phase-shifted fiber Bragg structures as intrinsic magnetic field sensors," *J. Lightwave Technol.* 28(18), 2667–2673 (2010).
511. P. Orr and P. Niewczas, "An optical fibre system design enabling simultaneous point measurement of magnetic field strength and temperature using low-birefringence FBGs," *Sens. Actuator. A Phys.* 163(1), 68–74, (2010).
512. J. C. Yong, S. H. Yun, M. L. Lee, and B. Y. Kim, "Frequency-division-multiplexed polarimetric fiber laser current-sensor array," *Opt. Lett.* 24, 1097–1099 (1999).
513. Z. Huang, H. Zhang, J. Jiang, et al., "A quasi-distributed all-fiber current sensor based on series structure," *Opt. Fiber Technol.* 32, 1–5 (2016).
514. L. Palmieri, D. Sarchi, and A. Galtarossa, "Distributed measurement of high electric current by means of polarimetric optical fiber sensor," *Opt. Express* 23, 11073–11079 (2015).
515. E. R. Perry, "Laser measures current," *Instrum. Contr. Syst.* 38(7), 121–124 (1965).
516. A. Braun and J. Zinkernagel, "Optoelectronic electricity meter for high-voltage lines," *IEEE Trans. Instrum. Meas.* 22(4), 394–399 (1973).
517. J. S. Subjak, "An EHV current transducer with feedback-controlled encoding," *IEEE Trans. Power Appar. Syst.* 94(6), 2124–2130 (1975).
518. L. E. Berkebile, S. Nilsson, and S. Sun, "Digital EHV current transducer," *IEEE Trans. Power Appar. Syst.* PAS-100(4), 1498–1504 (1981), doi: 10.1109/TPAS.1981.316497
519. R. Malewski, "High-voltage current transformers with optical signal transmission," *Opt. Eng.* 20(1) 200154 (1981), doi: 10.1117/12.7972662
520. M. B. Adolfsson, C. H. Einvall, P. Lindberg, J. Samuelsson, L. Ahlgren, and H. Edlund, "EHV series capacitor banks: a new approach to platform to grounds signalling, relay protection and supervision," *IEEE Trans. Power Deliv.* 4(2), 1369–1378 (1989), doi: 10.1109/61.25624
521. Y. N. Ning, T. Y. Liu, and D. A. Jackson, "Two low-cost robust electro-optic hybrid current sensors capable of operation at extremely high potential," *Rev. Sci. Instrum.* 63(12), 5771–5773 (1992), doi: 10.1063/1.1143361
522. N. A. Pilling, R. Holmes, and G. R. Jones, "Optically powered hybrid current measurement system," *Electron. Lett.* 29(12), 1049–1051 (1993).
523. A. Tardy, A. Derossis, and J. Dupraz, "A current censor remotely powered and monitored through an optical Fiber Link," *Opt. Fiber Technol.* 1, 181–185 (1995).
524. G. Zhang, S. Li, Y. Qin, and Z. Zhang, "A new electro-optic hybrid current-sensing scheme for current measurement at high voltage," *Rev. Sci. Instrum.* 70(9), 3755–3758 (1999). doi:10.1063/1.1149988
525. E. F. Donaldson, J. R. Gibson, G. R. Jones, N. A. Pilling and B. T. Taylor, "Hybrid optical current transformer with optical and power-line energisation," *IEE Proc. Gener. Transm. Distrib.*, 147(5), 304–309 (2000), doi: 10.1049/ip-gtd:20000604
526. G. Zhang, S. Li, Z. Zhang, and W. Cao, "A novel electro-optic hybrid current measurement instrument for high-voltage power lines," *IEEE Trans. Instrum. Meas.* 50(1) 59–62 (2001), doi: 10.1109/19.903878
527. M. M. Werneck and A. C. S. Abrantes, "Fiber-optic-based current and voltage measuring system for high-voltage distribution lines," *IEEE Trans. Power Deliv.* 19(3), 947–951 (2004), doi: 10.1109/TPWRD.2004.829916
528. Y. Wang, L. Zheng, P. Hou, and C. Hu, "Research on optically powered ultra current transformer," *Opt. Lasers Eng.* 43(10), 1145–1150 (2005).

529. Y. Wang, J. Yuan, H. Wang, and J. Liu, "Design and research of fiber optically powered Rogowski coil current transformer," *Procedia Eng.* 15, 886–890 (2011).
530. F. V. B. de Nazare and M. M. Werneck, "Hybrid optoelectronic sensor for current and temperature monitoring in overhead transmission lines," *IEEE Sens. J.* 12(5), 1193–1194 (2012), doi: 10.1109/JSEN.2011.2163709
531. F. R. Bassan, J. B. Rosolem, C. Floridia, et al. "Power-over-fiber LPIT for voltage and current measurements in the medium voltage distribution networks," *Sensors* 21, 547, (2021), doi: 10.3390/s21020547
532. *ABB Instrument Transformers: Application Guide*, Edition 4 (2015).
533. IEC 61869-11. Instrument Transformers—Part 11: Additional requirements for low power passive voltage transformers; International Electrotechnical Commission IEC: Geneva, Switzerland, 2017.
534. IEC 61869-10. Instrument Transformers—Part 10: Additional requirements for low power passive current transformers; International Electrotechnical Commission IEC: Geneva, Switzerland, 2017.
535. A. Erez, "Low-frequency electrical signal measurement by electrooptical methods," *IEEE Trans. Instrum. Meas.* 21(4), 358–360 (1972).
536. R. M. Ribeiro, L. Martins, and M. M. Werneck, "Wavelength demodulation of ultrabright green light-emitting diodes for electrical current sensing," *IEEE Sens. J.* 5(1), 38–47 (2005), doi: 10.1109/JSEN.2004.838663
537. J. L. Flores, G. A. Ayubi, J. Matias di Martino, et al., "Hybrid ac-current sensor based on the time modulation of an autonomous light source," *Optik* 152(1), 29–35 (2018).
538. B. Jaffe, W. R. Cook, and H. Jaffe, *Piezoelectric Ceramics* (Academic Press, New York, 1971).
539. Y. N. Ning, B. C. B. Chu, and D. A. Jackson, "Interrogation of a conventional current transformer by a fiber-optic interferometer," *Opt. Lett.* 16, 1448–1450 (1991).
540. C. M. Davis, "Phase-modulated fiber-optic current transformer/voltage transformer," *EPRI final report* EL-7421, Research Project 2734-3, Optical Technologies Inc., Herndon, VA, 1991.
541. C. McGarrity, Y. N. Ning, J. L. Santos, and D. A. Jackson, "A fiber-optic system for three-phase current sensing using a hybrid sensing technique," *Rev. Scient. Instrum.* 63, 2035–2039 (1992), doi: 10.1063/1.1143810
542. T. Wang, C. Luo, and S. Zheng, "A fiber-optic current sensor based on a differentiating Sagnac interferometer," *IEEE Trans. Instrum. Meas.* 50(3), 705–708 (2001), doi: 10.1109/19.930443
543. P. J. Henderson, N. E. Fisher, and D. A. Jackson, "Current metering using fibre-grating based interrogation of a conventional current transformer," in *12th Int. Conf. Optical Fiber Sensors*, Vol. 16 of 1997 OSA Technical Digest Series (Optica Publishing Group, 1997), paper OWC11.
544. N. E. Fisher, P. J. Henderson, and D. A. Jackson, "The interrogation of a conventional current transformer using an in-fibre Bragg grating," *Meas. Sci. Technol.* 8(10), 1080 (1997).
545. L. Dziuda, G. Fusiek, P. Niewczas, G. M. Burt, and J. R. McDonald, "Laboratory evaluation of a hybrid fiber-optic current sensor," *Sens. and Actuator. A Phys.* 136(1), 184–190 (2007).
546. G. Fusiek, "Hysteresis compensation for a piezoelectric fiber optic voltage sensor," *Opt. Eng.* 44(11), 114402 (2005).
547. G. Fusiek, Pawel Niewczas, James R. McDonald, "Improved method of hysteresis compensation for a piezoelectric fiber optic voltage sensor," *Opt. Eng.* 46(3), 034401 (2007), doi: 10.1117/1.2714931
548. P. Niewczas and J. R. McDonald, "Advanced optical sensors for power and energy systems applications," *IEEE Instrum. Meas. Mag.* 10(1), 18–28 (2007), doi: 10.1109/MIM.2007.339552
549. G. Fusiek, P. Niewczas, and J. McDonald, "Feasibility study of the application of optical voltage and current sensors and an arrayed waveguide grating for aero-electrical systems," *Sens. Actuator. A Phys.* 147(1), 177–182 (2008).
550. A. D. Kersey, et al., "Fiber grating sensors," *J. Lightwave Technol.* 15(8), 1442–1463 (1997), doi: 10.1109/50.618377

551. P. M. Cavaleiro, F. M. Araujo, and A. L. Ribeiro, "Metal-coated fibre Bragg grating sensor for electric current metering," *Electron. Lett.* 34(11), 1133–1135 (1998).
552. P. Orr, G. Fusiek, P. Niewczas, et al., "Distributed photonic instrumentation for power system protection and control," *IEEE Trans. Instrum. Meas.* 64(1), 19–26 (2015), doi: 10.1109/TIM.2014.2329740
553. P. Orr, P. Niewczas, C. Booth, G. Fusiek, A. Dysko, and F. Kawano, "An optically-interrogated Rogowski coil for passive multiplexable current measurement," *IEEE Sens. J.* 13(6), 2053–2054 (2013).
554. P. Orr, G. Fusiek, P. Niewczas, et al., "Distributed photonic instrumentation for power system protection and control," *IEEE Trans. Instrum. Meas.* 64(1), 19–26 (2015), doi: 10.1109/TIM.2014.2329740
555. D. Tzelepis A. Dysko, G. Fusiek, et al., "Single-ended differential protection in MTDC networks using optical sensors," *IEEE Trans. Power Deliv.* 32(3), 1605–1615 (2017), doi: 10.1109/TPWRD.2016.2645231
556. D. Tzelepis, A. Dysko, C. Booth, G. Fusiek, et al. "Distributed current sensing technology for protection and fault location applications in high-voltage direct current networks," *J. Eng.* 2018(15), 1169–1175 (2018).
557. P. Orr and P. Niewczas, "High-speed, solid state, interferometric interrogator and multiplexer for fiber Bragg grating sensors," *J. Lightwave Technol.* 29(22), 3387–3392 (2011).
558. P. Wei, C. Cheng, X. Wang, et al. "A high-performance hybrid current transformer based on a fast variable optical attenuator," *IEEE Trans. Power Deliv.* 29(6), 2656–2663 (2014).
559. P. Wei, C. Cheng, and T. Liu, "A photonic transducer-based optical current sensor using back-propagation neural network," *IEEE Photonics Technol. Lett.* 28(14), 1513–1516 (2016), doi: 10.1109/LPT.2016.2557339
560. P. Wei, C. Cheng, L. Deng, and H. Huang, "A hybrid electro-optic current transducer using a photodiode-based primary supply," *IEEE Sens. J.* 17(9), 2713–2717 (2017).
561. R. L. Heredero, R. F. de Caleya, H. Guerrero, et al., "Micromachined optical fiber current sensor," *Appl. Opt.* 38(25), 5298–5305 (1999).
562. R. L. Heredero, J. L. Santos, R. F. de Caleya, and H. Guerrero, "Micromachined low-finesse Fabry-Perot interferometer for the measurement of DC and AC electrical currents," *IEEE Sens. J.* 3(1), 13–18 (2003), doi: 10.1109/JSEN.2003.810112
563. J. Duplessis and J. Barker, "Intelligent measurement for grid management & control," in *USA Pacworld Conference* 2014, paper 0903.
564. N. A. Pilling, R. Holmes, and G. R. Jones, "Optical fibre current measurement system using liquid crystals and chromatic modulation," *IEE Proc.* C140(5), 351–356 (1993), doi: 10.1049/ip-c.1993.0052
565. J. D. Bull, N. A. F. Jaeger, and F. Rahmatian, "A new hybrid current sensor for high-voltage applications," *IEEE Trans. Power Deliv.* 20(1), 32–38 (2005), doi: 10.1109/TPWRD.2004.833889
566. J. P. Joule, "On a new class of magnetic forces," *Ann. Electr. Magn. Chem.* 8, 219–224 (1842).
567. E. Trémolet de Lacheisserie, *Magnetostriction – Theory and Applications of Magnetoelasticity* (CRC Press, 1993).
568. M. J. Dapino, "On magnetostrictive materials and their use in adaptive structures," *Struct. Eng. Mech.* 17(3–4), 303–330 (2004).
569. J. D. Livingston, "Magnetomechanical properties of amorphous metals," *Physica Status Solidi (a)* 70(2), 591–596 (1982).
570. L. Daniel, "An analytical model for the magnetostriction strain of ferromagnetic materials subjected to multiaxial stress," *Eur. Phys. J. Appl. Phys.* 83(3), article 30904 (2018).
571. A. Yariv and H. V. Windsor, "Proposal for detection of magnetostrictive perturbation of optical fibers," *Opt. Lett.* 5, 87 (1980).
572. A. Dandridge, A. B. Tveten, G. H. Sigel, E. J. West, and T. G. Giallorenzi, "Optical fiber magnetic field sensors," *Electron. Lett.* 17(15), 408–409 (1980).

573. J. Jarzynski, J. H. Cole, J. A. Bucaro, and C. M. Davis, "Magnetic field sensitivity of an optical fiber with magnetostrictive jacket," *Appl. Opt.* 19(22), 3746–3748 (1980).
574. S. C. Rashleigh, "Magnetic-field sensing with a single-mode fiber," *Opt. Lett.* 6(1), 19–21 (1981).
575. K. P. Koo and G. H. Sigel, "Characteristics of fiber-optic magnetic-field sensors employing metallic glasses," *Opt. Lett.* 7, 334–336 (1982).
576. K. Koo, A. Dandridge, A. Tveten, and G. Sigel, "A fiber-optic DC magnetometer," *J. Lightwave Technol.* 1(3), 524–525 (1983), doi: 10.1109/JLT.1983.1072141
577. A. D. Kersey, M. Corke, D. A. Jackson, and J. D. C. Jones, "Detection of DC and low frequency AC magnetic fields using an all single-mode fiber interferometer," *Electron. Lett.* 19, 469–471 (1983).
578. K. P. Koo and G. H. Sigel, "Detection scheme in a fiber-optic magnetic-field sensor free from ambiguity due to material magnetic hysteresis," *Opt. Lett.* 9, 257–259 (1984).
579. F. Bucholtz, K. Koo, G. Sigel, and A. Dandridge, "Optimization of the fiber/metallic glass bond in fiber-optic magnetic sensors," *J. Lightwave Technol.* 3(4), 814–817 (1985), doi: 10.1109/JLT.1985.1074286
580. A. Kersey, D. Jackson, and M. Corke, "Single-mode fibre-optic magnetometer with DC bias field stabilization," *J. Lightwave Technol.* 3(4), 836–840 (1985), doi: 10.1109/JLT.1985.1074285
581. K. Koo, A. Dandridge, F. Bucholtz, and A. Tveten, "An analysis of a fiber-optic magnetometer with magnetic feedback," *J. Lightwave Technol.* 5(12), 1680–1685 (1987).
582. F. Bucholtz, D. M. Dagenais, K. P. Koo, and S. Vohra, "High-frequency fibre-optic magnetometer with 70 fT/square root (Hz) resolution," *Electron. Lett.* 25, 1719 (1989).
583. K. P. Koo, F. Bucholtz, D. M. Dagenais, and A. Dandridge, "A compact fiber-optic magnetometer employing an amorphous metal wire transducer," *IEEE Photonics Technol. Lett.* 1(12), 464–466 (1989), doi: 10.1109/68.46051
584. D. Y. Kim, H. J. Kong, and B. Y. Kim, "Fiber-optic DC magnetic field sensor with balanced detection technique," *IEEE Photonics Technol. Lett.* 4(8), 945–948 (1992), doi: 10.1109/68.149918
585. M. Sedlar, I. Paulicka, and M. Sayer, "Optical fiber magnetic field sensors with ceramic magnetostrictive jackets," *Appl. Opt.* 35, 5340–5344 (1996).
586. N. Rajkumar, V. Jagadeesh Kumar, and P. Sankaran, "Fiber sensor for the simultaneous measurement of current and voltage in a high-voltage system," *Appl. Opt.* 32, 1225–1228 (1993).
587. K. D. Oh, J. Ranade, V. Arya, A. Wang, and R. O. Claus, "Optical fiber Fabry-Perot interferometric sensor for magnetic field measurement," *IEEE Photonics Technol. Lett.* 9(6), 797–799 (1997).
588. K. D. Oh, A. Wang, and R. O. Claus, "Fiber-optic extrinsic Fabry–Perot dc magnetic field sensor," *Opt. Lett.* 29, 2115–2117 (2004).
589. J. Mora, A. Diez, J. L. Cruz, and M. V. Andres, "A magnetostrictive sensor interrogated by fiber gratings for DC-current and temperature discrimination," *IEEE Photonics Technol. Lett.* 12(12), 1680–1682 (2000), doi: 10.1109/68.896347
590. K. S. Chiang, R. Kancheti, and V. Rastogi, "Temperature-compensated fiber-Bragg-grating-based magnetostrictive sensor for DC and AC currents," *Opt. Eng.* 42(7), 1906–1909 (2003).
591. D. Davino, C. Visone, C. Ambrosino, S. Campopiano, A. Cusano, A. Cutolo, "Compensation of hysteresis in magnetic field sensors employing fiber Bragg grating and magneto-elastic materials," *Sens. Actuator. A Phys.* 147(1), 127–136 (2008).
592. C. Ambrosino, S. Campopiano, A. Cutolo, and A. Cusano, "Sensitivity tuning in Terfenol-D based fiber Bragg grating magnetic sensors," *IEEE Sens. J.* 8(9), 1519–1520 (2008), doi: 10.1109/JSEN.2008.925159
593. X. Wang, S. Chen, Z. Du, et al., "Experimental study of some key issues on fiber-optic interferometric sensors detecting weak magnetic field," *IEEE Sens. J.* 8(7), 1173–1179 (2008).
594. X. Wang, X. Li, Z. Du, X. Wang, and J. Chen, "Experimental investigation on temperature dependence of the performance in a magnetostrictive fiber-optic interferometric magnetic field sensor," *IEEE Sens. J.* 9(10), 1234–1239 (2009).

595. M. Yang, J. Dai, C. Zhou, and D. Jiang, "Optical fiber magnetic field sensors with TbDyFe magnetostrictive thin films as sensing materials," *Opt. Express* 17(23), 20777–20782 (2009).
596. S. M. M. Quintero, A. M. B. Braga, H. I. Weber, A. C. Bruno, J. F. D. F. Araújo, "A magnetostrictive composite-fiber Bragg grating sensor," *Sensors* 10(9), 8119 (2010).
597. S. M. M. Quintero, C. Martelli, A. M. B. Braga, L. C. G. Valente, and C. C. Kato, "Magnetic field measurements based on Terfenol coated photonic crystal fibers," *Sensors* 11(12), 11103–11111 (2011).
598. G. Lanza, G. Breglio, M. Giordano, A. Gaddi, S. Buontempo, and A. Cusano, "Effect of the anisotropic magnetostriction on Terfenol-D based fiber Bragg grating magnetic sensors," *Sens. Actuator. A Phys.* 172(2), 420–427 (2011).
599. H. Liu, S. W. Or, and H. Y. Tam, "Magnetostrictive composite–fiber Bragg grating (MC–FBG) magnetic field sensor," *Sens. Actuator. A: Phys.* 173(1), 122–126 (2012).
600. Y. Du, T. Liu, Z. Ding, K. Liu, B. Feng, and J. Jiang, "Distributed magnetic field sensor based on magnetostriction using Rayleigh backscattering spectra shift in optical frequency-domain reflectometry," *Appl. Phys. Express* 8(1), 012401 (2014).
601. A. Masoudi and T. P. Newson, "Distributed optical fiber dynamic magnetic field sensor based on magnetostriction," *Appl. Opt.* 53, 2833–2838 (2014).
602. R. Amat, H. García-Miquel, D. Barrera, G.V. Kurlyandskaya, and S. Sales, "Magneto-optical sensor based on fiber Bragg gratings and a magnetostrictive material," *Key Eng. Mat.* 644, 232–235 (2015).
603. Y. Du, T. Liu, Z. Ding, et al., "Distributed magnetic field sensor based on magnetostriction using Rayleigh backscattering spectra shift in optical frequency-domain reflectometry," *Appl. Phys. Express* 8, 012401 (2015).
604. Z. Ding, Y. Du, T. Liu, K. Liu, B. Feng, and J. Jiang, "Distributed optical fiber current sensor based on magnetostriction in OFDR," *IEEE Photonics Technol. Lett.* 27(19), 2055–2058 (2015), doi: 10.1109/LPT.2015.2450237
605. I. M. Nascimento, J. M. Baptista, P. A. S. Jorge, J. L. Cruz, and M. V. Andrés, "Intensity-modulated optical fiber sensor for AC magnetic field detection," *IEEE Photonics Technol. Lett.* 27(23), 2461–2464 (2015), doi: 10.1109/LPT.2015.2470135
606. I. M. Nascimento, J. M. Baptista, P. A. S. Jorge, J. L. Cruz, and M. V. Andrés, "Passive interferometric interrogation of a magnetic field sensor using an erbium doped fiber optic laser with magnetostrictive transducer," *Sens. Actuator. A Phys.* 235, 227–233 (2015).
607. H. Yan, X. Thao C. Zhang, et al., "A novel current fiber sensor with magnetostrictive material based on the plasmon response," *Optik* 127(3), 1323–1325 (2016).
608. J. Han, H. Hu, H. Wang, et al., "Temperature-compensated magnetostrictive current sensor based on the configuration of dual fiber Bragg gratings," *J. Lightwave Technol.* 35, 4910–4915 (2017).
609. I. M. Nascimento, G. Chesini, J. M. Baptista, C. M. B. Cordeiro, and P. A. S. Jorge, "Vibration and magnetic field sensing using a long-period grating," *IEEE Sens. J.* 17(20), 6615–6621 (2017), doi: 10.1109/JSEN.2017.2743112
610. N. Kaplan, J. Jasenek, J. Červeňová, and M. Ušáková, "Magnetic optical FBG sensors using optical frequency-domain reflectometry," *IEEE Trans. Magn.* 55(1), art no. 4000704 (2019), doi: 10.1109/TMAG.2018.2873405
611. S. Xu, Q. Peng, F. Xing, et al., "A low-cost current sensor based on semi-cylindrical magnetostrictive composite," *Electronics* 9(11), 1833 (2020).
612. W. Li, X. Kewen, W. Ling, "Model and experimental study on optical fiber CT based on Terfenol-D," *Sensors* 20(8), 2255 (2020).
613. M. Zhou, Y. Zhao, G. Wang, et al., "Simultaneous AC and DC measurement based on an FBG-magnetostrictive fiber sensor," *Appl. Opt.* 60(24), 7131–7135 (2021).
614. D. Feng, Y. Gao, T. Zhu, M. Deng, X. Zhang, and L. Kai, "High-precision temperature-compensated magnetic field sensor based on optoelectronic oscillator," *J. Lightwave Technol.* 39, 2559–2564 (2021).

615. J. D. Lopez, A. Dante, C. C. Carvalho, R. Allil, and M. M. Werneck, "Simulation and experimental study of FBG-based magnetic field sensors with Terfenol-D composites in different geometric shapes," *Measurement* 172, 108893 (2021).
616. B. Zhan, T. Ning, Li Pei, et al., "Terfenol-D based magnetic field sensor with temperature independence incorporating dual fiber Bragg gratings structure," *IEEE Access* 9, 32713–32720 (2021).
617. R. Sun, L. Zhang, H. Wei, et al., "Quasi-distributed magnetic field fiber sensors integrated with magnetostrictive rod in OFDR system," *Electronics* 11(7), 1013 (2022).
618. D. Satpathi, J. A. Moore, and M. G. Ennis, "Design of a Terfenol-D based fiber-optic current transducer," *IEEE Sens. J.* 5(5), 1057–1065 (2005).
619. J. Mora, L. Martínez-León, A. Díez, J. L. Cruz, and M.V. Andrés, "Simultaneous temperature and ac-current measurements for high voltage lines using fiber Bragg gratings," *Sens. Actuator. A Phys.* 125, 313–316 (2006).
620. D. Reilly, A. J. Willshire, G. Fusiek, P. Niewczas, and J. R. McDonald, "A fiber-Bragg-grating-based sensor for simultaneous AC current and temperature measurement," *IEEE Sens. J.* 6(6), 1539–1542 (2006).
621. A. O. Cremonezi, E. C. Ferreira, A. J. B. Filho, and J. A. S. Dias, "A fiber Bragg grating RMS current transducer Based on the magnetostriction effect using a Terfenol-D toroidal-shaped modulator," *IEEE Sens. J.* 13(2), 683–690 (2013), doi: 10.1109/JSEN.2012.2226333
622. F. V. B. de Nazaré, M. M. Werneck, R. P. de Oliveira, et al., "Development of an optical sensor head for current and temperature measurements in power systems," *J. Sens.*, Art. no. 393406 (2013), doi: 10.1155/2013/393406
623. F. V. B. de Nazaré and M. M. Werneck, "Compact optomagnetic Bragg-grating-based current sensor for transmission lines," *IEEE Sens. J.* 15(1), 100–109 (2015), doi: 10.1109/JSEN.2014.2337518
624. A. Dante, J. D. Lopez, T. Trovão, R. W. Mok, C. C. Carvalho, R. C. da Silva Barros Allil, and M. M. Werneck, "A compact FBG-Based toroidal magnetostrictive current sensor with reduced mass of Terfenol-D," in *OSA Optical Sensors and Sensing Congress 2019* (Optical Society of America, Washington, DC, USA, 2019).
625. A. Dante, J. D. Lopez, C. C. Carvalho, R. C. da Silva Barros Allil, and M. M. Werneck, "A Compact FBG-based magnetostrictive optical current sensor with reduced mass of Terfenol-D," *IEEE Photonics Technol. Lett.* 31(17), 1461–1464 (2019), doi: 10.1109/LPT.2019.2932112
626. S. Wang, F. Wan, H. Zhao, W. Chen, W. Zhang, and Q. Zhou, "A sensitivity-enhanced fiber grating current sensor based on giant magnetostrictive material for large-current measurement," *Sensors* 19(8), 1755 (2019).
627. J. D. Lopez, A. Dante, R. M. Bacurau, et al., "Fiber-optic current sensor based on FBG and optimized magnetostrictive composite," *IEEE Photonics Technol. Lett.* 31(24), 1987–1990 (2019).
628. J. D. Lopez, A. Dante, A. O. Cremonezi, et al., "Fiber-optic current sensor based on FBG and Terfenol-D with magnetic flux concentration for enhanced sensitivity and linearity," *IEEE Sens. J.* 20(7), 3572–3578 (2020), doi: 10.1109/JSEN.2019.2959231
629. S. Xu, Q. Peng, C. Li, et al., "Optical fiber current sensors based on FBG and magnetostrictive composite materials," *Appl. Sci.* 11(1), 161 (2020).
630. E. Blums, A. Cebers, and M.M. Majorov, *Magnetic Fluids* (De Gruyter, Berlin 1996), ISBN 3-11-014390-9.
631. Y. Zhao, R. Q. Lv, H. Li, and Q. Wang, "Simulation and experimental measurement of magnetic fluid transmission characteristics subjected to the magnetic field," *IEEE Trans. Magn.* 50(5), 1–5 (2014).
632. Y. Zhao, D. Wu, R. Q. Lv, and Y. Ying, "Tunable characteristics and mechanism analysis of the magnetic fluid refractive index with applied magnetic field," *IEEE Trans. Magn.* 50(8), 1–5 (2014).

633. Z. Di, X. Chen, S. Pu, X. Hu, and Y. Xia, "Magnetic-field-induced birefringence and particle agglomeration in magnetic fluids," *Appl. Phys. Lett.* 89(21), art. no. 211106 (2006).
634. P. Zu et al., "Enhancement of the sensitivity of magneto-optical fiber sensor by magnifying the birefringence of magnetic fluid film with Loyt Sagnac interferometer," *Sens. Actuator. B Chem.* 191, 19–23 (2014).
635. M. Xu and P. J. Ridler, "Linear dichroism and birefringence effects in magnetic fluids," *J. Appl. Phys.* 82(1), 326–332 (1997).
636. J. Xia, Q. Wang, X. Liu, and H. Luo, "Fiber optic Fabry-Perot current sensor integrated with magnetic fluid using a fiber Bragg grating demodulation," *Sensors* 15, 16632–16641 (2015), doi: 10.3390/s150716632
637. R. Lv, Y. Zhao, D. Wang, and Q. Wang, "Magnetic fluid-filled optical fiber Fabry–Perot sensor for magnetic field measurement," *IEEE Photonics Technol. Lett.* 26(3), 217–219 (2014).
638. Y. Zhao, X. Liu, R.-Q. Lv, Y.-N. Zhang, and Q. Wang, "Review on optical fiber sensors based on the refractive index tunability of ferrofluid," *J. Lightwave Technol.* 35, 3406–3412 (2017), doi: 10.1109/JLT.2016.2573288
639. J. Dai, M. Yang, X. Li, et al., "Magnetic field sensor based on magnetic fluid clad etched fiber Bragg grating," *Opt. Fiber Technol.* 17(3), 210–213 (2011).
640. J. Zheng, X. Dong, P. Zu, et al., "Magnetic field sensor using tilted fiber grating interacting with magnetic fluid," *Opt. Express* 21(15), 17863–17868 (2013).
641. X. Li, X. Lyu, F. Lyu et al., "An all-fiber current sensor based on magnetic fluid clad microfiber knot resonator," *Int. J. Smart Sens. Intell. Syst.* 7(5) (2020).
642. L. Gao et al., "Long-period fiber grating within D-shaped fiber using magnetic fluid for magnetic-field detection," *IEEE Photonics J.* 4(6), 2095–2104, (2012), doi: 10.1109/JPHOT.2012.2226439
643. J. Luo et al., "A magnetic sensor based on a hybrid long-period fiber grating and a magnetic fluid," *IEEE Photonics Technol. Lett.* 27(9), 998–1001 (2015), doi: 10.1109/LPT.2015.2405079
644. Y. Chen, Q. Han, T. Liu, et al., "Optical fiber magnetic field sensor based on single-mode–multimode–single-mode structure and magnetic fluid," *Opt. Lett.* 38, 3999–4001 (2013).
645. H. Wang, S. Pu, N. Wang, S. Dong, and J. Huang, "Magnetic field sensing based on single-mode–multimode–singlemode fiber structures using magnetic fluids as cladding," *Opt. Lett.* 38, 3765–3768 (2013).
646. L. Li, Q. Han, Y. Chen, T. Liu, and R. Zhang, "An all-fiber optic current sensor based on ferrofluids and multimode interference," *IEEE Sens. J.* 14(6), 1749–1753 (2014), doi: 10.1109/JSEN.2014.2302812
647. C. Li, T. Ning, X. Wen, et al., "Magnetic field and temperature sensor based on a no-core fiber combined with a fiber Bragg grating," *Opt. Laser Technol.* 72, 104–107 (2015).
648. G.-H. Su, J. Shi, D.-G. Xu, et al., "Simultaneous magnetic field and temperature measurement based on no-core fiber coated with magnetic fluid," *IEEE Sens. J.* 16 (23), 8489–8493 (2016).
649. H. Tian, Y. Song, Y. Li, and H. Li, "Fiber-optic vector magnetic field sensor based on mode interference and magnetic fluid in a two-channel tapered structure," *IEEE Photonics J.* 11(6), 1–9, art no. 7104309 (2019), doi: 10.1109/JPHOT.2019.2944931
650. Y. Li, S. Pu, Y. Zhao, et al. "All-fiber-optic vector magnetic field sensor based on side-polished fiber and magnetic fluid," *Opt. Express* 27, 35182–35188 (2019).
651. X. Li and H. Ding, "Temperature insensitive magnetic field sensor based on ferrofluid clad microfiber resonator," *IEEE Photonics Technol. Lett.* 26(24), 2426–2429 (2014).
652. Z. Ma, Y. Miao, Y. Li, et al., "A highly sensitive magnetic field sensor based on a tapered microfiber," *IEEE Photonics J.* 10(4), 1–8, art no. 6803308 (2018), doi: 10.1109/JPHOT.2018.2862949
653. P. Li, H. Yan, Z. Xie, Y. Li, and X. Zhao, "An intensity-modulated and large bandwidth magnetic field sensor based on a tapered fiber Bragg grating," *Opt. Laser Technol.* 125, 105996 (2020).
654. M. Luo, Q. Yang, Y. He, and R. Liu, "Current sensor based on an integrated micro-ring resonator and superparamagnetic nanoparticles," *Opt. Express* 28, 5684–5691 (2020).

655. H. Ji, S. Pu, X. Wang, and G. Yu, "Magnetic field sensing based on V-shaped groove filled with magnetic fluids," *Appl. Opt.* 51(8), 1010–1020 (2012).
656. H. Ji, S. Pu, X. Wang, et al., "Magnetic field sensing based on capillary filled with magnetic fluids," *Appl. Opt.* 51(27), 6528–6538 (2012).
657. P. Zu, C. C. Chan, W. S. Lew, et al., "Magneto-optical fiber sensor based on magnetic fluid," *Opt. Lett.* 37, 398–400 (2012).
658. Y. Zhao, D. Wu, R. Lv, and J. Li, "Magnetic field measurement based on the Sagnac interferometer with a ferrofluid-filled high-birefringence photonic crystal fiber," *IEEE Trans. Instrum. Meas.* 65(6) 1503–1507, (2016), doi: 10.1109/TIM.2016.2519767
659. Q. Wang, J. Xia, X. Liu, Y. Zhao, J. Li, and H. Hu, "A novel current sensor based on magnetic fluid and fiber loop cavity ring-down technology," *IEEE Sens. J.* 15(11), 6192–6198 (2015), doi: 10.1109/JSEN.2015.2454451
660. R. Gao, Y. Jiang, and S. Abdelaziz, "All-fiber magnetic field sensors based on magnetic fluid-filled photonic crystal fibers," *Opt. Lett.* 38, 1539–1541 (2013).
661. J. Wang, L. Pei, J. Wang, et al., "Magnetic field and temperature dual-parameter sensor based on magnetic fluid materials filled photonic crystal fiber," *Opt. Express* 28, 1456–1471 (2020).
662. C. Liu, T. Shen, H. B. Wu, et al., "Applications of magneto-strictive, magneto-optical, magnetic fluid materials in optical fiber current sensors and optical fiber magnetic field sensors: a review," *Opt. Fiber Technol.* 65, 102634 (2021).
663. L. Martinez, F. Cecelja, and R. Rakowski, "A novel magneto-optic ferrofluid material for sensor applications," *Sens. Actuator. A Phys.* 123–124(23), 438–443 (2005).
664. N. Alberto, M. F. Domingues, C. Marques, P. André, and P. Antunes, "Optical fiber magnetic field sensors based on magnetic fluid: a review," *Sensors* 18, 4325 (2018), doi: 10.3390/s18124325
665. J. Lenz and A. S. Edelstein, "Magnetic sensors and their applications," *IEEE Sens. J.* 6(3), 631–649 (2006).
666. S. Ziegler, R. C. Woodward, H. H. C. Iu, and L. J. Borle, "Current sensing techniques: a review," *IEEE Sens. J.* 9(4), 354–376 (2009).
667. V. C. Gungor, B. Lu, and G. P. Hancke, "Opportunities and challenges of wireless sensor networks in smart grid," *IEEE Trans. Ind. Electron.* 57(10), 3557–3564 (2010).
668. K.-L. Chen, Y.-R. Chen, Y.-P. Tsai, and N. Chen, "A novel wireless multifunctional electronic current transformer based on ZigBee-based communication," *IEEE Trans. Smart Grid* 8(4), 1888–1897 (2017).
669. A. Bergström, "Vattenfall goes real time," *T&D World Magazine*, Nov. 2014.
670. J. H. Hines, A. T. Hines, D. Yaney, and D. Kirkpatrick, "Remotely powered line monitor," U.S. patent 10,620,238, filed Oct. 20, 2016.
671. T. Otto, S. Kurth, S. Voigt, et al., "Integrated microsystems for smart applications," *Sens. Mater.* 30(4), 767–778 (2018).
672. X. Sun, Q. Huang, Y. Hou, L. Jiang, and P. W. T. Pong, "Noncontact operation-state monitoring technology based on magnetic-field sensing for overhead high-voltage transmission lines," *IEEE Trans. Power Del.* 28(4), 2145–2153 (2013).
673. J. S. Boboski et al., "Calibrated single-contact voltage sensor for high-voltage monitoring applications," *IEEE Trans. Instrum. Meas.* 64, 923 (2015).
674. D. Lawrence et al., "Non-contact measurement of line voltage," *IEEE Sens. J.* 10(24), 8990 (2016).
675. Y. Cui et al., "Relative localization in wireless sensor networks for measurement of electric fields under HVDC transmission lines," *Sensors* 15, 3540 (2015).
676. K. Zhu, "Non-contact capacitive-coupling based and magnetic field-sensing assisted technique for monitoring voltage of overhead power transmission lines," *IEEE Sens. J.* 17, 1069 (2017).
677. Y. Xiang, K. Chen, Q. Xu, Z. Jiang, and Z. Hong, "A novel contactless current sensor for HVDC overhead transmission lines," *IEEE Sens. J.* 18(11), 4725–4732 (2018), doi: 10.1109/JSEN.2018.2828807

678. ISO 5725-1: "Accuracy (trueness and precision) of measurement methods and results - Part 1: General principles and definitions" (1994).
679. S. Shin, U. Sharma, H. Tu, W. Jung, and S. A. Boppart, "Characterization and analysis of relative intensity noise in broadband optical sources for optical coherence tomography," *IEEE Photonics Technol. Lett.* 22(14), 1057–1059 (2010), doi: 10.1109/LPT.2010.2050058
680. R. C. Rabelo, R. T. de Carvalho, and J. Blake, "SNR enhancement of intensity noise-limited FOGs", *J. Lightwave Technol.* 18(12), 2146–2150 (2000).
681. Telcordia Standard GR-1221-CORE-1999, "Generic reliability assurance requirements for passive optical components," Telcordia Technologies, 1999.
682. D. H. Stamatis, *Failure Mode and Effect Analysis: FMEA from Theory to Execution, Milwaukee* (ASQ Quality Press, WI, USA, 2003).
683. International Electrotechnical Commission, "Reliability of fiber optic interconnecting devices and passive components—part 2: quantitative assessment of reliability based on accelerated aging tests—temperature and humidity" (2001).
684. International Electrotechnical Commission, "Reliability of fiber optic interconnecting devices and passive components—part 3: relevant tests for evaluating modes and failure mechanisms for passive components" (2001).
685. J. Park and D. S. Shin, "Degradation of fiber optical communication devices under damp-heat aging," *IEEE Photonics Technol. Lett.* 15(8), 1106–1108, 2003.
686. H. Nagata, Y. Li, K. R. Voisine, and W. R. Bosenberg, "Reliability of nonhermetic bias-free LiNbO3 modulators," *IEEE Photonics Technol. Lett.* 16(11), 2457–2459 (2004).
687. A. Birolini, *Reliability Engineering* (Springer Verlag, Berlin, Germany, 2007).
688. G. Nicholson, "Reliability considerations: optical sensors for the control and measurement of power," in *Proc. of 2001 IEEE/PES Transmission and Distribution Conference and Exposition. Developing New Perspectives*, vol. 1, pp. 122–126 (2001), doi: 10.1109/TDC.2001.971220
689. A. A. Stolov, D. A. Simoff, and J. Li, "Thermal stability of specialty optical fibers," *J. Lightwave Technol.* 26(20), 3443–3451 (2008).
690. M. Lenner, L. Yang, A. Frank, and K. Bohnert, "Influence of optical fiber coatings on the long-term accuracy of interferometric fiber-optic current sensors," *Procedia Eng.* 168, 1735 (2016).
691. D. Yin, W. Yu, Z. Xiao, L. Han, and G. Wang, "Reliability assessment of an optical current sensor based on accelerated aging tests," *Appl. Opt.* 58, 5107–5114 (2019).
692. M. H. Samimi, A. A. S. Akmal, and H. Mohseni, "Optical current transducers and error sources in them: a review," *IEEE Sens. J.* 15(9), 4721–4728 (2015).
693. "Aluminum" Wikipedia, Wikimedia Foundation, 12 October, 2023, https://en.wikipedia.org/wiki/Aluminium
694. International Aluminium Institute https://international-aluminium.org/
695. F. Rahmatian and J. N. Blake, "Applications of high-voltage fiber optic current sensors," in *2006 IEEE Power Engineering Society General Meeting*, Montreal, QC, Canada, doi: 10.1109/PES.2006.1709517
696. T. Kumai, H. Nakabayashi, Y. Hirata, et al., "Field trial of optical current transformer using optical fiber as Faraday sensor," in *IEEE Power Engineering Society Summer Meeting*, 2002, vol. 2, pp. 920–925, doi: 10.1109/PESS.2002.1043498
697. K. Bohnert, A. Frank, T. Roininen, et al., "Fiber-optic current and voltage sensors as modern alternatives to conventional instruments transformers," in *Technical Digest of IEEMA TECH-IT 2014, Third International Conference on Instrument Transformers.*
698. F. J. M. Herrera, "Mixed line protection," *T&D World Magazine*, Feb. 2015, https://www.tdworld.com/overhead-transmission/article/20965194/mixedline-protection
699. Y. Barmenkov and F. Mendoza-Santoyo, "Faraday plasma current sensor with compensation for reciprocal birefringence induced by mechanical perturbations," *J. Appl. Res. Technol.* 1, 157–163 (2003).

700. F. Descamps, M. Aerssens, A. Gusarov, et al., "Simulation of vibration-induced effect on plasma current measurement using a fiber optic current sensor," *Opt. Express* 22, 14666–14680 (2014).
701. R. Motuz, W. Leysen, P. Moreau, et al., "Theoretical assessment of the OTDR detector noise on plasma current measurement in tokamaks," *Appl. Opt.* 58, 2795–2802 (2019).
702. M. M. Xue, D. L. Chen, B. Shen, et al., "Upgrade of poloidal field coils current measurement system on Experimental Advanced Superconducting Tokamak," *Fusion Eng. Des.* 148, 111264 (2019).
703. P. Dandu, A. Gusarov, P. Moreau, et al., "Plasma current measurement in ITER with a polarization-OTDR: impact of fiber bending and twisting on the measurement accuracy," *Appl. Opt.* 61, 2406–2416 (2022).
704. R. Malewski, "High-voltage current transformers with optical signal transmission," *Opt. Eng.* 20(1), 200154 (1981), doi: 10.1117/12.7972662
705. G. W. Day and A. H. Rose, "Faraday effect sensors: the state of art," *Proc. SPIE* 985, 138–150 (1988).
706. G. W. Day, M. N. Deeter and A. H. Rose, "Faraday effect sensors: a review of recent progress," *Proc. SPIE*, PM07, 11–26 (1992).
707. Y. N. Ning and D. A. Jackson, "Review of optical current sensors using bulk-glass sensing elements," *Sens. Actuator. A Phys.* 39(3), 219–224 (1993).
708. "Optical current transducers for power systems: a review," *IEEE Trans. Power Deliv.* 9(4), 1778–1788 (1994), doi: 10.1109/61.329511, paper jointly prepared by the *Emerging Technologies Working Group* of the Power Systems Instrumentation and Measurements Committee and the *Fiber Optics Sensors Working Group*, Fiber Optics Substation Committee, Power Systems Communications Committee.
709. Z. P. Wang, S. Q. Zhang, and L. B. Zhang, "Recent advances in optical current-sensing techniques," *Sens. Actuator. A Phys.* 50(3), 169–175 (1995).
710. "Optical Current Sensor Technology (Chapter 7)," in *Optical Fiber Sensor Technology Vol. 3 – Applications and Systems*, K. T. V. Grattan and B. T. Meggitt, Eds. (Kluwer Academic Publishers, London, 1998).
711. M. El-Hami and K. T. V. Grattan, "An overview of optical-fibre technology applications in electrical power systems," *Meas. Control* 33(10), 296–302 (2000).
712. A. H. Rose and G. W. Day, "Optical fiber current and voltage sensors for the electric power industry," in *Handbook of Optical Fibre Sensing Technology*, Jose Miguel Lopez-Higuera, Ed. (Wiley & Sons, Ltd., West Sussex, England, 2002), pp. 569–618, Chapter 27.
713. S. Q. Xu, S. X. Dai, et al., "Recent progress of all-fiber optic current sensors," *Laser Optoelectron. Prog.* 41(1), 41–45 (2004).
714. P. Niewczas and J. R. McDonald, "Advanced optical sensors for power and energy systems applications," *IEEE Instrum. Meas. Mag.* 10(1), 18–28 (2007), doi: 10.1109/MIM.2007.339552
715. P. Ripka, "Electric current sensors: a review," *Meas. Sci. Technol.* 21(11), 112001 (2010).
716. R. M. Silva, H. Martins, I. Nascimento, et al., "Optical current sensors for high power systems: a review," *Appl. Sci.* 2(3), 602–628 (2012).
717. R. Wang, S. Xu, W. Li, et al., "Optical fiber current sensor research: review and outlook," *Opt. Quantum Electron* 48(9), article 422 (2016), doi: 10.1007/s11082-016-0719-3
718. P. Mihailovic and S. Petricevic, "Fiber-optic sensors based on the Faraday effect," *Sensors* 21, 6564 (2021), doi: 10.3390/s21196564

PART III

Electric Field and Voltage Sensors

CHAPTER 13

Electro-Optic Effects

13.1 POCKELS EFFECT AND KERR EFFECT

Many optical electric field and voltage sensors make use of the linear electro-optic effect, also called Pockels effect after the German physicist Friedrich C. A. Pockels, who first described the phenomenon in 1893. The quadratic electro-optic effect, or Kerr effect, discovered in 1875 by the Scottish physicist John Kerr, is much weaker and has largely been reserved for the measurement of high impulse fields. The Pockels and Kerr effects result from anharmonic terms in the excursion of bound electrons in a medium under the influence of an applied electric field. At given optical frequency ϖ, the medium's polarization vector $\boldsymbol{P}(\varpi, \boldsymbol{E})$ in the presence of static or low frequency field $\boldsymbol{E} = \boldsymbol{E}(\Omega)$ of frequency $\Omega \sim 0$, expanded in a power series of $\boldsymbol{E}$, is [1]

$$\boldsymbol{P}(\varpi,\boldsymbol{E}) = \varepsilon_0\left[\chi^{(1)}(\varpi)\cdot\boldsymbol{E}_{\mathrm{opt}}(\varpi) + \chi^{(2)}(\varpi):\boldsymbol{E}\boldsymbol{E}_{\mathrm{opt}}(\varpi) + \chi^{(3)}(\varpi)\vdots\boldsymbol{E}\boldsymbol{E}\boldsymbol{E}_{\mathrm{opt}}(\varpi) + \ldots\right], \tag{13.1}$$

where e_0 is the vacuum permeability, $\chi^{(m)}$ ($m = 1, 2, \ldots$) is the mth order susceptibility, and $\boldsymbol{E}_{\mathrm{opt}}(\varpi)$ is the optical field; $\chi^{(m)}$ is a tensor of rank $m + 1$. More explicitly, the ith component of $\boldsymbol{P}$ at frequency ϖ is

$$P_i = \varepsilon_0\left[\sum_{j=1}^{3}\chi_{ij}^{(1)}E_{j,\mathrm{opt}} + \sum_{j=1}^{3}\sum_{k=1}^{3}\chi_{ijk}^{(2)}E_jE_{k,\mathrm{opt}} + \sum_{j=1}^{3}\sum_{k=1}^{3}\sum_{l=1}^{3}\chi_{ijkl}^{(3)}E_jE_kE_{l,\mathrm{opt}} + \ldots\right]. \tag{13.2}$$

Eq. (13.1) implies a field-dependent dielectric displacement $\boldsymbol{D} = \varepsilon_0\varepsilon_r\boldsymbol{E}$ and, consequently, a field-dependent permittivity (dielectric constant) [2]:

$$\varepsilon_r(\varpi,\boldsymbol{E}) = \left[\varepsilon_r^{(1)}(\varpi) + \varepsilon_r^{(2)}(\varpi)\boldsymbol{E} + \varepsilon_r^{(3)}(\varpi)\boldsymbol{E}\boldsymbol{E} + \ldots\right]. \tag{13.3}$$

DOI: 10.1201/9781003100324-17

The first term in (13.3), $\boldsymbol{\varepsilon}_r^{(1)}$, defines the refractive indices of the medium ($n = \sqrt{\varepsilon_r}$), while the higher order terms represent field-induced index changes and corresponding linear birefringence. The term linear in $\boldsymbol{E}$ describes the Pockels effect, and the term quadratic in $\boldsymbol{E}$ is the dc Kerr effect. (Note that in nonlinear optics, the $\chi^{(2)}$-term represents second-order nonlinearities such as optical frequency doubling, and the $\chi^{(3)}$-term represents third order effects like four-wave mixing, self-focusing, and optical-field-induced birefringence, that is, the ac Kerr effect. All fields in [13.1] are then optical fields.) For symmetry reasons, the Pockels effect only occurs in media without inversion symmetry. Out of the 32 crystallographic point groups, 20 groups exhibit the Pockels effect [2, 3]. By contrast, the dc Kerr effect is always present but commonly only of interest in media with a center of symmetry such as glasses, liquids, or gases, where it is not masked by the Pockels effect. Even though the index changes are relatively small, they can easily be exploited for electric field sensing.

The refractive optical properties of a crystal are often represented by the index ellipsoid or indicatrix (see Section 2.3):

$$\frac{x^2}{n_x^2} + \frac{y^2}{n_y^2} + \frac{z^2}{n_z^2} = 1. \tag{13.4}$$

The directions x, y, and z are the principal dielectric axes, i.e., the directions along which $\boldsymbol{D}$ and $\boldsymbol{E}$ are parallel, and the axis lengths correspond to the principal refractive indices n_i.

More generally, the index ellipsoid is written as [2]

$$B_{ij} x_i x_j = 1, \tag{13.5}$$

where B_{ij} represents the impermeability tensor: $B_{ij} = \varepsilon_0 \delta E_i / \delta D_j$ with D_j being the electric displacement. The elements of B_{ij} correspond to the inverse squares of the refractive indices. In expanded form and with $x_i = x, y, z$, (13.5) reads

$$\frac{x^2}{n_{xx}^2} + \frac{y^2}{n_{yy}^2} + \frac{z^2}{n_{zz}^2} + \frac{2yz}{n_{yz}^2} + \frac{2xz}{n_{zx}^2} + \frac{2xy}{n_{xy}^2} = 1. \tag{13.6}$$

Eq. (13.6) reduces to (13.4), if x, y, and z are chosen parallel to the principal dielectric axes of the crystal. An electric field $\boldsymbol{E} = (E_x, E_y, E_z)$ alters the size, shape, and/or orientation of the ellipsoid. The modification can be represented by a power series expansion of the index terms $n_{ij}^{-2}(\boldsymbol{E})$ [2, 3]:

$$\frac{1}{n_{ij}^2(\boldsymbol{E})} = \left(\frac{1}{n_{ij}^2(\boldsymbol{E}=0)}\right) + \sum_k r_{ijk} E_k + \sum_{k,l} q_{ijkl} E_k E_l + \ldots. \tag{13.7}$$

The coefficients r_{ijk} constitute the linear electro-optic tensor (Pockels effect), and the coefficients q_{ijkl} constitute the quadratic electro-optic tensor (dc Kerr effect); $i, j, k, l = 1, 2, 3$.

In the following, we restrict ourselves to the Pockels effect. Kerr-effect-based sensors will be discussed in Section 14.7. Since $B_{ij} = B_{ji}$ and $n_{ij} = n_{ji}$, it follows that $r_{ijk} = r_{jik}$ [2]. This reduces the number of independent coefficients r_{ijk} from 27 to 18. The crystal symmetry determines how many of the 18 coefficients r_{ij} ($i = 1, 2, \ldots 6, j = 1, 2, 3$) are non-zero and the relationships between them. With

$$\Delta\left(\frac{1}{n^2}\right)_i = \left(\frac{1}{n^2(E)}\right)_i - \left(\frac{1}{n^2(E=0)}\right)_i, \quad i = 1,2,\ldots.6 \tag{13.8}$$

(13.7) then becomes

$$\Delta\left(\frac{1}{n^2}\right)_i = \sum_{j=1}^{3} r_{ij} E_j, \tag{13.9}$$

or written in matrix form

$$\begin{pmatrix} \Delta\left(\frac{1}{n^2}\right)_1 \\ \Delta\left(\frac{1}{n^2}\right)_2 \\ \Delta\left(\frac{1}{n^2}\right)_3 \\ \Delta\left(\frac{1}{n^2}\right)_4 \\ \Delta\left(\frac{1}{n^2}\right)_5 \\ \Delta\left(\frac{1}{n^2}\right)_6 \end{pmatrix} = \begin{pmatrix} r_{11} & r_{12} & r_{13} \\ r_{21} & r_{22} & r_{23} \\ r_{31} & r_{32} & r_{33} \\ r_{41} & r_{42} & r_{43} \\ r_{51} & r_{52} & r_{53} \\ r_{61} & r_{62} & r_{63} \end{pmatrix} \begin{pmatrix} E_x \\ E_y \\ E_z \end{pmatrix}. \tag{13.10}$$

In the presence of an electric, the index ellipsoid field then reads [2, 3]

$$\frac{x^2}{n_1^2 + \Delta\left(\frac{1}{n^2}\right)_1} + \frac{y^2}{n_2^2 + \Delta\left(\frac{1}{n^2}\right)_2} + \frac{z^2}{n_3^2 + \Delta\left(\frac{1}{n^2}\right)_3} + \frac{2yz}{\Delta\left(\frac{1}{n^2}\right)_4} + \frac{2zx}{\Delta\left(\frac{1}{n^2}\right)_5} + \frac{2xy}{\Delta\left(\frac{1}{n^2}\right)_6} = 1, \tag{13.11}$$

where n_1, n_2, n_3, are the indices at $\boldsymbol{E} = 0$. In general, the principal axes x', y', z' of the new ellipsoid do not coincide with the principal axes x, y, z of ellipsoid without applied field. The orientation and lengths of the new axes follow from the eigenvalues of the matrix [3–4]

$$\begin{pmatrix} n_1^2+\Delta\left(\frac{1}{n^2}\right)_1 & \Delta\left(\frac{1}{n^2}\right)_6 & \Delta\left(\frac{1}{n^2}\right)_5 \\ \Delta\left(\frac{1}{n^2}\right)_6 & n_2^2+\Delta\left(\frac{1}{n^2}\right)_2 & \Delta\left(\frac{1}{n^2}\right)_4 \\ \Delta\left(\frac{1}{n^2}\right)_5 & \Delta\left(\frac{1}{n^2}\right)_4 & n_3^2+\Delta\left(\frac{1}{n^2}\right)_3 \end{pmatrix}. \tag{13.12}$$

For further illustration, we consider the electro-optic effect in $Bi_4Ge_3O_{12}$ (BGO), a material frequently used in electro-optic voltage sensors. BGO belongs to the cubic crystal class $\bar{4}3m$ (zincblende structure) and has a 4-fold inversion rotation axis (this is by convention the z-direction), four 3-fold axes (corresponding to the body diagonals of the cubic lattice), and a mirror plane. Crystals of this class are free of natural birefringence ($n_x = n_y = n_z = n_0$), and the matrix of electro-optic coefficients has the simple form [2]

$$r_{ij} = \begin{pmatrix} 0 & 0 & 0 \\ 0 & 0 & 0 \\ 0 & 0 & 0 \\ r_{41} & 0 & 0 \\ 0 & r_{41} & 0 \\ 0 & 0 & r_{41} \end{pmatrix}, \tag{13.13}$$

that is, the only nonvanishing elements are r_{41}, r_{52}, and r_{63}, having identical values r_{41}. With (13.10) and (13.11), we obtain the index ellipsoid in the presence of an electric field as follows:

$$\frac{x^2}{n_0^2}+\frac{y^2}{n_0^2}+\frac{z^2}{n_0^2}+2r_{41}\left(E_x yz+E_y xz+E_z xy\right)=1. \tag{13.14}$$

If we choose the field parallel to z, i.e., $\mathbf{E} = (0, 0, E_z)$, (13.14) becomes

$$\frac{x^2}{n_0^2}+\frac{y^2}{n_0^2}+\frac{z^2}{n_0^2}+2r_{41}E_z xy=1. \tag{13.15}$$

The mixed terms yz, xz, and xy in (13.14) indicate that the major axes of the index ellipsoid are no longer parallel to x, y, and z. In the new coordinate system with axes x', y', z' parallel to the major axes of the modified ellipsoid, (13.14) reduces to

$$\frac{x'2}{n_{x'}^2}+\frac{y'2}{n_{y'}^2}+\frac{z'2}{n_{z'}^2}=1. \tag{13.16}$$

With $\boldsymbol{E} = (0, 0, E_z)$, (13.16) can be transformed into [3]

$$\left(\frac{1}{n_0^2}+r_{41}E_z\right)x'^2+\left(\frac{1}{n_0^2}-r_{41}E_z\right)y'^2+\frac{z^2}{n_0^2}=1. \tag{13.17}$$

The z'-axis of the new coordinate system coincides with the z-axis of the field-free crystal, whereas x' and y' are rotated against x and y by 45° in counterclockwise direction. Finally, assuming $r_{41}E_z \ll n_0^{-2}$ and with $dn = -(n^3/2)d(2/n^2)$, we obtain the new principal indices as [3]

$$n_{x'} = n_0 - \frac{n_0^3}{2} r_{41} E_z, \tag{13.18}$$

$$n_{y'} = n_0 + \frac{n_0^3}{2} r_{41} E_z, \tag{13.19}$$

$$n_{z'} = n_z = n_0. \tag{13.20}$$

Figure 13.1 shows the corresponding BGO crystal orientation and a cross-section of the index ellipsoid in the (x, y)-plane. The light propagates in z-direction ($\bar{4}$-fold crystal axis, [001]-direction, perpendicular to [001]-crystal plane) and is linearly polarized along x ([100]-direction); for the designation of crystal lattice directions and planes, see [5]. In the absence of an electric field, the propagation velocity is determined by the index n_0, independent of the polarization direction. An electric field parallel to the propagation direction reduces the refractive index $n_{x'}$ for the optical field component parallel to x' ([$\bar{1}$10]-direction) and enhances the refractive index $n_{y'}$ for the optical field component parallel to y' ([110]-direction) according to (13.18) and (13.19). (A reversal of the field direction changes the signs of the index changes.) Hence, the two orthogonal polarization components parallel to x' and y' accumulate a phase difference $\Delta\phi$ given by

$$\Delta\phi = \left(2\pi/\lambda\right) n_0^3 r_{41} E_z l \tag{13.21}$$

or

$$\Delta\phi = \left(2\pi/\lambda\right) n_0^3 r_{41} V, \tag{13.22}$$

where l is the optical wavelength, l is the crystal length, and $V = E_z/l$ corresponds to the applied voltage. The product $n_0^3 r_{41}$ serves as a figure of merit for the sensitivity of an electro-optic material. Note that for the longitudinal electro-optic effect (the electric field is parallel to the propagation direction), a given voltage always produces the same phase shift independent of the crystal length. In the configuration Figure 13.1a, transverse field components (E_x, E_y) leave the relative phase of the orthogonal SOP unchanged, i.e., a corresponding

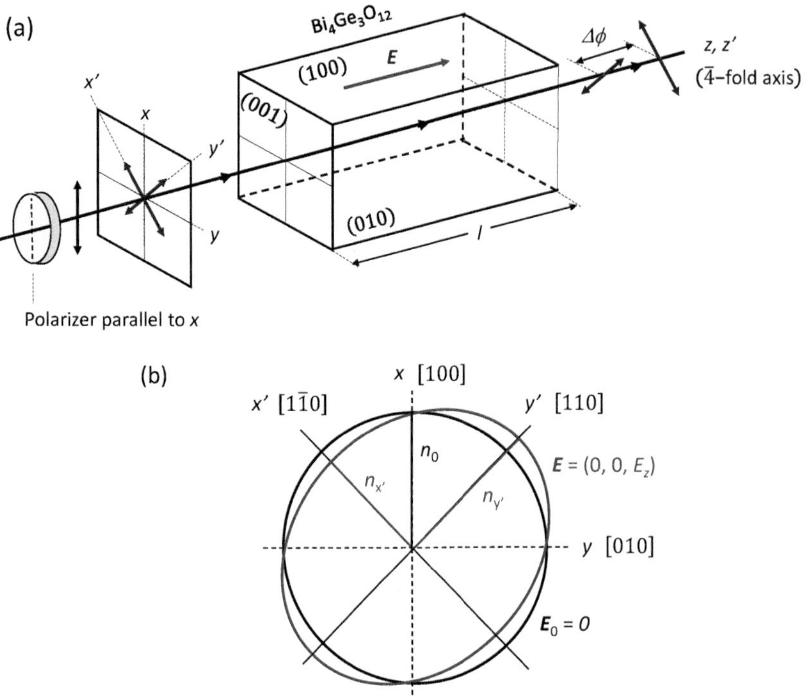

FIGURE 13.1 (a) Longitudinal electro-optic effect in crystals of class $\bar{4}3m$; polarization components parallel to x', y' of light propagating in z-direction accumulate an electro-optic phase difference $\Delta\phi$; (b) cross-sections of the index ellipsoid in the (x', y')-plane without electric field E (black) and with electric field (blue).

sensor measures the line integral $\int \mathbf{E} \cdot d\mathbf{s}$ along light propagation path z [4, 6]. No phase retardations occur for x or y propagation directions.

Figure 13.2 shows a common geometry of the transverse electro-optic effect. The light propagates in the $\left[\bar{1}10\right]$-direction, and the field is in the [110]-direction, orthogonal to the optical path. Both directions are perpendicular to z. The x and y directions lie in the (001)-plane at angles of ±45° to the (110)-plane. The electric field has components $\mathbf{E} = (E_x, E_y, 0)$ with $E_x = E_y = (1/\sqrt{2})E$. The new principal axes x', y', z' are obtained by rotating the axes x, y, z by 90° about the field direction. z' is parallel to the optical path, and x' and y' are parallel to the $(\bar{1}10)$-facet and at ±45° to the (110)-plane [7, 8]. The differential electro-optic phase retardation of the polarization components parallel to x' and y' is

$$\Delta\phi = \frac{2\pi}{\lambda}\frac{l}{d} n_0^3 r_{41} V, \tag{13.23}$$

where d is the crystal thickness in field direction and $V = Ed$. Note that the retardation now increases with the crystal length l. Again, there is no phase retardation for orthogonal beam directions, and only the field component in [110]-direction causes a retardation [4]. In another possible transverse configuration, the electric field is perpendicular to a (111)-plane, but the phase retardation is then reduced by a factor $(1/2)\sqrt{3}$ [7, 9].

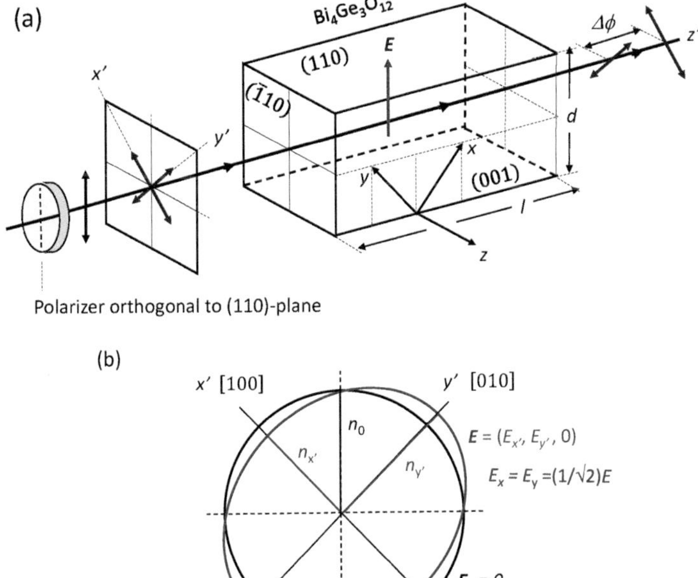

FIGURE 13.2 (a) Transverse electro-optic effect in crystals of class $\bar{4}3m$; (b) cross-sections of the index ellipsoid in the (x', y')-plane without electric field E (black) and with electric field (blue).

Similarly, as in Figure 3.1, a combination of a quarter-wave retarder and a Wollaston prism (alternatively a beam splitter and two orthogonal polarizers) convert the phase retardation *Df* into two anti-phase signals I_1, I_2 proportional to the sine of $\Delta\phi$. The voltage follows from the signal $S = (I_1 - I_2) / (I_1 + I_2)$:

$$S = \sin(\Delta\phi). \tag{13.24}$$

Here, we have again assumed a perfect fringe contrast ($K = 1$). A frequently given specification of a sensor is its half-wave voltage V_p, that is, the voltage at $\Delta\phi = p$. V_p of the longitudinal electro-optic effect of $B_4Ge_3O_{12}$ is about 74 kV at 1310 nm (48 kV at 850 nm). Obviously, S is proportional to $\Delta\phi$ only at voltages $V \ll V_p$ and becomes ambiguous beyond V_p. An unambiguous detection technique for voltages $V > V_p$ will be discussed further below. As noted, the phase retardation of the longitudinal electro-optic effect in $B_4Ge_3O_{12}$ (and any other crystal of zincblende structure, $\bar{4}3$m) corresponds to the line integral $\int \boldsymbol{E} \cdot \boldsymbol{ds}$. Other crystal classes suited for field integration include $\bar{4}2m$, 23, $\bar{6}2m$, and $2mm$. Table 13.1 lists the corresponding orientations and gives examples of materials [6]. Crystals of classes $\bar{6}2m$ and $2mm$ have natural birefringence, but the principal axes of the natural and electro-optic birefringence coincide and the corresponding phase retardations simply add up. Crystal classes with natural birefringence for which this is not the case have been omitted in the Table. A study on the measurement of field components, choice of crystals, and their orientation has also been reported by L. Duvillaret et al. [4].

TABLE 13.1 Some Crystal Classes for Line Integration of the Electric Field with a Polarimetric Pockels Sensor

Crystallographic Point Group	Crystal Axis Parallel to Light Propagation	Direction of Electro-Optic Axes	Materials
$\bar{4}3m$	$\bar{4}$-fold axis [001]	[110], [1$\bar{1}$0]	$Bi_4Ge_3O_{12}$, $Bi_4Si_3O_{12}$
23	2-fold axis [001]	[110], [1$\bar{1}$0]	ZnTe, ZnSe,
$\bar{4}2m$	$\bar{4}$-fold axis [001]	[110], [1$\bar{1}$0]	GaAs
$\bar{6}2m$	2-fold axis [001]	Parallel and orthogonal to 3-fold axis	$Bi_{12}GeO_{20}$, $Bi_{12}SiO_{20}$, $NaClO_3$
2*mm*	2-fold axis [001]	[100], [010]	KH_2PO_4 (KDP)$LiNaCO_3$ $MgBaF_4$

The linear electro-optic effect is accompanied by the converse piezoelectric effect, i.e., an applied electric field also induces piezo-electric strains or stresses in the crystal (Chapter 16). Alternating fields at frequencies well above the acoustic resonances of the crystal (e.g., in the microwave range) prevent the crystal from deforming, that is, the crystal is in a state of constant strain but periodic piezoelectric stress. The crystal is then said to be clamped. The additional stress-induced piezoelectric field adds to the external field and enhances the electro-optic effect. By contrast, unclamped crystals are free to deform and are in a state of constant stress. The latter applies to high voltage sensors operated at power line frequencies of 50 Hz or 60 Hz. For distinction, the electro-optic coefficients of unclamped and clamped crystals, r_{ij}^T and r_{ij}^S, are commonly marked with superscripts T and S, respectively. Very roughly, the coefficients r_{ij}^S exceed the coefficients r_{ij}^T by a factor of 2.

Finally, it should be mentioned that whereas electro-optic voltage sensors mostly operate as polarimetric sensors, sensors for electric field mapping often measure the Pockels effect in an integrated-optic Mach-Zehnder interferometer or a Fabry-Pérot resonator [10–12].

13.2 ELECTRO-OPTIC MATERIALS

Many criteria should be considered in the selection of an adequate electro-optic material for voltage sensing:

- Obviously, the material should be available in good quality, with adequate dimensions, and at a reasonable price.
- Other general prerequisites are mechanical stability and chemical inertness under the respective conditions of operation.
- Optical loss at the wavelength of operation, typically in the near infrared, should be small. (Most inorganic electro-optic crystals are transparent from the visible well into the infrared.)
- The sensitivity, i.e., the term $(n^3/2)r$, should be of sufficient magnitude.

- The temperature dependence of the electro-optic effect should be small.
- Preferably, the material should be free of natural birefringence to avoid a temperature-dependent polarimetric phase bias, an effect particularly adverse in dc voltage sensing. But poor quality (imperfections, built-in stress) may cause birefringence also in isotropic crystals [13].
- For similar reasons, materials exhibiting optical activity and electro-gyration (rotation of linear polarization in a transverse electric field) can be less suited.
- A small dielectric constant e_r with little sensitivity to temperature is advantageous if the sensing element is used without electrodes. This applies to sensors for electric field mapping but also to many voltage sensors. A large permittivity screens the electric field and reduces the sensitivity.
- In the absence of electrodes, the shape of the sensing element determines the depolarization factor and together with e_r the strength of the internal field [14–16].
- A small e_r also helps to limit field distortions and reduce the risk of partial discharge and dielectric breakdown at high voltages.
- If exposed to high voltages (hundreds of kilovolts), the material's dark resistivity should be sufficiently high to avoid excessive ohmic heating by resistive currents. As the free carrier concentration increases exponentially with temperature, resistive heating may initiate a thermal runaway effect and dielectric breakdown. Therefore, semiconductor materials such as GaAs or ZnTe should be considered with some caution in high voltage sensing [17]. In photorefractive crystals such as BGO, optical excitation of free carriers may give rise to photoconductivity, especially at short wavelengths [18].
- Dielectric loss (loss factor tan d) should be small to minimize response delays and internal heating.

Many reported electro-optic voltage sensors have used the ternary materials bismuth germanate (BGO) [8, 13, 17–38], or bismuth silicate (BSO) [39–45]. BGO has modifications $Bi_4Ge_3O_{12}$ and $Bi_{12}GeO_{20}$, and BSO has modifications $Bi_4Si_3O_{12}$ and $Bi_{12}SiO_{20}$. BGO and BSO also serve as scintillators in high energy physics or nuclear medicine and are therefore readily available. $Bi_{12}GeO_{20}$ and $Bi_{12}SiO_{20}$ belong to crystal class 23, and $Bi_4Ge_3O_{12}$ and $Bi_4Si_3O_{12}$ to class $\bar{4}3$m. Both classes have cubic symmetry. $Bi_{12}GeO_{20}$ and $Bi_{12}SiO_{20}$ give significantly larger electro-optic phase retardations than $Bi_4Ge_3O_{12}$ and $Bi_4Si_3O_{12}$ but are optically active and of higher permittivity. Both crystal classes are free of natural birefringence, and their electro-optic tensors have identical forms; Eq. (13.13). Therefore, the orientations of Figure 13.1 apply all four BGO and BSO modifications with the difference that the z-direction corresponds to the $\bar{4}$-fold axis in class $\bar{4}3m$ crystals and to the 2-fold axis in class 23 crystals. Among the four materials, $Bi_4Ge_3O_{12}$ [18, 46] has arguably become the most widely used transducer material in high voltage sensing. Typically grown by the

TABLE 13.2 Selected Electro-Optic Materials and Their Properties; The Optical Parameters Are Valid at l = 632.8 nm, If Not Stated Otherwise; r_{ij}^T and r_{ij}^S Are Unclamped and Clamped Electro-Optic Coefficients, Respectively; [4, 18, 31, 45, 52, 71–73] and References Therein.

Material	Crystallographic Point Group	Relative Permittivity e_r	Refractive Index	Electro-Optic Coefficients (pm/V)	Optical Activity	Dark Resistivity (Ωcm)
BGO ($Bi_4Ge_3O_{12}$)	$\bar{4}3m$ Isotropic	16	2.1	$r_{41} = 0.96$ $r_{41}^T = 0.95$	NA	5.2×10^{14}
BGO ($Bi_{12}GeO_{20}$)	23 Isotropic	38	2.548	$r_{41} = 3.67$	9.6°/mm, 850 nm	10^{11} $- 10^{13}$
BSO ($Bi_4Si_3O_{12}$)	$\bar{4}$ $3m$ Isotropic	-	2.05	$r_{41}^T = 0.54$	NA	-
BSO ($Bi_{12}SiO_{20}$)	23 Isotropic	56	2.54	$r_{41}^T = 4.25$	10.5°/mm, 870 nm	0.2×10^{13}
BTO ($Bi_{12}TiO_{20}$)	23 Isotropic	-	2.56	$r_{41}^T = 5.75r_{41} = 3.81$	2.8°/mm, 870 nm	-
ZnTe	$\bar{4}3$m Isotropic	10.1	2.894	$r_{41}^S = 4.3$	NA	-
Lithium niobate ($LiNbO_3$)	$3m$ Uniaxial	$e_{r,o} = 43$ $e_{r,e} = 28$	$n_o = 2.286$ $n_e = 2.200$	$r_{13}^S = 8.6$ $r_{33}^S = 30.8$ $r_{51}^S = 28$ $r_{22}^S = 3.4$	NA	-
Lithium tantalate ($LiTaO_3$)	$3m$ Uniaxial	$e_{r,o} = 41$ $e_{r,e} = 43$	$n_o = 2.176$ $n_e = 2.180$	$r_{13}^S = 7.5$ $r_{33}^S = 33$ $r_{51}^S = 20$ $r_{22}^S = 1.0$	NA	-
KD*P (KD_2PO_4)	$\bar{4}$ $2m$ Uniaxial	$e_{r,o} = 58$ $e_{r,e} = 48$	$n_o = 1.503$ $n_e = 1.465$	$r_{41} = 1.4$ $r_{41}^S = 8.8$ $r_{51} = 3.3$ $r_{63}^S = -24.3$	NA	-
Quartz (SiO_2)	32 Uniaxial	$e_{r,o} = 4.514$ $e_{r,e} = 4.634$	$n_o = 1.534$ $n_e = 1.543$	$r_{11} = -0.45$ $r_{41} = 0.19$	17.3°/mm, 656 nm	~10^{16}
Poled polymer (example)	∞mm Uniaxial	4	$n_o = 1.66$ $n_e = 1.74$	$r_{13} = 23$ $r_{33} = 70$	NA	-

Czochralski method [47, 48], $Bi_4Ge_3O_{12}$ has favorable mechanical and electric properties, is optically transparent from 350 nm to 4 μm, free of natural birefringence, non-hydroscopic, and not pyroelectric. Moreover, the temperature dependence of the electro-optic effect is relatively small: $(1/T)\delta[(n^3/2)r]/\delta T = 1.54 \times 10^{-4}\ °C^{-1}$ [46].

Other materials explored for high voltage sensing include bismuth titanate ($Bi_{12}TiO_{20}$, BTO) [49–51], lithium niobate ($LiNbO_3$), a material frequently used in electro-optic modulators [52–60], potassium dihydrogen phosphate (KH_2PO_4, KDP and KD_2PO_4, KD*P), known from nonlinear optics [14, 61, 62], *a*-quartz [63–68], and others [69]. However, KDP and KD*P are hydroscopic, while lithium niobate is pyroelectric and its electro-optic effect has a large temperature dependence, $Bi_{12}TiO_{20}$ is like $Bi_{12}GeO_{20}$ optically active. Quartz has many excellent properties but exhibits a comparatively small electro-optic effect; also, optical activity and electro-gyration complicate matters. A further option is poled polymers [70]. They have the benefit of large electro-optic coefficients and small permittivity, but typically exhibit larger temperature dependence, and the poling-induced electro-optic effect tends to decay over time, especially at elevated temperatures. Table 13.2 summarizes properties of selected electro-optic materials.

CHAPTER 14

Electro-Optic Voltage Sensors

14.1 VOLTAGE SENSING METHODS

Figure 14.1 illustrates some basic configurations for high voltage sensing that subsequent sections will discuss in more detail [74]. First, Figure 14.1a shows an ideal albeit not existing voltage sensor that measures the path integral $\int \boldsymbol{E} \cdot d\boldsymbol{s}$ between ground and high voltage potential by means of an electro-optic fiber. A field integrating sensor is appealing in that it is immune to crosstalk from neighboring power lines and perturbations of the electric field, e.g., by nearby structures or weather. By contrast, an electric field sensor as depicted in Figure 14.1b that only measures the field at a single point near the power line provides a more indirect measure for the line voltage and is sensitive to variations in the field distribution [75, 76]. As an alternative, the sensor may be arranged in a capacitive voltage divider (Figure 14.1c), where it measures a small but well-defined fraction of the primary voltage V ($V = V_1 + V_2$ with $V_2 \ll V_1$). Corresponding solutions have been demonstrated for both air-insulated outdoor substations (AIS) [77–79] and indoor SF_6 gas-insulated switchgear (GIS) [80–81]. In Figure 14.1d, a cylinder-shaped transducer crystal is exposed to the full line voltage. The relatively short crystal lengths (typically 100–250 mm) result in high electric field strengths that require a careful dielectric design of the sensor to prevent breakdown, especially at over-voltages. But with proper crystal symmetry and orientation, the sensor measures the path integral $\int \boldsymbol{E} \cdot d\boldsymbol{s}$, i.e., the true line voltage [17, 82, 83]. Finally, the system of Figure 14.1e consists of several local field sensors that approach the path integral by a discrete summation of local fields [37, 38].

14.2 VOLTAGE SENSING WITH LOCAL ELECTRIC FIELD SENSORS

Many of the early works on electro-optic voltage sensors from the 1970s to the 1990s investigated basic polarimetric Pockels cells, implemented with a variety of electro-optic materials. Typically, the researchers applied voltages up to a few kilovolts, mostly in transverse geometry, in order to study sensitivity, linearity, frequency response, or temperature dependence [13, 14, 19, 21, 24, 30, 31, 33, 34, 39, 40–43, 45, 50–66], as well as adverse effects due to linear birefringence [59], optical activity and electro-gyration [63, 65, 84], photorefractivity [85], or beam misalignment [54], etc. Methods for temperature compensation

DOI: 10.1201/9781003100324-18

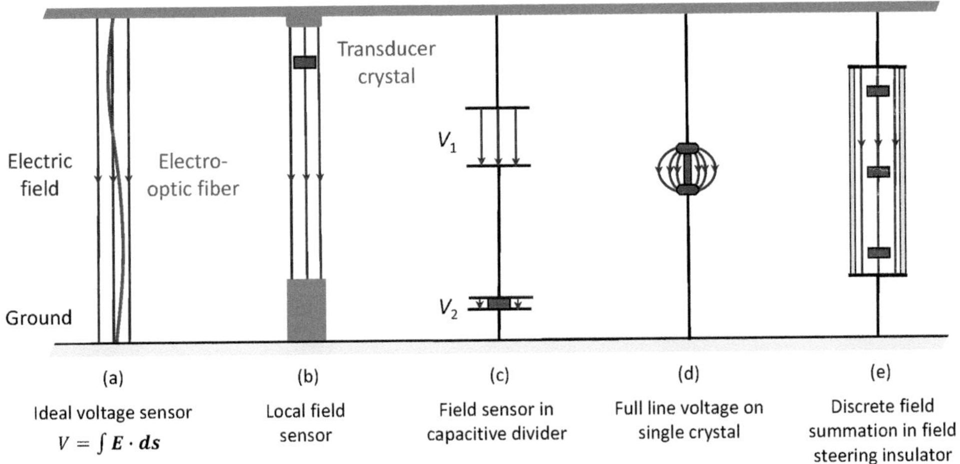

FIGURE 14.1 Basic methods for sensing of power line voltages [74].

were investigated in [21, 35, 43] and methods for birefringence compensation in [13, 86, 87]. Typical crystal lengths were around 10–30 mm. In the same period, first demonstrations and installations of such devices at high voltage substations were reported. Massey et al. mounted KD*P electric field sensors at the top of hollow insulator columns in 500 kV ac and 400 kV dc facilities according to the scheme of Figure 14.1b. Free-space laser beams through the insulators interrogated the crystals [75, 76]. Sasano et al. [88] and Rogers [89] reported similar implementations.

Christensen demonstrated a 132-kV voltage sensor, where, on top of an insulator, an electro-optic field probe ($Bi_4Ge_3O_{12}$) with multimode fiber leads resided in a recess of an elevated ball-shaped ground electrode mounted inside a dome-shaped high voltage electrode as schematically illustrated in Figure 14.2a. The design resulted in a relatively stable field distribution and enhanced field strength. Sulphur hexafluoride gas (SF_6) between the electrodes provided sufficient dielectric strength. The signal varied linearly with voltage, and the temperature-dependent error was within ±1.3% between −40°C and 70°C [25].

Sensors for gas-insulated switchgear (GIS) often employed a capacitive voltage divider and measured only a small but defined fraction of the total voltage. GIS substations are much more compact than air-insulated stations (AIS) and, therefore, often the solution of choice in densely populated areas. Figure 14.2b illustrates a sensor configuration for single phase encapsulated GIS [81]. The current carrying high voltage busbar runs through a concentric SF_6-gas-filled enclosure at ground potential. The voltage between a concentric electric electrode at floating potential and ground is applied to the sensor. (The voltage seen by the sensor was 1.6 kV in [81] at a phase-to-ground voltage of 220 kV/$\sqrt{3}$.) A capacitive-divider-based voltage sensor for 300 kV 3-phase encapsulated GIS was reported in [60]. From 2% to 110% of the rated voltage, amplitude errors at constant temperature were within ±1% and phase errors within ±40 min. Between 20°C to 60°C, the signal stayed within ±1%. A divider-based sensor for 132 kV air-insulated substations was reported by

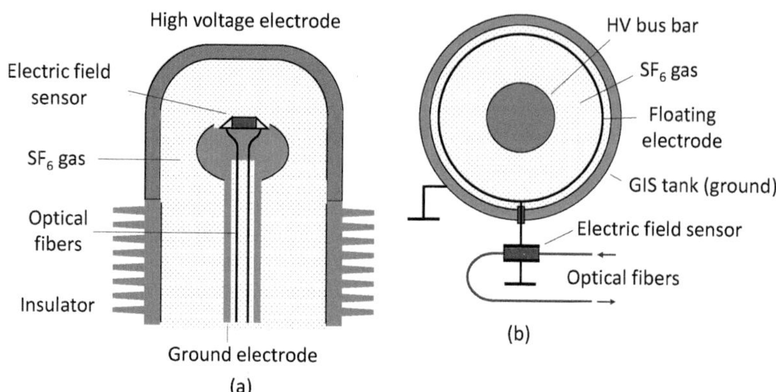

FIGURE 14.2 Examples of early electro-optic voltage transducer: (a) OVT for air-insulated substations (according to [25]); (b) OVT for SF6-gas-insulated switchgear (according to [81]).

Tonnesen et al. [77]. Today, the direct conversion of the stepped-down voltage into a digital optical output by means of digital electronics and an optical transmitter is often a more economical alternative [90].

14.3 ELECTRIC FIELD INTEGRATING SENSORS

14.3.1 OVT Measuring True Line Integral of Electric Field

Figure 14.3 depicts an electro-optic voltage transducer (EOVT) from ABB Ltd [17, 91]. Led by R. C. Miller, its development began at Westinghouse in the 1980s, and continued at ABB after ABB's purchase of Westinghouse's T&D business in 1989. The ABB EOVT has probably become the most widely deployed optical voltage transducer for transmission grids with rated voltages between 110 kV and 550 kV. On special request, some 800 kV EOVT were also built. The total line voltage is longitudinally applied to a rod-shaped $Bi_4Ge_3O_{12}$ crystal according to the scheme of Figure 14.1d. Operated in reflective mode, the crystal is

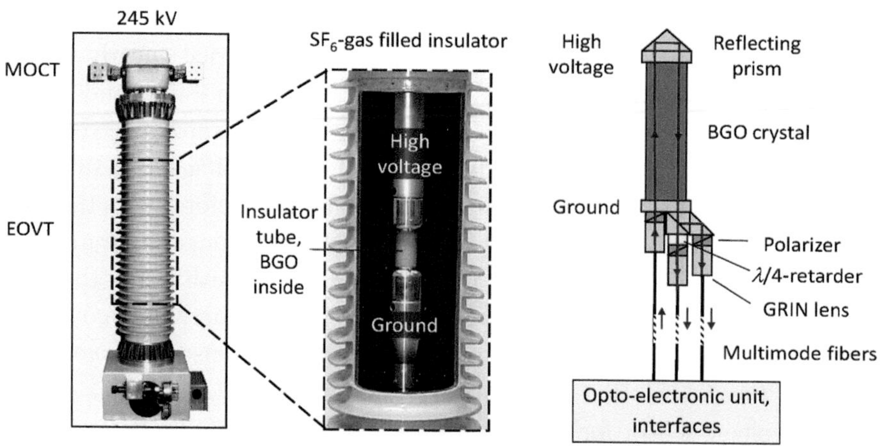

FIGURE 14.3 245-kV optical metering unit (OMU), consisting of an electro-optic voltage transducer (EOVT) and a magneto-optic current transducer (MOCT, left); EOVT assembly with electrodes in SF_6 gas-filled hollow insulator (center); EOVT optical assembly (right) [91].

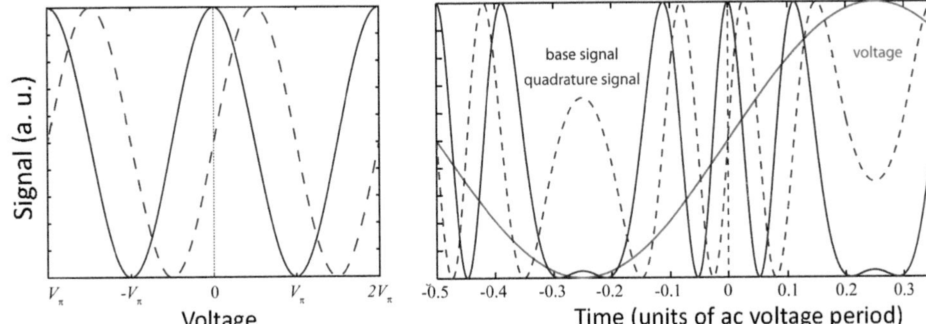

FIGURE 14.4 Simulated base and quadrature signals as a function of applied voltage in units of V_p (left); voltage waveform and corresponding base and quadrature signals as a function time for a peak-to-peak voltage of about 150 kV and V_p = 24 kV (right) (© 2013 IEEE. Reprinted with permission from [93].)

oriented as shown in Figure 13.1a. Hence, the sensor measures the path integral $\int \mathbf{E} \cdot d\mathbf{s}$ and is unaffected by field perturbations, stray fields, thermal changes in dielectric constants, thermal expansion, etc. The crystal length is between 100 mm and 250 mm, depending on the rated voltage, and the diameter is 20 mm or less. (The maximum length of good quality Czochralski-grown crystals is about 300 mm.) Polarized at 45° to the electro-optic axes x', y', 850-nm light from an LED enters the crystal at the ground potential side, makes a round-trip through the crystal and finally splits into two exit channels, each with an analyzer parallel to the entrance polarizer. A quarter-wave retarder in one of the two channels introduces a 90°-phase shift between the two polarimetric signals. Multimode fibers guide the light to and from the sensor. The electro-optic phase shift Df between two vertices of the voltage results in multiple interference fringes (Figure 14.4); the half-wave voltage (V_p) is 24 kV in this configuration. But appropriate processing of the two signals (called base and quadrature signals in Figure 14.4) unambiguously returns the voltage amplitude and waveform [92, 93].

The high field strengths in the vicinity of the crystal require careful control of the field distribution by appropriately designed and positioned electrodes to prevent partial discharge and dielectric breakdown. The sensor must be able to withstand not only the rated voltage but over-voltages as well as lightning and switching impulse voltages according to industry standards [94]. For example, at a rated voltage of 420 kV (phase-to-ground voltage of 420 kV/$\sqrt{3}$), the sensor must withstand an over-voltage of 520 kV for 1 min as well as 15 consecutive lightning and switching impulses of 1425 kV-peak and 1050 kV-peak, respectively, of positive and negative polarity. Specified rise and decay times are 1.2 ms and 50 ms, respectively, for lightning impulses, and 250 ms and 2500 ms for switching impulses. The rise time is defined as time to peak; the decay as time to half value. The switching impulse tests must also be performed at wet conditions, i.e., while the transducer is being sprinkled with water. At the peak of a 1425 kV impulse, the average field along a 200-mm long crystal is about 7.1 kV/mm. For comparison, the breakdown field of air is about 3 kV/mm.

Particularly crucial are so-called triple points, i.e., points or lines where three media with different dielectric constants meet each other, for example, the BGO crystal, optics components, and the surrounding insulation gas. To lower the field strengths in those zones, the crystal is equipped with a pair of tube-like inner electrodes that protrude over the crystal ends (not shown in Figure 14.3). The electrodes are in electric contact with transparent and conducting indium tin oxide (ITO) layers on the respective crystal end faces. The optical assembly including the inner electrodes is mounted in a fiber reinforced epoxy tube. Layers of foam mechanically decouple the assembly from the remainder of the transducer. The epoxy tube connects a pair of outer electrodes that also protrude over the crystal ends (Figure 14.3, center); see [17] for further details.

The whole assembly resides in a hollow core insulator column filled with SF_6 gas at a pressure around 600 kPa. The SF_6 breakdown field is 9 kV/mm/100 kPa. The diameter of the insulator column is such that the electric field on the outer surface remains well below the breakdown field of air. The transducer meets class-0.2 accuracy in outdoor temperature ranges. An optical temperature sensor serves to compensate the temperature dependence of the Pockels effect. Often, the EOVT has been combined with the magneto-optic current transducer (MOCT) of Figure 6.5 to an optical metering unit (OMU); Figure 14.3 left. Figure 14.5 shows an OMU retrofit installation in a HV substation. Scarce space made it necessary to mount the OMU hanging upside down from an existing line entrance tower which hardly would have been possible for conventional VT and CT [91].

In further work, ABB researchers developed an EOVT version with all-solid-state insulation. The modifications make the transducer maintenance-free (no gas pressure monitoring) and result in considerably slimmer insulators. Figure 14.6 depicts a 300 kV demonstrator [95–96]. The insulator builds on technology adopted from high voltage bushings [97]. The BGO crystal with a length of 200 mm and diameter of 5 mm resides in

FIGURE 14.5 Optical metering units (OMU) in HV substation [91].

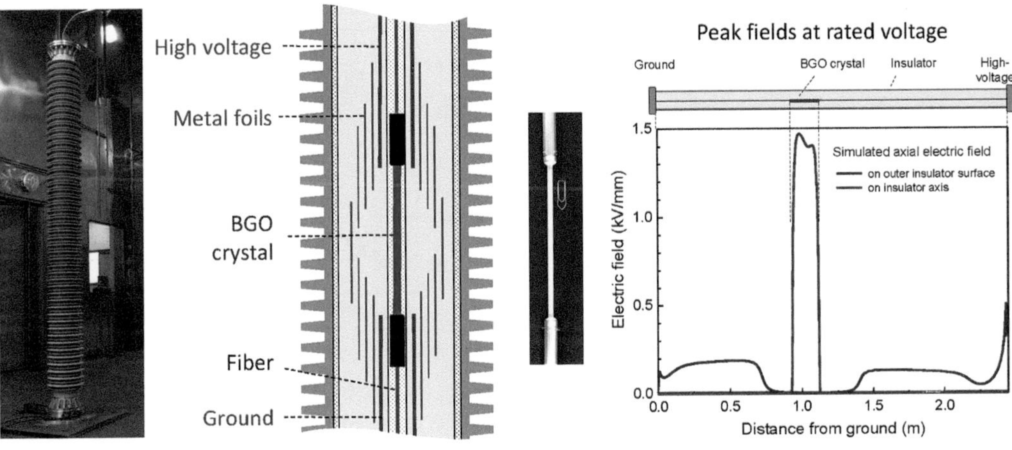

FIGURE 14.6 Electro-optic voltage transducer (300 kV) with solid state insulation (left) [74], BGO crystal in hollow-core insulator with embedded field steering metal foils, photograph of crystal with electrodes (center); simulated electric field strengths on insulator axis (red) and along outer insulator surface (blue) at 250 kV phase-to-ground (corresponding to the peak voltages at the rated alternating voltage of 300 kV) (© 2013 IEEE. Reprinted with permission from [95]).

the 25-mm diameter bore of a 2.44-m-long condenser core made of resin impregnated paper (RIP). The crystal is operated in transmission-mode (at 1310 nm) to minimize the lateral dimensions of the optical assembly. An outer 2.7-m-long insulator column of fiber-reinforced epoxy provides sufficient mechanical strength (Figure 14.6, center). (The insulator column can be omitted if the core is made of resin impregnated fiber instead of RIP.)

The core contains two sets of concentric and appropriately staggered aluminum foils. The two innermost foils are at ground and high-voltage potentials, respectively, whereas the other foils are on floating potential. The foils steer the planes of equipotential such that on the outer insulator surface the voltage drops sufficiently uniformly over the full insulator length keeping the field strengths within safe bounds for air, whereas inside the insulator the voltage drops over a 170-mm-long cavity containing the crystal. The graph in Figure 14.6 shows the field on the insulator axis (red) and outer insulator surface (blue) at a phase-to-ground voltage of 250 kV-peak; that voltage corresponds to the voltage vertices at the rated voltage of 300 kV: (300 kV/$\sqrt{3}$) × $\sqrt{2}$. The field on the insulator axis is about 1.5 kV/mm at the crystal location and essentially zero elsewhere. The field on the outer insulator surface stays below 0.25 kV/mm apart from the HV end, where it rises towards 0.5 kV/mm. The crystal is again equipped with tube-type electrodes that are in contact with the respective ground and HV potentials and house the collimator and polarizer optics. The remaining bore volume is filled with compressible silicone foam.

The sensor passed high voltage stress tests according to the relevant industry standards. The signal remained undisturbed by water or ice on the silicone sheds, stray fields from neighbor phases, or severe distortions of the external field. Figure 14.7a shows the relative error between 1 kV and 545 kV. The sensor meets the accuracy requirements of IEC metering class 0.2 and protection class 3P. The respective voltage ranges (80–120% and 5–190%

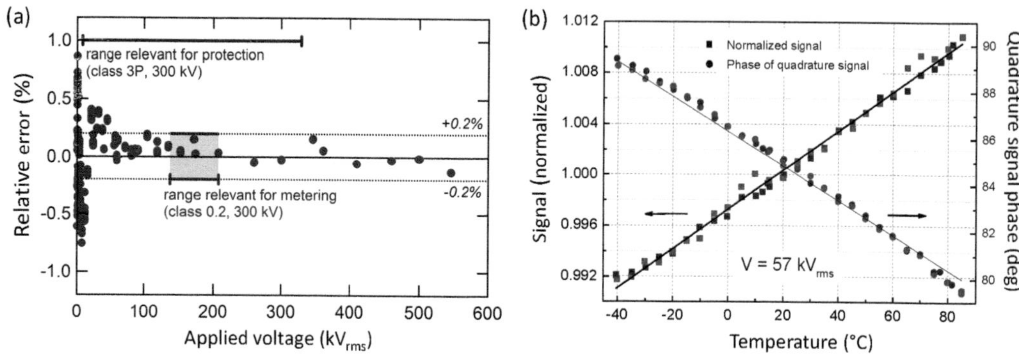

FIGURE 14.7 (a) Measurement error vs. applied voltage (phase-to-ground). The voltage ranges relevant for metering (80–120% of $V_r/\sqrt{3}$, green) and protection (5–190% of $V_r/\sqrt{3}$, blue) are indicated for a rated voltage of 300 kV (phase-to-phase) (© 2013 IEEE. Reprinted with permission from [95]). (b) Signal at constant voltage (57 kV rms) vs. temperature and quadrature signal phase (relative to base signal) for temperature compensation; blue and green symbols correspond to two consecutive temperature ramps [74].)

of the rated voltage) are indicated. The variation of the signal with temperature (1.54×10^{-4}°C^{-1}) between −40°C and 85°C (Figure 14.7b, squares, left-hand scale) is consistent with the temperature dependence of the Pockels effect in BGO [46]. Using a multiple-order quarter-wave retarder, the thermal variation of the differential phase between the base and quadrature signals (Figure 14.7b, dots, right-hand scale) can be used for temperature compensation.

As mentioned, two optical signals at quadrature serve to resolve the EOVT signal ambiguity at large phase shifts. As an alternative, Santos et al. developed a multisegmented sensing element consisting of thin BGO slices separated by transparent dielectric spacers, so that the total electro-optic phase shift remained within ±p/2 [26, 27]. Eight BGO slices, each with a thickness of 1 mm, and 7 spacers with thicknesses of 14.6 mm totaled a length of 110 mm. The segmentation raised the halfwave voltage at 1310 nm by a factor of 198 to about 14 MV. Note that in this configuration, the relatively high dielectric constant of BGO (e_r =16) significantly reduces the effective field strength. Measurement errors (at constant temperature) were within ±0.1% for alternating voltages between 40 kV and 170 kV and ±0.2% for dc voltages between 50 kV and 230 kV [27]. Potential errors of multisegmented OVT and their mitigation by additional field-controlling elements were investigated in [98, 99]. Finally, Kumada et al. reported a transducer, where the voltage (up to 220 kV peak) was applied to eight serial BGO rods, each with a length of 120 mm resulting in a total length of 960 mm. The sensor was operated at two different wavelengths (1300 nm and 1550 nm) for removing the signal ambiguity [30].

14.3.2 OVT with Discrete Field Summation

A team at the University of British Columbia and subsequently at NxtPhase Corporation developed an electro-optic voltage transducer according to the concept of Figure 14.1e [38, 100, 101]. Here, an insulator column with electrode plates at the top and bottom ends contains N Pockels cells placed at position z_i on the longitudinal insulator axis (Figure 14.8a).

The sum of the weighted fields $a_i E_z(z_i)$ at the locations z_i serves as a measure for the applied voltage V:

$$V \cong \sum_{i=1}^{N} \alpha_i E_z(z_i), \tag{14.1}$$

where a_i is the weight of the ith field sensor. The number of required sensors depends on the expected worst-case perturbation of the field distribution; obviously, a single sensor would suffice in a perturbation-free situation. Moreover, the insulator contains a concentric shielding tube of a slightly conductive material of high permittivity that is to minimize field perturbations inside the insulator when the external field distribution changes [37, 102]. The remaining field perturbation along the insulator axis z can be expressed in terms of a polynomial $r(z)$ that relates the perturbed field $E_z^{(p)}(z)$ to the unperturbed field $E(z)$: $E_z^{(p)}(z) = r(z)E(z)$. For the assumed worst-case scenario, the so-called quadrature method then yields the number of required sensors, the sensor positions z_i, and weight factors a_i [100, 101].

In the specific case of a 230 kV transformer, three miniature Pockels cells resided in an insulator column with a height of 2245 mm, an outer diameter of 312 mm, and filled with dry nitrogen. The Pockels crystals were 20 mm long, had a diameter of 3.5 mm, and measured the z-components of the field. The longitudinal crystal axes and light propagation directions were parallel to z. The fiber-pigtailed field sensors were mounted in a fiber glass tube (not shown in Figure 14.8) with an inner diameter of 198 mm at distances from the ground electrode of $z_1 = 231$ mm, $z_2 = 1230$ mm, and $z_3 = 1998$ mm, z_2 being the location with the smallest field perturbation. The corresponding weight factors were $a_1 = 0.60$, $a_2 = 1.00$, and $a_3 = 0.41$.

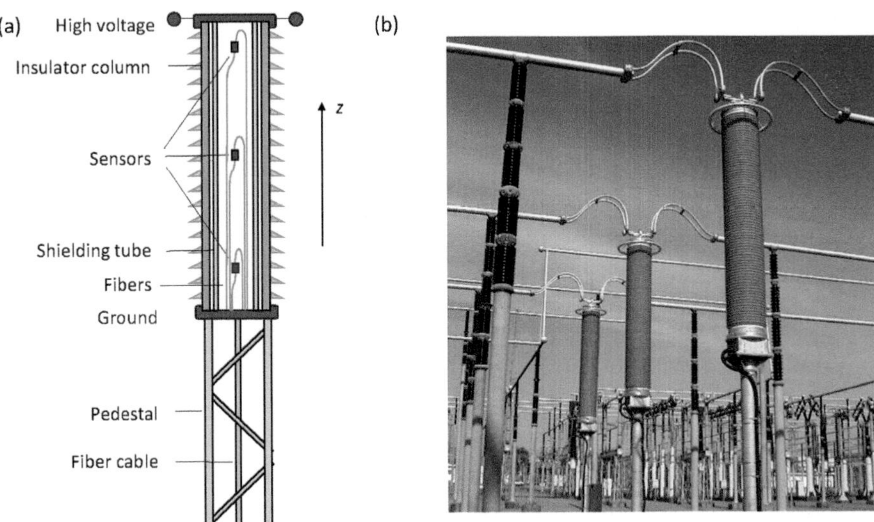

FIGURE 14.8 (a) Schematic setup of NxtPhase electro-optic voltage transducer (adapted from US patent 6,252,388B1 [100]); (b) field installation of combined 550 kV optical voltage and current transducers (Reprinted with permission from CIGRE, ©2004 [109]).

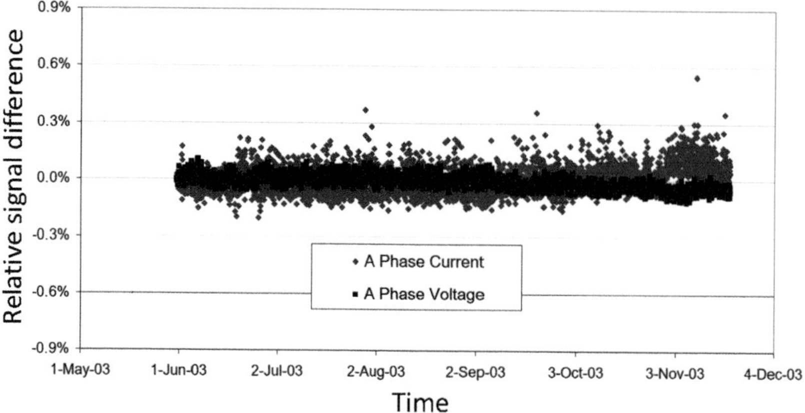

FIGURE 14.9 Field trial of NxtPhase 550 kV voltage and current transducers at a BC Hydro substation near Vancouver: relative signal difference between optical and conventional voltage and current measurements for phase A. (Reprinted with permission from CIGRE, ©2004 [109].)

NxtPhase, in collaboration with the University of British Colombia, reported multiple field installations with rated voltages between 138 kV and 550 kV (Figure 14.8b) [103–109]. It was shown that the transducers were compliant with IEC accuracy class 0.2 and ANSI/IEEE class 0.3 over wide temperature ranges. Figure 14.9 depicts corresponding data measured during an extended field trial of 550 kV voltage and current transducers. The accuracy was well maintained in the presence of severe field distortions and pollution. Even so, subsequent owners have further pursued the concept.

14.4 DC VOLTAGE SENSORS

DC voltage transformers have applications, for example, in high voltage direct current (HVDC) power transmission and corresponding ac/dc power converters. Optical dc voltage transducers are an alternative to resistive capacitive voltage dividers. Integrated in HVDC switchgear on converter platforms of offshore wind parks, for instance, optical sensors can help to save weight and space [110]. Electro-optic dc voltage measurement poses several challenges, however. Dark currents and space charge accumulation at the crystal surfaces (of electrode-less sensors) cause the field distribution in the crystal to change with time and temperature [23, 111–115]. Charge migration can also lead to local field enhancements in the vicinity of the crystal and result in partial discharge and a reduced breakdown voltage. And without further measures, electro-optic dc phase changes are indistinguishable from thermally or stress-induced phase shifts. For BGO in a transverse electric field, Kurosawa et al. reported signal drift due to space charge accumulation of up to 10% of the saturated signal over time periods of some tens of minutes after dc voltages of a few hundred volts were applied [23]. The electro-optic signal of a bare $LiNbO_3$ crystal was observed to exponentially decay on a time scale of a few hours [115].

In 1985, Robertson et al. showed for cylinder-shaped $LiNbO_3$ and $Bi_4Ge_3O_{12}$ crystals in transverse dc fields up to 100 kV/m that signal drift could be eliminated by rotating the crystals around their longitudinal axes at a frequency of 20 Hz that was high enough to prevent charge migration [52]. Already in 1982, Hidaka et al. had applied the technique to the measurement of discharge fields [116]. More recently, new studies on a rotating BGO crystal sensor were reported by Li et al. [117]. Instead of rotating the sensor, Zhang et al. rotated one of the electrodes of an integrated-optic lithium niobate field sensor to circumvent the problem of fiber pigtailing a rotating sensor [118]. Kurosawa et al. measured dc voltages of up to ±187.5 kV by chopping the down-stepped voltage from a resistive divider before transversely applying the voltage to a $Bi_4Ge_3O_{12}$ crystal. Errors were well within 1% at voltages above 20 kV and smaller than 5% below. The device was tested in a 1-year field trial at a converter station of the Tokyo Electric Power Company (TEPCO) [23].

Signal drifts due to space charges and dark currents can effectively be suppressed by operating the sensor in a longitudinal configuration. In the development towards a 320 kV dc voltage transducer for gas-insulated HVDC switchgear, Gu et al. exposed 5-mm-thick BGO samples ($Bi_4Ge_3O_{12}$) to a constant dc voltage of 5 kV and measured the electrical conductivity as well as the longitudinal electro-optic phase shift over a period of 11 days. The results showed that, while a slow conductivity decay reflected a redistribution of the electric field inside the crystal (consistent with dielectric spectroscopy measurements), the electro-optic phase shift exhibited no signs of drift [110]. Kumada et al. also observed stable operation of longitudinally operated BGO crystals under dc voltage, albeit over shorter time periods [30].

14.5 INTEGRATED-OPTIC ELECTRIC FIELD SENSORS

Integrated-optic electric field and voltage sensors have been developed since the 1980s in parallel with integrated-optic modulators. Applications include electro-magnetic compatibility, microwave fields, electromagnetic impulses, accelerators for high energy physics, and electric power transmission. The literature on the subject is extensive, and this section can only give a short overview. Mostly based on lithium niobate (LN), the devices are either directly immersed into the electric field, or a voltage is applied to electrodes parallel to the waveguides. Device lengths range from a few mm to some tens of mm. Figure 14.10 illustrates some common designs. Often, the waveguides form a Mach-Zehnder interferometer (MZI) with electrodes as schematically illustrated in Figure 14.10a. Bulmer et al. and Kuwabara et al. were among the pioneers of integrated-optic MZI [12, 119–123]. The waveguides are commonly generated by titanium in-diffusion into the LN substrate and support TM and TE modes with polarizations orthogonal and parallel to the plane of the substrate, respectively. A SiO_2 buffer layer separates the electrodes from the substrate to minimize optical loss. In Figure 14.10a, the substrate is x-cut (the crystallographic x-axis is orthogonal to the substrate), the waveguides are along y, and the voltage is applied in $\pm z$-directions. Hence, the MZI operates in a push-pull fashion, that is, the electro-optic phase shifts in the two interferometer arms have opposite signs. The probe light arrives through a polarization-maintaining fiber (PMF) and excites the TE mode. The interference intensity at the output reads in normalized units [12]

$$I = \left(\frac{1}{2}\right)\left[1 + \cos\left(\theta_0 + \pi\frac{V}{V_\pi}\right)\right]. \tag{14.2}$$

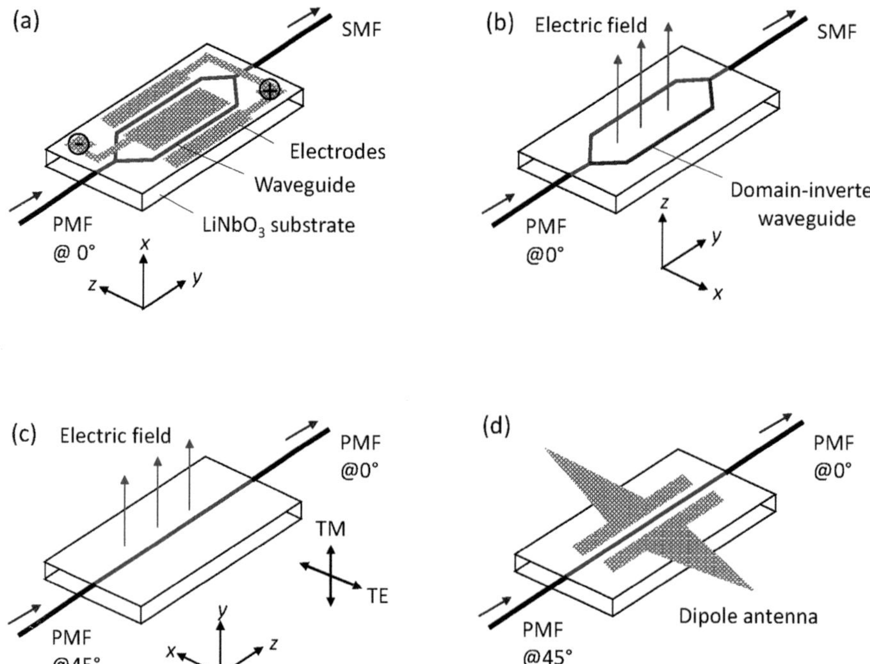

FIGURE 14.10 Integrated electro-optic lithium niobate electric field and voltage sensors: (a) Mach-Zehnder interferometer (MZI) with electrodes for push-pull operation; (b) MZI with a domain-inverted interferometer arm for direct immersion into an electric field; (c) single waveguide sensor utilizing the electro-optic phase shift between TM and TE modes; (d) sensor with dipole antenna for electric field mapping.

The halfwave voltage V_p is given by

$$V_\pi = \frac{\lambda g}{2\delta n^3 r_{33} L}, \tag{14.3}$$

where l is the wavelength, δ is an efficiency factor, n is the relevant refractive index, r_{33} is the relevant electro-optic coefficient, and L and g are the electrode length and gap width, respectively. The phase biasθ_0 is

$$\theta_0 = (2\pi n \Delta l)/\lambda. \tag{14.4}$$

Here, Dl is the length difference between the interferometer arms. For operation in the linear range, θ_0 must be tuned to p/2 (modulo p), e.g., by an added dc voltage, asymmetric arm lengths [12], laser ablation in one interferometer arm [120, 121], or applied stress [123].

The sensor of Figure 14.10b is without electrodes and directly immersed into the electric field to be measured. One of the two interferometer arms runs through a region of the z-cut LN substrate with inverted ferroelectric polarization. The z-directions in the inverted and non-inverted zones are anti-parallel, which results in opposite electro-optic phase shifts at a common field direction [124–127]. Alternatives to domain inversion are MZI with asymmetric arm lengths [12] or unequal effective mode indices in the two arms [128]. A sensor as in Figure 14.10b has comparatively low sensitivity, since the high permittivity of LN causes significant field screening, but this is commonly not an issue in high field applications. Figure 14.10c shows another example of an electrode-less sensor [129]. The sensor consists of a single waveguide in *y*-cut lithium niobate. The input light is polarized at 45° to *y* and excites both orthogonal modes, TM and TE, propagating in *z*-direction [130–133]. A field along *y* increases the refractive index of the TE-mode and decreases the index of the TM-mode by $\pm(1/2)n_0^3 r_{22}E_y$, respectively. In *z*-cut substrates, the difference of r_{33} and r_{13} governs the phase shift [134]. Electrodes equipped with an antenna or antenna array, as schematically illustrated in Figure 14.10d, enhance the sensitivity for electric field mapping in selected frequency ranges [123, 129, 133, 135, 136]. Besides the sensor types of Figure 14.10, sensors based on integrated optic 1×2-directional couplers were also studied [137, 138], and Gutierrez-Martinez et al. reported sensor interrogation by low coherence interferometry, which circumvented the need for bias tuning [134, 139, 140].

The following gives a few selected examples of sensor demonstrations and achieved performances: Bulmer et al. demonstrated MZI according to Figure 14.10a with 1.1-mV sensitivity and an 84-dB linear range for a detection bandwidth of 3 kHz. With 14-mm long electrodes, the *Vp*-voltage was 2 V at 1310 nm [12]. Jaeger et al. measured with a similar sensor the voltage between the concentric ground and high voltage conductors of gas-insulated HV switchgear [78]. To this end, a simple voltage divider down-stepped the GIS voltage of 50 kV to a voltage of 9.5 V that was applied to the *y*-cut MZI. The MZI was intentionally made with a large V_p of 50 V at 633 nm. The same team also demonstrated an electrode-less single waveguide sensor as in Figure 14.10c for direct sensing of HV electric fields. Here, the *y*-cut, z-propagating waveguide was 28.5-mm long and had an *Ep*-field of 146 V/mm at 670 nm [130]. For a similar sensor, Gutierrez-Martinez et al. reported a measurement range from 18 V/mm to 350 V/mm at frequencies up to 20 kHz [134]. Sun et al. reported a MZI device with an antenna array capable of measuring high frequency fields from 10 kHz to 18 GHz with a minimum detectable field of 0.4 V/m at 1 GHz [136]. Sensors for intense pulsed electric fields were reported, for example, in [141–143], and voltage measurement on overhead transmission lines was investigated in [144]. Finally, it is worth mentioning that compact micro-ring waveguide resonators evanescently coupled to electro-optic thin films have also been demonstrated for electric field sensing. Chen et al. coupled a thin lithium niobate layer to a (non-electro-optic) silicon ring resonator with a radius of 20 mm. In spite of the small size, the device reached sensitivity of 4.5 $(\mathrm{V/m})/\sqrt{\mathrm{Hz}}$ at radio frequency fields of 1.86 GHz due to a quality factor of 13,000 [145].

14.6 FIBER TIP ELECTRIC FIELD SENSORS

Reflective fiber tip electro-optic sensors are of particular interest for electric field mapping, for example, on antennas and high-speed electronic devices. They are less intrusive than metallic probes and can provide spatial resolutions as small as 10 mm and bandwidths up to tens of GHz [146]. Often, the sensors apply a modulation technique for controlling the operating point and down-converting the high frequency signal into a lower frequency regime [147–149]. Early sensors were commonly configured as polarimetric devices with a small electro-optic crystal mounted in a dielectric ferrule at the distant

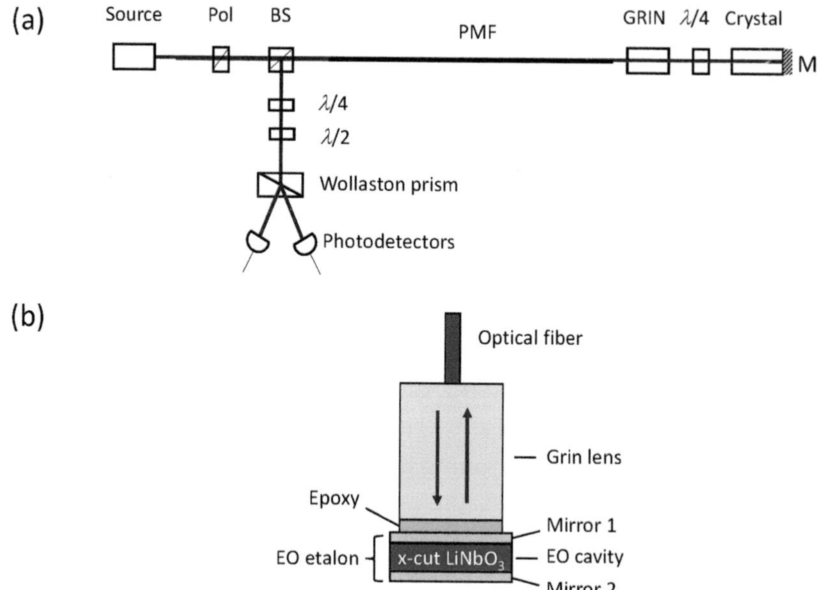

FIGURE 14.11 Fiber tip electro-optic sensors for electric field mapping: (a) polarimetric sensor; (b) Fabry-Pérot etalon sensor (Adapted/reprinted with permission from [150, 155] © The Optical Society). Pol: polarizer, BS: beam splitter, PMF: polarization-maintaining fiber, GRIN: graded index lens, M: mirror.

end of an optical fiber and the analyzer optics positioned at the source side end of the fiber (Figure 14.11a) [147–152]. Typical crystal diameters were 1–2 mm and crystal lengths several mm. Stationary devices without a fiber lead have been developed, for example, for monitoring the fabrication quality of complex radio frequency devices [149]. Others have configured the electro-optic element as a fiber tip Fabry-Pérot etalon. This has resulted in a dramatic miniaturization and higher spatial resolution at still adequate sensitivity [147, 153–156]. Lee et al., for example, reported lithium niobate (LN) etalons with thicknesses of 0.5 mm and 0.35 mm and calculated Q-factors of 1.1×10^4 and 2.0×10^5, respectively (Figure 14.11b) [155]. They demonstrated field mapping with a spatial resolution of 50 mm at field strengths between 1 V/m and 10^4 V/m and frequencies up to 3 GHz.

Calero et al. reduced the fiber tip device to a circular thin film with a diameter of 125 μm and a thickness of 700 nm. A photonic crystal structure confined the light within the LN film and resulted in a spatial resolution of under 20 mm and a frequency range from 1 Hz to tens of GHz. (The theoretical bandwidth was 10^{12} Hz.) A Fano resonance increased the effective electro-optic phase shift by a factor of 400 with respect to bulk LN and gave a field detection threshold of 32 (V/m)/$\sqrt{\text{Hz}}$. By comparison, typical thresholds of much longer bulk crystals are on the order of 100 (mV/m)/$\sqrt{\text{Hz}}$ [146, 157]. Besides lithium niobate, probe materials have included $LiTaO_3$ [147, 154, 156], $Bi_{12}SiO_{20}$ [158], and semiconductors like GaAs [159], AlGaAs [153], and ZnTe [152]. Three-dimensional vectorial field measurement was demonstrated, for example, in [160–162], and Poynting vector sensors were reported in [163, 164]. There is extensive further literature on field mapping sensors and their applications. However, a more detailed review is beyond the scope of this book.

14.7 KERR EFFECT-BASED SENSORS

In 1875, the Scottish physicist John Kerr discovered that an isotropic transparent substance becomes birefringent when exposed to an electric field perpendicular to the propagation direction of the light. The refractive index difference for light waves polarized parallel and perpendicular to the field is

$$\Delta n = n_{\parallel} - n_{\perp} = \lambda_0 K E^2. \quad (14.5)$$

The Kerr constant K is proportional to the difference of the elements χ_{1111} and χ_{1122} of the susceptibility tensor $\chi^{(3)}$ in eq. (13.2). K is largest in polar liquids such as nitrobenzene ($C_6H_5NO_2$), nitrotoluene ($C_7H_7NO_2$), or propylene carbonate ($C_4H_6O_3$), whereas nonpolar liquids and glasses exhibit only a small Kerr effect (Table 14.1). Other than nitrobenzene and nitrotoluene, propylene carbonate has the benefit of being nontoxic. At least since the 1960s, Kerr cells have been routinely used in laboratory environments for the recording of ac voltages and short HV pulses up to hundreds of kilovolts [165–169]. The Kerr effect has also served to study electric field distributions, space charge, and charge transport phenomena both in liquids and gases, e.g., in the development of electric power equipment [170–174]. Kerr effect-based optical fiber sensors were also reported.

In 1981, Rogers proposed an optical fiber voltage sensor comprising a single-mode fiber arranged on a helical path between ground and high voltage potentials. Polarization-optical time domain reflectometry was to spatially resolve the Kerr optical phase shift in the fiber. Appropriate signal processing would then yield the line integral of the field [175]. However, no experimental demonstration was given. Han et al. reported a silica micro-wire electric field sensor. The micro-wire — 65-mm long, with 1-mm radius, and immersed in propylene carbonate — formed one arm of a fiber Mach-Zehnder interferometer. The evanescent wave of the micro-fiber saw the field-induced refractive index change of the liquid, which led to a shift of the MZI interference fringes. The minimum and maximum detectable fields were 59 V/cm and 14 kV/cm, respectively, measured for 50 Hz ac fields and 200-ms impulses [176]. Hou et al. presented a propylene-carbonate-coated microfiber resonator for electric fields from 200 V/cm to 4000 V/cm [177].

TABLE 14.1 Kerr Constants of Selected Liquids and Solids at Room Temperature

Substance	Kerr Constant K (10^{-14} mV^{-2})	Wavelength (nm)	References
Nitrobenzene	245	589	[178]
Nitrotoluene	137	589	[178]
Propylene carbonate	110	589	[172]
Water	5.2	589	[178]
Transformer oil	0.25	633	[170]
Flint glass SF-57	0.11	633	[179]
Fused silica fiber	0.053	633	[180]

14.8 ALTERNATIVE INTERROGATION TECHNIQUES

The sensors of the previous sections employed polarimetric detection schemes, integrated-optic Mach-Zehnder interferometers, or Fabry-Pérot cavities to measure the electro-optic phase shift. The adaptation of the polarization-rotated reflection technique from interferometric current sensors (Figure 8.10) to voltage sensing is a further alternative [181]. Wildermuth et al. first demonstrated the method for a piezo-optic voltage transducer [182, 183]. Subsequently, numerous further researchers studied the method in the context of electro-optic voltage sensors, e.g., [87, 110, 184–191]. Figure 14.12 illustrates the basic setup. Whereas the current sensor works with circular states of polarizations (SOP), the electro-optic sensor works with orthogonal linear SOP. Accordingly, a 45°-Faraday rotator replaces the fiber quarter-wave retarder of the current sensor and provides the necessary round-trip polarization rotation of 90°. The method is attractive in that the same interrogation technique can be used for current and voltage, particularly if the voltage sensing element is a fiber like in poled fiber sensors or in the piezo-optic sensors presented below. In case of bulk electro-optic crystals, however, the need to combine single-mode fibers with bulk optics and added cost tend to outweigh such advantages. Optical voltage transducers with polarization-rotated reflection and passive polarimetric detection were reported in [151, 192, 193]. Still other detection methods for Pockels sensors can be found in [194–196].

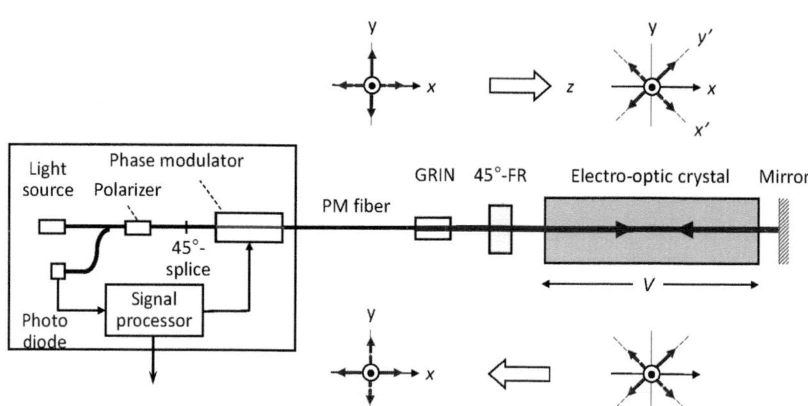

FIGURE 14.12 Electro-optic voltage sensor interrogated by a polarization-rotated reflection technique (Adapted from [110, 181]). GRIN: graded index lens, FR: Faraday rotator.

CHAPTER 15

Electro-Optic Fibers

15.1 POLED FIBERS

15.1.1 Thermal Fiber Poling

Electric poling of glass leads to a permanent frozen-in electric field in the glass matrix. The field lifts the centro-symmetry of the structure and thus induces a non-zero second-order nonlinearity $\chi^{(2)}$ and linear electro-optic effect. Poling techniques include temperature-assisted poling [197], UV-assisted poling [198, 199], electron beam irradiation [200], corona discharge poling [201], induction poling [202, 203], and others. Optical fiber poling is commonly done with temperature-assisted electric poling or thermal poling in short; for reviews on poling techniques and poled fiber devices, see [203, 204]. Thermal poling of bulk fused silica was first demonstrated by Myers et al. in 1991, achieving an induced second-order nonlinearity of $\chi^{(2)} \approx 1$ pm/V [197]. The corresponding electro-optic coefficient is about $r \approx 2\chi^{(2)}/n_0^4$, where n_0 is the refractive index of the glass. With $n_0 \approx 1.45$, r is roughly $\chi^{(2)}/2$ in silica-based glasses and fibers [198, 205]. In 1994, Kazansky et al. demonstrated the first thermally poled fibers with an effective $\chi^{(2)}$ of 0.2 pm/V. The germanosilicate fibers had a D-shaped cladding, and the poling field was applied via external electrodes [206]. The following year, Fujiwara et al. reported poling of a germanosilicate fiber with electrodes inserted in hollow channels in the fiber cladding parallel to the core achieving about 1 pm/V [207]. Numerous further publications on poled fibers followed, motivated by the perspective of all-fiber devices such as modulators or switches for optical communication. Direct measurement of the Pockels effect in poled glass and fibers was reported in [198, 208–212]. Moreover, noteworthy work has been done on poled polymers and poled polymer fibers; see, for example, [213, 214].

Explanations of the poling-induced $\chi^{(2)}$ and linear electro-optic effect have included dipole alignment, buildup of spatially separated space charges in the fiber, and charge injection from the environmental air under the influence of the poling voltage [215–218]. Experimental evidence strongly points to space charge fields as the dominating cause [204, 216, 218–223]. Figure 15.1 schematically illustrates thermal poling of a twin hole fiber. The poling voltage (typically a few kilovolts) is applied to the two electrode wires (e.g., from tungsten) inserted into the twin holes through side-polished openings in the fiber

DOI: 10.1201/9781003100324-19

cladding. The anode wire is commonly close to the core (within about 5 micrometers) for maximum effect. Upon heating the fiber to the poling temperature (250–325°C), positively charged alkali ions such as K^+, Li^+, and Na^+, attached to defects in the glass matrix become mobile and drift under the influence of the poling field towards the cathode wire, leaving behind a negative space charge. As a result, an electric screening field builds up between the separated positive and negative space charges. The width of the layer depleted of impurity ions is relatively narrow and on the order of 10 mm [197]. After dwelling at the poling temperature for about 1–2 hours, the fiber is led to cool down with the poling voltage still applied. After removal of the voltage, the space charges and recorded space charge field E_{rec} remain frozen-in and ascertain a persisting, non-zero $\chi^{(2)}$ (Figure 15.1c). In 2009, Margulis et al. introduced a modified, cathode-less thermal poling process with both electrodes at the same anodic potential. This reduced the risk of dielectric breakdown during poling [224]. A more detailed description of the poling mechanism can be found in [203, 204]. It is worth mentioning that, not surprisingly, only small second-order nonlinearities (<0.1 pm/V) were achieved in pure silica fiber [225, 226].

The diameter of the electrode wires is on the order of 20–50 μm; the twin hole separation is typically 10–20 μm [227], and the lengths of the inserted wires may range from a few centimeters to some tens of centimeters. Electrode lengths of up to 200 m were obtained by co-drawing a mono-hole fiber together with an electrode wire and coating the fiber with conductive polyimide that served as the second electrode; but only a modest electro-optic coefficient of 0.085 pm/V was measured [204, 228]. An alternative to solid electrodes are conducting liquids, e.g., mercury, gallium, or alloys such as AuSn or BiSn [229, 230]. Liquid electrodes are easier to insert over longer distances and give a more reproducible pattern of the poling field. Fokine et al. reported BiSn electrodes with lengths of up to 22 m [229].

Even though the poling-induced second-order nonlinearity commonly exhibits good long-term stability at room temperature, it is subject to decay as a result of reversed charge migration and recombination, especially at elevated temperatures [211, 231–236]. Usually, there is an initial fast decay, followed by a slower decay — both governed by Arrhenius's law. The different decay times are associated with different mobilities and/or activation energies of the involved cations and therefore strongly depend on the glass composition. Janos et al. reported a lifetime of the induced electro-optic effect of only 15 days at room

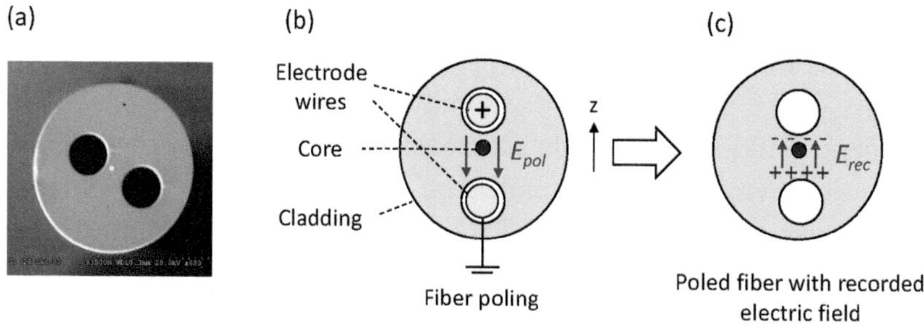

FIGURE 15.1 (a) SEM image of the cross section of a 125-μm twin hole fiber for poling. (Reprinted with permission, from [230] © The Optical Society); (b) thermal poling of a twin hole fiber; (c) poled fiber with recorded space charge field E_{rec}.

temperature in boron co-doped germanosilicate fiber [234]. Xu et al. observed a decay rate in a twin hole fiber with a pure silica cladding of 2.6% per day at 110°C after the initial fast decay. The rate was reduced to 0.6% per day in a fiber with a borosilicate layer on the inside of the anode hole [211]. Much longer lifetimes were reported by Sugihara et al. for thin film waveguides of poled germanosilicate glass on a silica substrate. The team extrapolated a room temperature lifetime of 310 years. Deparis et al. found strongly enhanced $\chi^{(2)}$-stability in aluminosilicate and aluminoborosilicate glasses with a lifetime increase by at least five orders of magnitude with respect to silica [236].

15.1.2 Electro-Optic Effect in Poled Fibers

The nonlinear polarization of eq. (13.1) is associated with a field-dependent refractive index. With the intrinsic susceptibility $\chi^{(2)} = 0$, the refractive index of a poled fiber is given in a simplified scalar form as [237]

$$n = n_0 + \frac{3\chi^{(3)}}{2n_0}\left(E_{\text{rec}} + E_{\text{app}}\right)^2, \tag{15.1}$$

$$n = n_0 + \frac{3\chi^{(3)}}{2n_0}\left(E_{\text{rec}}^2 + E_{\text{app}}^2 + 2E_{\text{rec}}E_{\text{app}}\right). \tag{15.2}$$

Here, E_{rec} is the recorded field, and E_{app} is the applied field to be measured. The quadratic terms in (15.2) represent the dc Kerr effect; the third term is proportional to the applied field and represents the poling-induced linear electro-optic effect. The recorded field thus results in an effective second-order susceptibility proportional to E_{rec}:

$$\chi_{\text{eff}}^{(2)} = 3\chi^{(3)}E_{\text{rec}}. \tag{15.3}$$

The matrix of electro-optic coefficients of poled glass is equivalent to the one of the crystal classes 4 mm and 6 mm [205] and given by

$$r_{ij} = \begin{pmatrix} 0 & 0 & r_{13} \\ 0 & 0 & r_{13} \\ 0 & 0 & r_{33} \\ 0 & r_{13} & 0 \\ r_{13} & 0 & 0 \\ 0 & 0 & 0 \end{pmatrix}. \tag{15.4}$$

Here, it assumed that the poling direction is parallel to z. An applied field in that direction induces electro-optic index changes for optical polarizations parallel and orthogonal to z. The modified indices read [205]:

$$n_{x'} = n_{y'} = n_0 - \frac{n_0^3}{2}r_{13}E_z, \tag{15.5}$$

$$n_{z'} = n_0 - \frac{n_0^3}{2} r_{33} E_z. \tag{15.6}$$

Hence, the polarimetric electro-optic phase shift in a poled fiber with an applied electric field parallel to the poling direction is

$$\Delta\phi = \frac{\pi}{\lambda} n_0^3 \left(r_{33} - r_{13}\right) E_{\text{app}} l, \tag{15.7}$$

where l is the length of the poled fiber section. Fields orthogonal to the poling direction have no effect. In particular, a field parallel to the propagation direction of the fiber will not produce an electro-optic phase shift. For symmetry reasons, the theoretical ratio of r_{33}:r_{11} is 3:1 [205, 238]. Experimental data for poled bulk glass were found to be well in line with that ratio [205]. By contrast, poled fibers have exhibited somewhat smaller ratios. Long et al., for instance, reported a ratio of 2.4:1 [210], Fujiwara et al. of (2–3):1 [198], and Michie et al. a ratio as small as 1.15:1 [204, 212]. The causes for the lower ratios in fibers were still speculative at the time.

15.1.3 Poled Fiber Electric Field and Voltage Sensors

A poled optical fiber can be employed to line integration of the electric field. To this end, the fiber (with the electrodes removed) is wound as a helix with constant pitch and radius onto an insulator rod between ground and high voltage potentials (Figure 15.2a). The poling direction must be aligned parallel to the longitudinal helix axis — or, more precisely, parallel to the binormal of the helix, that is, the direction normal to the radius and tangent vectors. Another prerequisite is that the electro-optic coefficients are constant along the fiber. Given the twin hole fiber structure, the fiber is polarization maintaining and can be polarimetrically interrogated.

A team at the Optical Fiber Technology Centre (OFTC) at the University of Sydney developed, with support from industry, a corresponding voltage transducer for demonstration

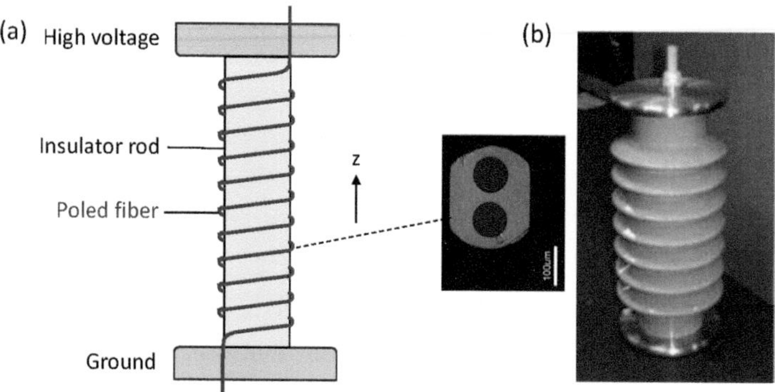

FIGURE 15.2 (a) Schematic illustration of a poled fiber voltage sensor with helical fiber path; (b) poled fiber voltage transducer developed at OFTC (Sydney) with self-aligning fiber; from [204].

purposes (Figure 15.2b) [204, 239–241]. The twin hole fiber with a linear electro-optic effect of ≈0.16 pm/V had two flat cladding surfaces for self-alignment [242] and was wound on an insulator rod with a diameter of 95 mm and a length of 500 mm. Low coherence interferometry was applied to determine the voltage-induced differential group delay between the orthogonal modes. The transducer showed good linearity at 50-Hz voltages between 2.5 and 12 kV, but dielectric breakdown occurred at the latter voltage. In other work, Magalhães et al. proposed distributed electric field sensing with a poled fiber interrogated by time domain reflectometry (OTDR) [237], and optical fiber electric field sensors based on poled polymers were reported in [243, 244].

As a final note, a high voltage transducer according to Figure 15.2 appears to be an attractive all-fiber solution. Nonetheless, to the author's knowledge, no such sensor was ever tested in the field. The finite lifetime of the electro-optic effect, particularly at higher temperature, has remained a concern. Further work towards poled fibers with significantly enhanced stability would still be necessary. Low stress and high voltage proof packaging of the fiber is another challenge. Also, optical phase shifts due to shock and vibration may require compensation measures [245].

15.2 CRYSTALLINE FIBERS

A crystalline electro-optic fiber with proper crystal symmetry and orientation enables electric field integration and voltage measurement according to the concept of Figure 14.1a and arguably might be considered an ideal high voltage transducer. Other than a poled fiber, a crystalline fiber can bridge the distance between ground and high voltage potentials in a straight line, i.e., there is no need for helical fiber winding and axis orientation. As a result, the fiber can be packaged in a slim and flexible high voltage proof fiber cable. Also, the electro-optic effect is inherently long-term stable. Moreover, a fiber of appropriate length significantly lowers the average field strength compared to field integration with shorter bulk crystals. Alas, such a fiber does not exist at present. However, there have been a few attempts towards crystalline electro-optic fibers, some of which we will consider in the following.

15.2.1 Micro-Pulling-Down Growth of BGO Fibers

Bulk crystals such as BGO are typically grown by the conventional Czochralski method. That method, however, is not suited to the growth of sub-millimeter thick fibers. Techniques for crystalline fiber growth include the laser-heated pedestal growth technique [246–248] and the micro-pulling-down (mPD) technique [249–255]. Both techniques have been applied to the growth of single crystal fibers such as Nd:YAG or Yb:YAG fibers for fiber lasers [248, 250] or $LiNbO_3$ fibers for nonlinear optical and electro-optical devices [247, 251]. mPD-grown crystal fibers have also included fibers from $KNbO_3$, $Tb_3Ga_5O_{12}$ (TGG), and others [254]. Of particular interest has been the mPD-growth of BGO ($Bi_4Ge_3O_{12}$) and BSO ($Bi_4Si_3O_{12}$) fibers [252–255]. As already noted, BGO and BSO are scintillating materials. Grown as fibers, they can be readily cut to small-sized pixels for spatially resolved radiation measurements. FiberCryst S.A.S., Villeurbanne (France), a

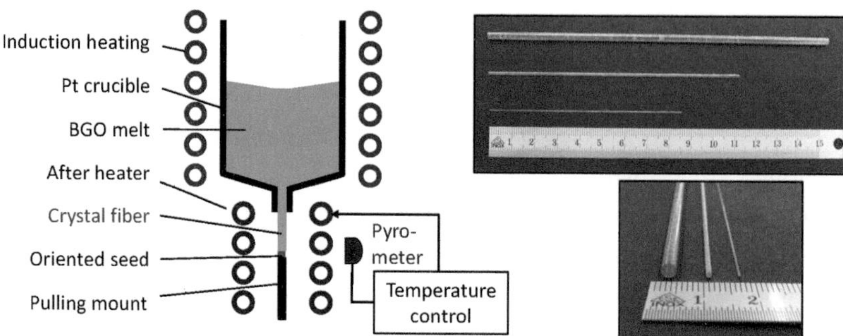

FIGURE 15.3 Micro-pulling-down apparatus for BGO fiber growth (left) and images of grown fiber samples (right) [256].

spin-off from the University of Lyon and one of the pioneers in the field, in collaboration with ABB has endeavored to grow $Bi_4Ge_3O_{12}$ fibers with optical quality and crystal orientation adequate for high voltage sensing [256]. Figure 15.3 schematically illustrates the fiber growth apparatus. Czochralski-grown BGO material was molten in an induction-heated platinum crucible. The melt flowed through a micro-capillary at the bottom of the crucible and crystallized at an appropriately oriented seed crystal beneath the exit nozzle. The seed crystal was then slowly pulled downwards at a speed of typically 0.5–1 mm/min. For stable growth, the temperature of the nozzle zone had to be kept in a narrow range of about 5°C around 1115°C. (The BGO melting point is at 1050°C.) To this end, an after-heater with closed-loop control served to stabilize the temperature. It was also important to suppress Bi_2O_3 evaporation from the melt due to its high vapor pressure to maintain the correct melt stoichiometry [254].

BGO fiber samples with lengths of up to 850 mm (limited by the apparatus) and diameters between 0.3 mm and 3 mm were grown [256]. Due to thermal issues, the fiber sections free of buckles or bends were only 150–200 mm long, however (Figure 15.3). A visual inspection indicated good optical quality without major defects. The absorption coefficients at 633 nm of 18 samples from three different growth runs were between <0.005 cm^{-1} and 0.045 cm^{-1}, with most samples near 0.01 cm^{-1}. For comparison, absorption coefficients between 0.0005 cm^{-1} and 0.03 cm^{-1} have been reported for conventionally grown bulk BGO [257]. The birefringent phase retardation of a 150-mm-long sample with a diameter of 3 mm was 0.6 rad (34°) at 1310 nm compared to 0.1–0.2 rad of good-quality Czochralski-grown crystals of similar lengths. The crystal orientation, determined by x-ray diffraction, was essentially constant along the sample length, but the [001]-direction was off from the desired direction (longitudinal axis) by about 20°, which was attributed to a misaligned seed crystal. Measurements with a longitudinally applied 60-Hz voltage showed a linear electro-optic response up to the maximum voltage of 30 kV. The BGO fiber samples did not yet have a core/cladding structure for guided wave propagation. Two concentric crucibles with appropriately doped BGO melts might enable the growth of fibers with a core/cladding index step [258]. But the success probability was judged as too uncertain, and no further work was attempted.

15.2.2 Molten Core Fabrication of BGO Fibers

Molten core crystal fiber fabrication has been pioneered, among others, by J. Ballato and coworkers at Clemson University. The initial focus was on semiconductor core fibers [259–261], but the fabrication of crystalline oxide core fibers was also investigated [262]. The fibers or canes are drawn on a regular drawing tower from a preform that typically consists of a cladding glass tube containing the core material, either in form of a crystal rod derived from a Czochralski-grown single crystal boule or in powder form. The core material melts during drawing but remains encapsulated in the viscous glass cladding. Compared to other crystal growth methods, the molten core method enables higher growth speed, longer sample lengths, and provides a core/cladding structure for optical waveguiding. On initiative from ABB, the Clemson group started a project towards molten core fabrication of $Bi_4Ge_3O_{12}$ core fibers [263]. The ultimate vision was a fiber with single crystalline core of defined orientation over a few meters of fiber. The investigations provided important new insights, but high voltage sensing-based molten core fabricated fibers remained elusive at the time; see [263] for details.

15.2.3 Crystallized Glass Fibers

Thermal treatment of glasses at sufficiently high temperatures leads to devitrification, that is, the formation of crystallites in the glass matrix. In special glass ceramics, the crystallites have a non-centrosymmetric structure and exhibit strong second-order nonlinearities and a linear electro-optic effect. Researchers in Japan have explored fresnoite-type glass ceramics, particularly $BaO\text{-}TiO_2\text{-}SiO_2$ and $BaO\text{-}TiO_2\text{-}GeO_2$ (BTGS), and optical fibers drawn from these materials as candidates for active optical devices such as switches or modulators [264–269]. In appropriately prepared transparent ceramics, they found highly oriented crystallized layers of long-range order which ascertained a macroscopic electro-optic effect. For a ceramic composition of $30BaO\text{-}20TiO_2\text{-}50SiO_2$, Iwafuchi et al. reported electro-optic coefficients r_{13} and r_{33} of 3.15 and 1.00 pm/V, respectively, at 633 nm and a field frequency of 100 Hz, which roughly amounted to 60% of the coefficients of corresponding fresnoite single crystals [266]. Ohara et al. demonstrated the electro-optic effect in BTGS-based double clad fibers [268, 269]. Laser irradiation space-selectively crystallized the inner cladding around the fiber core over lengths of 1 cm and 2 cm. The crystallized structures were found to be oriented with the *c*-axis in radial direction. A transverse electric field produced an electro-optic phase shift between polarizations parallel and orthogonal to the field of 0.1 rad/cm at 1.55 mm and a field strength of 12.5 kV/mm. The propagation loss was 1.8 dB/cm.

15.3 D-FIBERS COUPLED TO ELECTRO-OPTIC CRYSTAL SLABS

Slab-coupled electro-optic field sensors (SCOS) have been extensively investigated by a team at Brigham Young University (Utah) [70, 270–276]. The sensors consist of a fiber with D-shaped cross-section attached to a thin electro-optic slab of refractive index n_0 and thickness t (some tens of micrometers) (Figure 15.4a). The close proximity of the fiber core to the slab results in resonant coupling and power transfer between the excited polarization

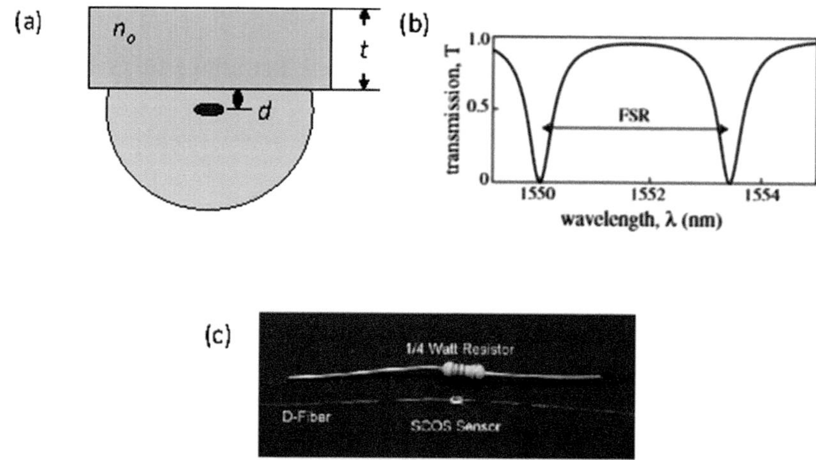

FIGURE 15.4 (a) Cross-section of an SCOS electric field sensor consisting of a D-fiber coupled to an electro-optic slab; (b) SCOS transmission spectrum; (c) Image of an SCOS sensor. (Adapted/reprinted with permission from [70, 271] © The Optical Society.)

mode of fiber and the multiple modes of the slab. (Note that a D-fiber is polarization-maintaining with eigen-polarizations parallel and perpendicular to the flat cladding surface.) To this end, the distance d between the core and flat of the pristine fiber is reduced by wet etching, e.g., from $d = 13$ μm to $d = 1.5$ μm [270]. The resonance wavelengths appear as dips in the fiber's transmission spectrum (Figure 15.4b) and are given by [70]

$$\lambda_m = \frac{2t}{m}\sqrt{n_0^2 - n^2}, \tag{15.8}$$

where n is the effective refractive index of excited fiber mode and m is the slab mode number. An applied electric field changes the slab index n_0 and shifts the resonance wavelengths. The shift can be measured as a change in the transmitted optical power at an appropriately chosen wavelength of operation; see [70] for a detailed transmission analysis. Typically, the sensor is designed to measure the field component perpendicular to the flat fiber surface. SCOS can be made very small with a cross-sectional diameter around 0.2 mm and a length of a few mm or less (Figure 15.4c). Slab materials have included poled polymers [70, 272] and inorganic crystals, in particular, lithium niobate ($LiNbO_3$) [270, 271, 274–276]. Poled polymers have the advantage of a small permittivity and large electro-optic coefficients, $e_r = 4.0$ and $r_{33} = 70$ in [70], but suffer from limited long-term stability. Sensors with lithium niobate slabs have been employed, for example, to the measurement of pulsed electric fields from 1.5 kV/mm to 18 kV/mm [274] and alternating and pulsed voltages in coaxial high voltage cables [276]. Reported accuracies were 0.7% for a 100 kV voltage pulse, and 28% and 128% for 2-kHz alternating voltages of 500 V_{pp} and 100 V_{pp}, respectively. Slab-coupled electric field sensors were also reported in [277, 278].

CHAPTER 16

Piezo-Optic Voltage Sensors

16.1 THE PIEZOELECTRIC EFFECT

The piezoelectric effect, discovered in 1880 by the French brothers Jacques and Pierre Curie, is the phenomenon that occurs when mechanical stress applied to certain crystalline materials such as quartz leads to an electric polarization of the materials proportional to the stress [279]. The stress alters the separation between positive and negative charges in each crystal unit cell, resulting in net surface charges and a macroscopic electric field across the material samples. Conversely, when an electric field is applied to a piezoelectric material, the material deforms slightly. This is known as the converse or inverse piezoelectric effect. Like the Pockels effect, the piezoelectric effect occurs in crystals with no inversion symmetry. The direct and converse piezoelectric effects have found numerous applications, e.g., in actuators, ultrasonic transducers, motors, inkjet printing, ignitors, energy harvesting, and sensors, including optical electric field and voltage transducers.

The piezoelectric strains S_{jk} induced by the components E_i of an applied electric field are given by [2]

$$S_{jk} = d_{ijk} E_i, \tag{16.1}$$

where d_{ijk} is a third rank tensor of piezoelectric coefficients. Like the tensor r_{ijk} of the Pockels coefficients, d_{ijk} can be reduced to a 2×2 matrix. Eq. (16.1) then reads

$$S_j = d_{ij} E_i, \quad (i = 1, 2, 3; j = 1, 2, \ldots, 6). \tag{16.2}$$

Here, S_1, S_2, S_3 are tensile strains in x, y, and z directions, respectively, and S_4, S_5, S_6 are shear strains about x, y, z. Optical electric field and voltage sensors convert a piezoelectric strain into an optical signal, for example, by means of a fiber interferometer, a Fabry-Pérot cavity, a fiber Bragg grating, or a displacement sensor.

DOI: 10.1201/9781003100324-20

16.2 PIEZOELECTRIC MATERIALS

Crystalline quartz and tourmaline are two classical piezoelectric materials. Quartz is best known for its use as a high-Q oscillator in watches. The most widespread piezo materials are lead zirconate titanate ceramics, in short PZT ($PbZr_xTi_{1-x}O_3$) [280]. These polycrystalline ferroelectrics of perovskite crystal structure obtain their macroscopic piezoelectric properties by electric poling of the sintered ceramics at elevated temperature. Their piezoelectric coefficients d_{ij}, also called charge constants, are 2–3 orders of magnitude larger than in quartz. The electric and mechanical properties of PZT can be adapted within wide ranges to specific applications. In particular, one distinguishes so-called hard PZT and soft PZT, as already mentioned in Section 9.2.

Soft PZT have comparatively large piezoelectric coefficients and permittivity but poorer linearity and more pronounced hysteresis, higher dielectric loss, and are more susceptible to depolarization due to smaller Curie temperatures and coercive field strengths. The opposite is generally true for hard PZT. PZT are available in many different shapes such as hollow tubes, disks, plates, or hemispheres. The high piezoelectric coefficients have also made PZT a material of choice for voltage sensing. However, PZT materials have several properties that may hamper the performance and stability of a sensor and must be taken into consideration when selecting an adequate material:

- The poling decays logarithmically over time (Arrhenius's law). Accordingly, parameters such as the charge constants d_{ij} decrease according to $d_{ij} = d_{ij}(t_0) - a(T)\log(t)$, where t_0 is the starting time, and a is a temperature-dependent aging rate coefficient (a increases with temperature) [281]. Even though the rate of change slows over time and can be partially forestalled by pre-aging, the effect should be borne in mind, especially in hot climates. Commonly, soft PZT age faster than hard PZT.
- The piezoelectric coefficients d_{ij} and dielectric constants e_{ij} have comparatively high temperature dependence, and the temperature coefficients themselves are often functions of temperature. The temperature coefficients of the d_{ij} are roughly in the range $(2–10) \times 10^{-4}\ K^{-1}$; those of the e_{ij} can even be larger [282, 283].
- Dielectric losses are relatively high and cause corresponding phase delays between input voltage and output signal. The loss factor tan δ of PZT-5A, for instance, is around 0.02 at 100 Hz and 20°C. For comparison, quartz has a loss factor of $\approx 10^{-5}$.
- The coercive field strengths E_c are typically between 0.5 and 1.5 kV/mm. The maximum field strength a sensor is exposed to should be kept sufficiently far below E_c at all conditions of operation, including potential over-voltages, to prevent depolarization and damage.
- Given the above criteria, hard PZT are commonly preferential for voltage sensing [284].

Polyvinylidene fluoride (PVDF) is another interesting piezoelectric material, as it can be produced in thin flexible foils. It is made piezoelectric by stretching and subsequent poling [285]. Table 16.1 summarizes properties of selected piezoelectric materials.

TABLE 16.1 Selected Piezoelectric Materials and Their Properties

Material	Crystallographic Point Group	Piezoelectric Coefficients d_{ij} (10^{-12} C/N)	Temperature Coefficient $(1/d_{ij})\, \partial d_{ij}/\partial T$ (10^{-4} K^{-1})	Relative Permittivity (Unclamped) e_r	Temperature Coefficient $(1/e_r)\, \partial e_r/\partial T$ (10^{-4} K^{-1})	Dielectric Loss tan δ	References
Quartz	32	$d_{11} = 2.31$ $d_{14} = -0.72$	Td_{11}: −2.15 Td_{14}: 12.9	$e_{11} = 4.52$ $e_{33} = 4.64$	Te_{11} : 0.28 Te_{33} : 0.39	10^{-5}	[286, 282]
$LiNbO_3$	3m	$d_{31} = -0.85$ $d_{33} = 6$ $d_{22} = 20.8$ $d_{15} = 69.2$	Td_{11}: −8.9 Td_{33}: 3.0 Td_{22}: 2.3 Td_{15}: 2.6	$e_{11} = 99.5$ $e_{33} = 38.5$	Te_{11} : 4.11 Te_{33} : 9.4	10^{-4}	[282, 287, 288]
ZnO	6mm	$d_{31} = -5.12$ $d_{33} = 12.3$ $d_{15} = 8.3$		$e_{11} = 8.67$ $e_{33} = 11.3$			[282, 289]
PZT - 5A	∞mm	$d_{31} = -171$ $d_{33} = 374$ $d_{15} = 584$	See text	$e_{33} = 1700$	See text	0.02 (1 kHz)	[282, 283]
PZT - 5H	∞mm	$d_{31} = -274$ $d_{33} = 593$ $d_{15} = 741$	See text	$e_{13} = 3130$ $e_{13} = 3400$	See text	0.02 (1 kHz)	[282, 283]
PVDF	2mm	$d_{31} = 18$ $d_{33} = -30$		$e_r = 12$		0.05 (100 Hz)	[282]

16.3 INTERFEROMETRIC PIEZOELECTRIC VOLTAGE SENSORS

16.3.1 Quartz Transducer with Dual-Mode Sensing Fiber

Bohnert et al. at ABB have developed interferometric high voltage sensors based on the converse piezoelectric effect in quartz. In spite of its relatively small piezoelectric coefficients, quartz offers superior material properties (see Table 16.1), including low permittivity and hence little field distortion, relatively small temperature dependence of the piezoelectric and dielectric constants, high resistivity (7.5×10^{17} Ωm) and negligible dielectric loss, chemical inertness and excellent long-term stability [290]. Quartz belongs to the crystallographic point group 32 and possesses a threefold crystal axis, conventionally referred to as the z-axis, and three twofold axes (x-axes). The matrix of piezoelectric coefficients reads [2]

$$d_{ij} = \begin{pmatrix} d_{11} & -d_{11} & 0 & d_{14} & 0 & 0 \\ 0 & 0 & 0 & 0 & -d_{14} & -2d_{11} \\ 0 & 0 & 0 & 0 & 0 & 0 \end{pmatrix}. \tag{16.3}$$

The basic transducer element is a quartz disk or cylinder with a polyimide-coated optical fiber wrapped with constant pitch and defined tension onto the circumferential surface (Figure 16.1a). The longitudinal transducer axis coincides with a twofold crystal axis (x-axis). An alternating electric field E_x applied along that axis produces alternating piezoelectric strains along y and x and hence a modulation of the transducer's circumferential length l. The relative length change, given by $Dl/l = -(1/2)d_{11}E_x$, is transmitted to the fiber and measured interferometrically. The chosen crystal orientation is selected because the electric field's components in the plane of the disk (E_y and E_z) to a first order,

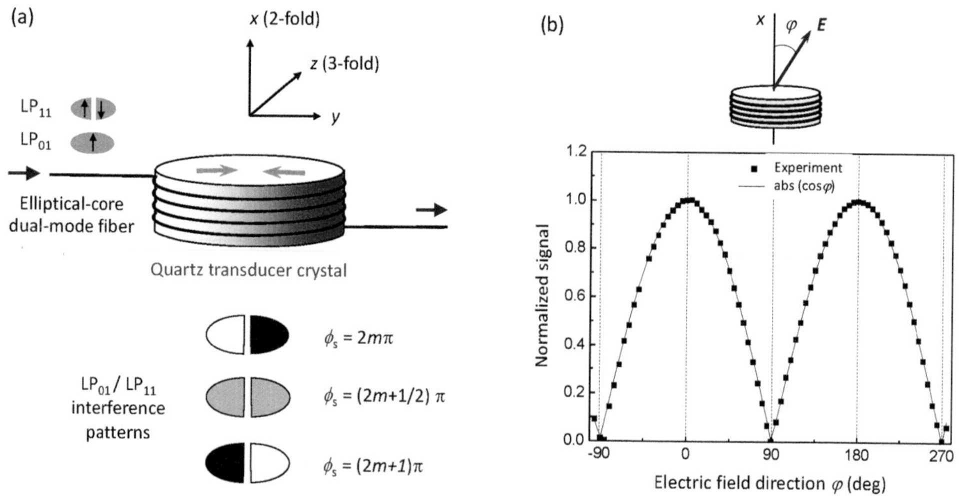

FIGURE 16.1 (a) Quartz transducer crystal with attached elliptical-core dual-mode fiber and LP_{01}/LP_{11} modal interference patterns (in the near field) at characteristic phase differences ϕ_s. An electric field in positive x-direction causes compressive strain along y (green arrows). (b) Sensor signal (absolute value) vs. angle φ between electric field and transducer normal (2-fold axis) (Adapted/reprinted with permission from [291] © The Optical Society.)

do not contribute to the signal. Hence, the transducer is only sensitive to E_x (Figure 16.1b) [291]. Hence, a sensor comprising a series arrangement of multiple quartz elements (with optional dielectric spacers between the elements) or a long solid quartz cylinder measures the line integral of the field, as demonstrated in [292]. Alternatives to quartz disks and cylinders are quartz slabs with the fiber attached in a straight line. Further crystal classes suitable for field integration and corresponding transducer/fiber orientations are listed in [291], for both disk- and slab-shaped transducers.

The sensing fiber has an elliptical-core and supports at the wavelength of operation two spatial modes, namely the fundamental LP_{01} mode and even LP_{11} higher order mode [293, 294]. The modes are polarized either along the major or minor core axis. The alternating fiber strain results in a modulation of the differential phase of the two modes and is detected by low coherence interferometry (Figure 16.2) [83, 295, 296]. To this end, a second dual-mode fiber of the same type, being part of the detection system, acts as a receiver interferometer. The light source is a low-coherent 780-nm multimode laser diode with 4-mW rated output power. Insensitive single-mode elliptical-core fibers guide the light to and from the remote sensor head. The different fiber sections are joined by fusion splices made with lateral offsets such that the LP_{01} mode of the single-mode fibers couples approximately equally to the LP_{01} and the LP_{11} modes of the dual-mode fibers, or vice versa. The two modes of the sensing fiber propagate with different group velocities and accumulate an optical path difference DL_s, which falls into a minimum of the multiple-peak fringe visibility function of the laser diode [297, 298], that is, the modes are incoherent at the sensing fiber end. The relative modal delay DL_r in the receiver fiber matches DL_s, which restores partial coherence at the end of the receiver fiber. Per meter of fiber, the modes accumulate delays $DL_{s,r}$ of about 3–4 mm depending on the fiber parameters [83, 296]. Two photodiodes detect the two anti-phase lobes of the modal interference pattern with intensities $I_{1,2}$:

$$I_{1,2} = \frac{I_0}{2}\left[1 \pm K\cos\left(\phi_r - \phi_s\right)\right], \tag{16.4}$$

where I_0 is proportional to the light source power, K is the fringe visibility of the partially coherent light ($K = 1/2$ at ideal conditions), and ϕ_r, ϕ_s are the relative modal phases in the

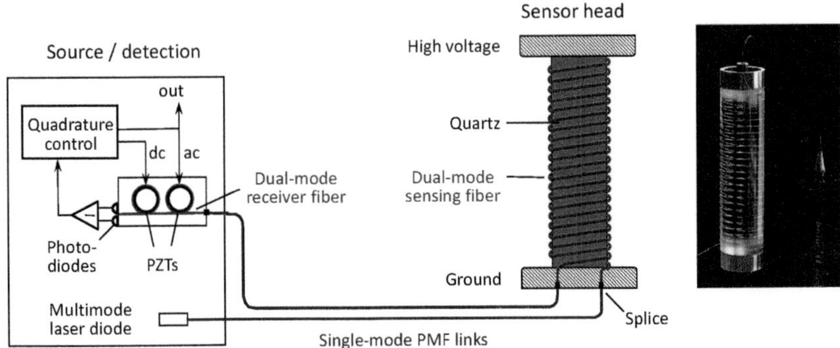

FIGURE 16.2 Piezo-optic voltage transducer; PMF: polarization maintaining fiber. (Adapted with permission from [83] © The Optical Society.)

two dual-mode fiber sections. Phase ϕ_s is composed of a quasi-static term θ_s that varies with temperature and an alternating term $\phi_{0s}\sin\varpi t$ with amplitude ϕ_{0s} produced by the applied ac voltage:

$$\phi_s(t) = \theta_s(t) + \phi_{0s}\sin\varpi t. \tag{16.5}$$

The receiver dual-mode fiber is wrapped onto two tube-type PZT transducers that are part of a closed-loop homodyne phase tracking system [299]. The phase tracker controls the receiver fiber phase ϕ_r such that the intensities I_1 and intensities I_2 of the two interference lobes are kept balanced. The combined phase difference $\phi_r - \phi_s$ is then locked at quadrature, and the feedback voltage from the phase tracker images the phase modulation ϕ_s(t) of the sensor fiber. Large quasi-static phase excursions due to temperature, $\theta_s(T)$ or $\theta_r(T)$, initiate 2p-phase resets. It should be mentioned that in [183, 300, 301], adapted versions of non-reciprocal phase modulation, equivalent to the one in Figure 14.12, were presented as an alternative to coherence-tuned interrogation.

16.3.2 170 kV POVT for Gas-Insulated Switchgear

Figure 16.3 shows a piezo-optic voltage transducer (POVT) for 170 kV gas-insulated switchgear (GIS) [83]. The full phase-to-ground voltage (170 kV/$\sqrt{3}$) is applied to a quartz crystal with a length of 100 mm, a diameter of 30 mm, and 24 bifilarly wound loops of sensing fiber (see photograph in Figure 16.2). The crystal resides in a compact housing that can be plugged into accordingly prepared GIS. An insulator tube from fiber-reinforced epoxy connects the high voltage and ground potential parts of the housing. Pieces of silicone foam help to insulate the crystal from mechanical shock and vibration. (A method for canceling mechanical noise has been disclosed in [301].) The tube interior stands in gas exchange with the GIS system. The housing and the electrodes are designed such that the field strength at a 750-kV lightning impulse voltage (the highest test voltage according to IEC standards [94]) does not exceed an empirically determined safety limit of 23 kV/mm at any location in the

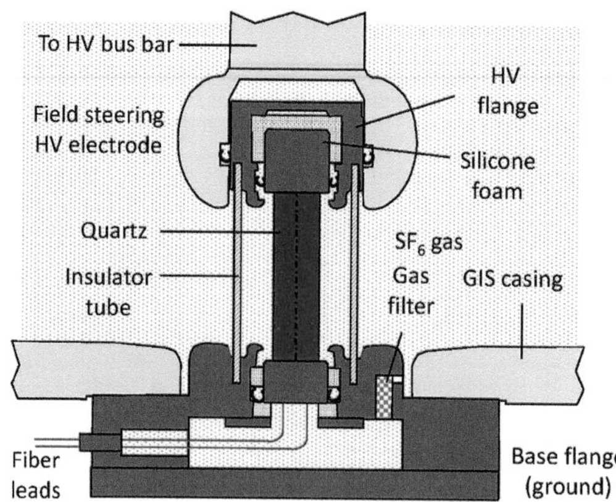

FIGURE 16.3 Piezo-optic voltage transducer for 170 kV gas-insulated switchgear (sensing fiber not shown). The overall height is 260 mm. (Reprinted with permission from [83] © The Optical Society.)

SF6 gas volume and of 15 kV/mm within the epoxy insulator. For instance, the maximum field strengths at 750 kV are 22.7 kV/mm at the HV electrode, 9.6 kV/mm on the quartz surface, and 10.7 kV/mm at the insulator tube. The fiber causes local field enhancements by a factor of 1.64. The transducer passed the required HV tests according to IEC standards that included an over-voltage withstand test (50 Hz ac voltage of 365 kV applied for 2 min) and lightning impulse tests (15 repetitive 750-kV impulses of positive and negative polarity). It was also confirmed that the transducer was free of partial discharge (PD) at the 1.1-fold of the rated voltage before and after the tests. PD onset and breakoff were above 200 kV.

The transducer was successfully tested in laboratory experiments and an extended field trial at a GIS substation in Germany (Figure 16.4) [83]. The minimum detectable voltage corresponded to 11.3 V/$\sqrt{\text{Hz}}$ at 60 Hz and was equivalent to a modal phase modulation ϕ_s of 2×10^{-5} rad/$\sqrt{\text{Hz}}$ at 60 Hz. Above 1 kV, the deviations from a reference voltage probe were within ±0.2% (Figure 16.4a,b). The phase modulation amplitude at the rated phase-to-ground voltage (98.1 kV) was 0.173 rad, corresponding to an rms fiber length change δl_{2p} of 93 mm, here for the polarization parallel to the major core axis. The signal was essentially frequency independent up to the first piezoelectric resonances of the quartz transducer above 10 kHz. Crosstalk between two neighboring sensors was below 5×10^{-4}. The signal at constant voltage decreased with temperature at a rate of -2.12×10^{-4} K^{-1} (Figure 16.4c), governed by the temperature dependence of the d_{11} piezoelectric coefficient of quartz (Table 16.1), i.e., there was no significant contribution from the polyimide-coated sensing

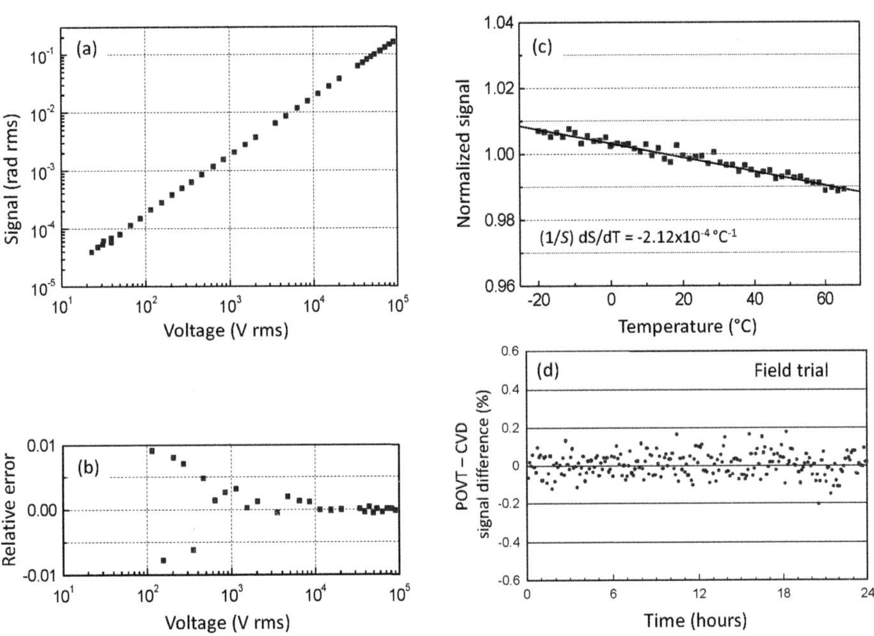

FIGURE 16.4 170 kV POVT test data: (a) Signal vs. voltage; (b) Relative error vs. voltage; (c) Signal at constant voltage vs. temperature; (d) Relative signal difference between optical and conventional voltage transducers versus time during a field test in 110 kV GIS (63.5 kV phase-to-ground). (Reprinted with permission from [83] © The Optical Society.)

fiber. Temperature can be accounted for either by a temperature measurement or by making use of the temperature-dependent interference fringe contrast [302]. The field trial confirmed accuracy to within ±0.2% (Figure 16.4d).

16.3.3 420 kV POVT for Air-Insulated Substations

Figure 16.5 depicts a 420-kV POVT for outdoor air-insulated substations with maintenance-free solid-state insulation [303]. Here, the line voltage (242.5 kV phase-to-ground) drops across a series of four quartz crystals, each with a length of 300 mm and a diameter of 30 mm. The crystals share a common sensing fiber with 22 fiber loops per crystal. Electrodes at the crystal ends provide a relatively homogeneous field along the crystals. Aluminum tubes with flexible joints establish electric contact between the crystals and to ground and high voltage potentials. Partitioning of the voltage on several crystals reduces the electric field strengths and makes it possible to mount the whole assembly in a slender and light insulator tube with a length of 3.2 m and inner and outer diameters of 80 mm and 92 mm, respectively. Soft polyurethane resin fills the remaining hollow volume in the insulator. The sum of the individual voltages V_i across the quartzes corresponds to the total line voltage V_0. The individual voltage fractions V_i can be controlled via the quartz positions on the insulator axis and the geometry of the corona ring depicted in Figure 16.5. Ideally, all V_i are equal to $V_0/4$. From top to bottom, the actual voltage drops are $V_1 = 0.30V_0$, $V_2 = 0.282V_0$, $V_3 = 0.201V_0$, and $V_4 = 0.217V_0$. For the given number of quartzes, the insulator dimensions have been chosen such that the electric field strength at the peak of a 1425 kV impulse voltage (IEC test requirement) does not exceed an empiric limit of 3.5 kV/mm at any surface exposed to air; for details, see [303]. The transducer passed the IEC-required high voltage tests (320 kV ac overvoltage, ±1425-kV lightning impulses, and ±1050-kV switching

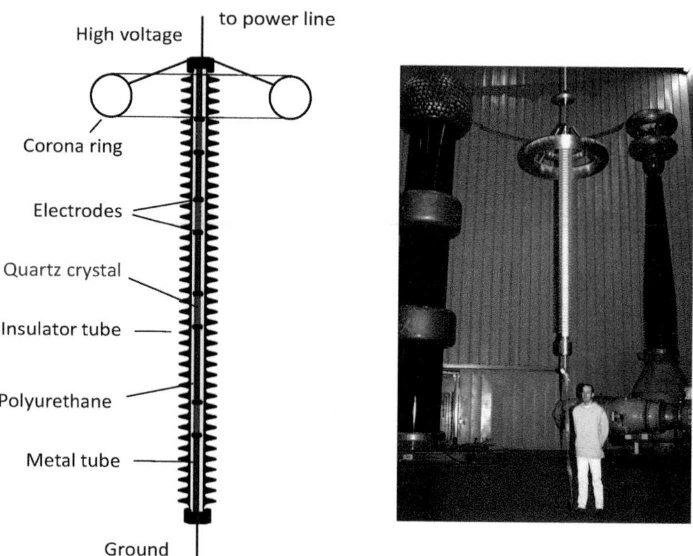

FIGURE 16.5 Piezo-optic voltage transducer for 420 kV outdoor substations [303].

impulses at dry and wet conditions). It should be mentioned, that even though successfully tested, the POVT concept was eventually abandoned in favor of ABB's Electro-Optic Voltage Transducer (EOVT) described in Section 14.3.1, as the EOVT was already more established at the time. PZT-type sensors with interferometric readout were studied, for example, in [304–306].

16.4 PIEZOELECTRIC VOLTAGE SENSORS WITH FBG INTERROGATION

Fiber Bragg gratings (FBG), for the most part, replaced interferometric techniques for piezo-optic voltage transducers after gratings became readily available in the 1990s. Pacheco et al., in 1999, were among the first to demonstrate a PZT/FBG transducer [284]. The FBG was radially attached to a flat surface of a 6.33-mm-thick PZT disk, which resulted in a linear output up to 6.3 kV with a resolution of 150 V. Extensive further work towards PZT/FBG-based current and voltage transducers for electric power networks has been carried out at Strathclyde University, Glasgow, and its spin-off, Synaptec Ltd, for example, [307–313]; see also Section 9.2 on corresponding current sensors. The voltage is either directly applied to the transducer [311, 312] or, at higher voltages, the transducer is operated in conjunction with a capacitive voltage divider (CVD). Figure 16.6 shows a transducer module for combination with a CVD [313]. The CVD output is applied to a rod-shaped PZT (PIC181 from Physik Instrumente; a hard PZT material) with a length of 20 mm and a diameter of 5 mm, a piezoelectric charge constant d_{33} of 265 pm/V, and a maximum permissible field strength of 2.5 kV/mm. The FBG is suspended between two ceramic holders attached to the electrodes protruding over the PZT end faces. At the rated line voltage (here, 76 kV phase-to-ground, 132 kV phase-to-phase), the voltage seen by the transducer is 1 kV and results in peak-to-peak Bragg wavelength shifts of ±39 pm, measured with a commercial FBG interrogator. Laboratory tests at room temperature showed that the transducer module

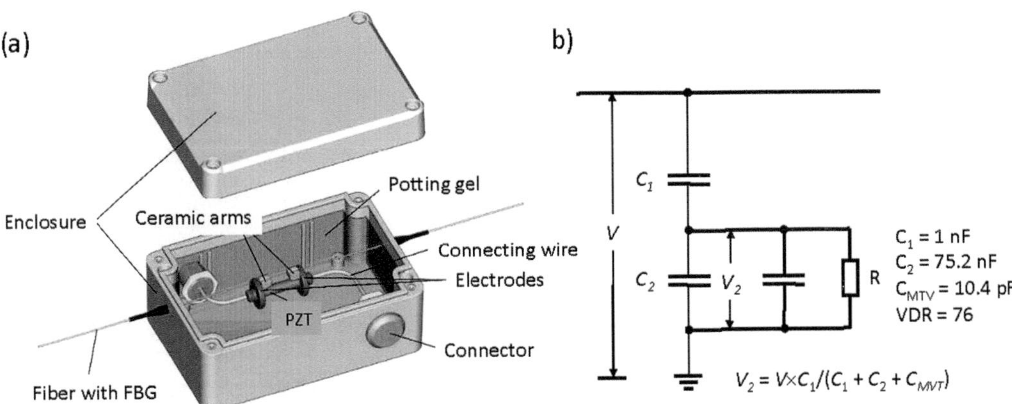

FIGURE 16.6 (a) PZT/FBG-transducer module for combination with a capacitive voltage divider (CVD); (b) CVD parameters for a voltage divider ratio (VDR) of 76. The output voltage V_2 applied to the transducer is 1 kV at the rated phase-to-ground voltage V of 76 kV. (Adapted from [313].)

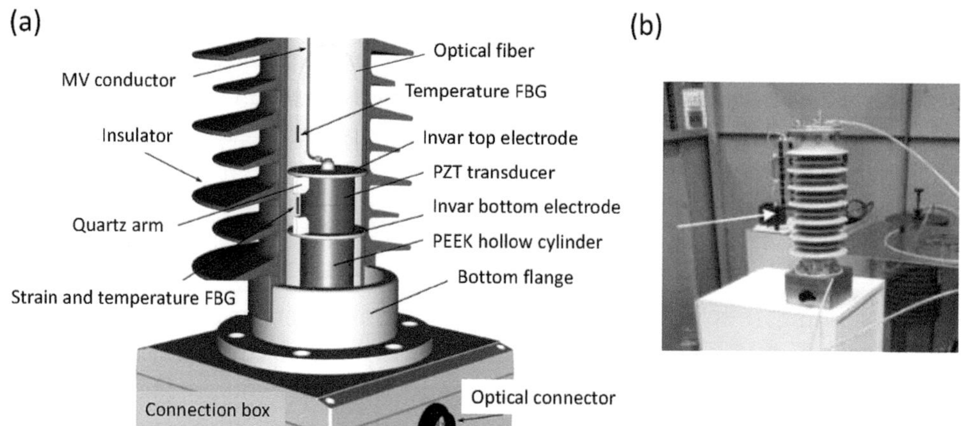

FIGURE 16.7 (a) PZT transducer with attached FBG in hollow core medium voltage insulator; (b) photograph of the device. (Adapted from [312].)

can meet the IEC requirements of metering class 0.1 and protection class 1P, as specified in [94]. Hysteresis of the hard PZT material was found to be minimal.

Figure 16.7 depicts a medium voltage transducer from the same team, where a 40-mm-thick PZT cylinder (PIC181) resides in a medium voltage hollow core insulator, and the line voltage (6.35 kV phase-to-ground, 11 kV phase-to-phase) is directly applied to the PZT [311]. The FBG is mounted at the PZT, as in Figure 16.6. An improved version with PZT protection against lightning impulses was reported in [312].

Similar work towards PZT/FBG voltage sensors for MV power distribution was also reported by Werneck et al. at the Federal University of Rio de Janeiro [314–316]. As an example, Figure 16.8 depicts a transducer for power quality monitoring in 13.8 kV

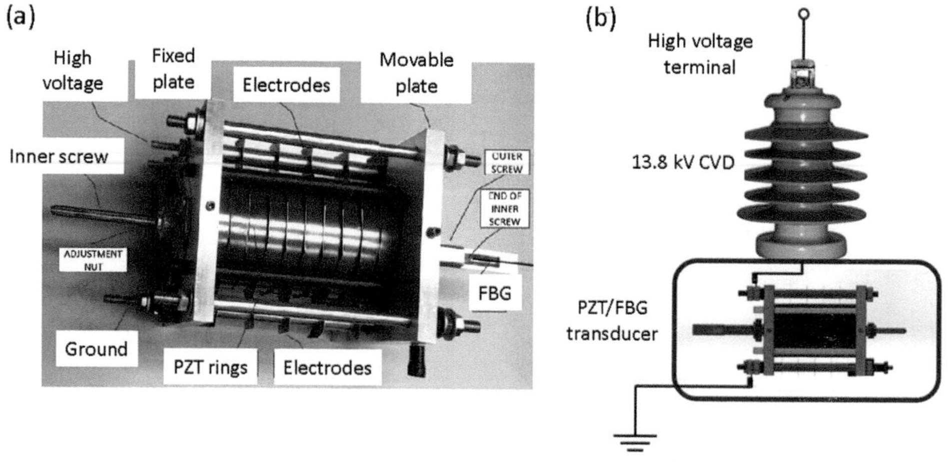

FIGURE 16.8 (a) PZT/FBG voltage transducer comprising a stack of 10 PZT rings; (b) combination of the transducer with a 13.8 kV capacitive voltage divider (CVD). (Adapted from [316].)

networks, to detect voltage sags or swells, for instance, unbalances between phases, or line interruption [316]. The transducer consists of a stack of 10 PZT rings with inner and outer diameters of 38.0 and 50.8 mm, respectively (PZT-4, again a hard PZT material with d_{33} = 300 pm/V). The total height of the stack is 80 mm. The PZT rings are electrically connected in parallel, that is, each ring is exposed to the external voltage. The fiber with the FBG runs along the longitudinal axis of the stack. A special mechanism transmits the PZT strains to the FBG with a sensitivity of 232 pm/kV. The maximum allowed transducer voltage is 4 kV. At 13.8 kV, the transducer is therefore combined with a CVD, as illustrated in Figure 16.8b. The experimental evaluation showed that the transducer was capable of reproducing transient voltage variations up to the 41st harmonic without significative distortion and impulsive surges up to 2.5 kHz. In still other work, He et al. reported multiplexed PZT/FBG voltage transducers [317–319]. Similar work is also known from Law et al. [320].

16.5 LIGHT INTENSITY MODULATING VOLTAGE SENSORS

Here, the PZT vibration caused by an applied ac voltage modulates the optical power transmitted by the transducer. In the example of Figure 16.9, reported by Lukens et al., light emitted from the central fiber of a multimode fiber bundle reflects off from a PZT surface (PZT-4 with dimensions 12 × 1.5 × 0.5 mm^3) and is collected by six fibers surrounding the emitting fiber [321]. The collected light power is a function of the gap width between the PZT and fiber bundle and thus varies as a function of the voltage — Section 9.3 presented an analog sensor for current measurement. With the steady state gap width adjusted to 280 mm, the minimum detectable displacement was 1.5 nm at 100 Hz. The maximum displacement was chosen as 3 mm and corresponded to an applied voltage of 20 V. Higher voltages again require a CVD. The frequency response was flat from 10 Hz to 500 Hz and had a resonance at 2 kHz. Other PZT-based intensity modulating voltage sensors were reported, for example, in [322, 323].

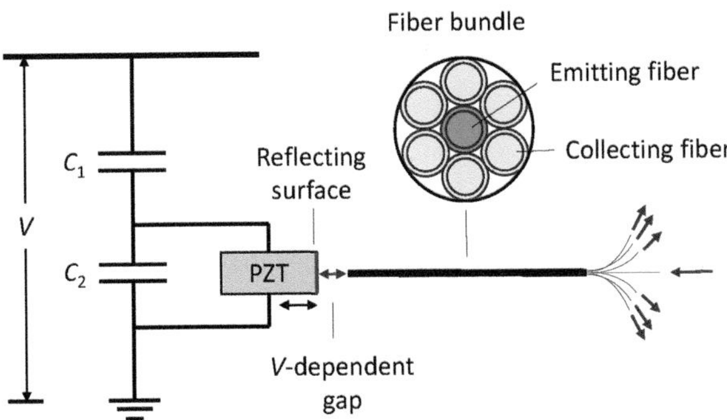

FIGURE 16.9 Intensity modulating voltage sensor in combination with a capacitive voltage divider, according to [321].

16.6 PIEZOELECTRIC POLYMER-JACKETED FIBERS

In the 1980s, Koo et al. and others explored fibers attached to poled PVDF films or with poled PVDF jackets for electric field sensing and optical phase modulation [324–328]. The fibers were interrogated in a Mach-Zehnder interferometer and sensitive to a field component perpendicular to the fiber. Imai et al. reported a field-induced phase shift of 1.5×10^{-7} rad/(V/m) per meter of vinylidene fluoride/trifluoroethylene-jacketed fiber [328], and Mermelstein et al. estimated a minimum detectable field of 33 mV/m for a 1-km-long fiber on PVDF film [327]. The field became dormant in later years, however.

CHAPTER 17

Further Electric Field and Voltage Sensors

The following sections are short overviews of further methods for optical electric field and voltage sensing. For details, the reader is referred to the cited literature.

17.1 ELECTROSTRICTIVE SENSORS

The electric field-induced strains in a medium commonly include, in addition to the first-order terms associated with the converse piezoelectric effect (if present), terms that also vary in proportion to the square of the field. The latter phenomenon is referred to as electrostriction. In a sense, electrostriction is the electric analog to magnetostriction (Section 10.1). Whereas the piezoelectric effect only occurs in materials with no inversion symmetry, electrostriction occurs in all crystal classes as well as in isotropic media such as glasses and liquids. In order to include electrostriction, eq. (16.1) of the converse piezoelectric effect must be supplemented as follows [2]:

$$S_{jk} = d_{ijk} E_i + \gamma_{iljk} E_i E_l, \tag{17.1}$$

where γ_{iljk} is a fourth rank tensor symmetrical in i and l and in j and k. In isotropic materials, electrostriction results in an elongation in the direction of the field (independent of the sign of the field) and contraction in orthogonal directions according to the Poisson ratio. The relative elongation is $Dl/l = \gamma E^2$, where γ is the relevant electrostrictive coefficient. While electrostriction is rather small in most materials, the electrostriction of so-called relaxor ferroelectrics that include, e.g., lead magnesium niobate (PMN) or lead lanthanum zirconate titanate (PLZT) ceramics, is sufficiently large for practical applications [329].

Vohra et al. at the Naval Research Laboratory, Washington, D.C., have developed interferometric dc and low frequency field and voltage sensors, as well as ac sensors based on electrostrictive ceramics [330–333]. As in the case of magnetostrictive sensors (Section 10.1), a carrier field E_w of high frequency ϖ is superimposed to the low frequency field E_Ω to be measured. Due to the nonlinearity, the low frequency field can be detected as

DOI: 10.1201/9781003100324-21

sidebands ($\varpi \pm \Omega$) of the carrier in a frequency regime far away from low frequency noise, which strongly improves the strain resolution. The strain-induced optical phase shifts $f_{w \pm \Omega}$ at the sideband frequencies are proportional to the product $E_w E_\Omega$. DC fields E_{dc} generate a phase shift f_w at the carrier frequency proportional to $E_w E_{dc}$. Operating the transducers in a fiber Mach-Zehnder interferometer, Fabiny et al. measured voltage resolutions of 40 nV/$\sqrt{\text{Hz}}$ at 1 Hz and 129 nV/$\sqrt{\text{Hz}}$ at 0.03 Hz [332]. Vohra et al. demonstrated a field resolution of 7 (mV/m)/$\sqrt{\text{Hz}}$ for an ac electric field sensor operated at a transducer resonance (32.5 kHz) and biased with 100 Vdc. The equivalent voltage resolution was 1 nV/$\sqrt{\text{Hz}}$ [331].

Tomic et al. reported an electrostrictive sensor for 50-Hz voltages comprising a multilayer PMN ceramic rod with a length of 10 mm. The researchers applied voltages up to 50 V parallel to the longitudinal rod axis (without a bias field) and measured the electrostrictive rod elongations with a displacement sensor based on a 3 × 3 optical fiber coupler operated at 1310 nm. They achieved a voltage resolution of about 10 mV (corresponding to a rod elongation of 1 nm); the temperature dependence was 2.5%/°C [334]. Electrostrictive field sensors employing FBG for readout were reported in [335, 336].

17.2 LIQUID-CRYSTAL-BASED SENSORS

Liquid crystals (LC) [337] are best known for their use in flat panel displays. However, the variation of the optical anisotropy of LC as a function of an applied electric field also lends itself to electric field and voltage sensing. LC electric field sensors mostly work with nematic LC (NLC). (Other well-known LC phases are the smectic and cholesteric phases.) The rod-shaped NLC molecules lack positional order but tend to align in a common direction (called the director) over macroscopic scales. Most nematic phases are uniaxial (others are biaxial), i.e., they have ordinary and extraordinary refractive indices, n_o and n_e, with the director representing the optical axis (see indicatrix of Figure 2.6). In an electric field, the LC molecules tend to align in the direction of the field. Probe light experiences the realignment as a change in the effective NLC index or birefringence. It should be mentioned, however, that the optical response often exhibits hysteresis, strongly depends on temperature, and is comparatively slow. Early LC electric field sensors comprised simple LC cells with fiber pigtails, for instance, for monitoring the presence of voltage in HV equipment [338, 339]. Other researchers have dispersed LC micro-droplets in an isotropic polymer matrix (PDLC) [340–342]. Without an applied field, the orientations of the LC molecules vary from droplet to droplet. Refractive index mismatches between the LC droplets and the polymer matrix then attenuate the transmitted light by scattering. An applied field parallel to the light propagation direction forces the molecules to align in this direction. Hence, the effective indices of the droplets approach the ordinary index n_o, which matches the polymer index. As a result, the PDLC transmission increases. PDLCs have been designed as thin films [340, 342] or used as a cladding material on short sections of fiber [341]. Scherschener et al., for instance, reported a PDLC thin film sensor (10-mm thick, indium tin oxide electrodes) for 50-Hz ac voltages between 0 and 10 V [342]. A dc offset-voltage of 40 V moved the operating point into the linear regime of the sensor. The measured sensor response at 633 nm exhibited good linearity, but the slope strongly depended on temperature. The response time was about 0.5–1 s.

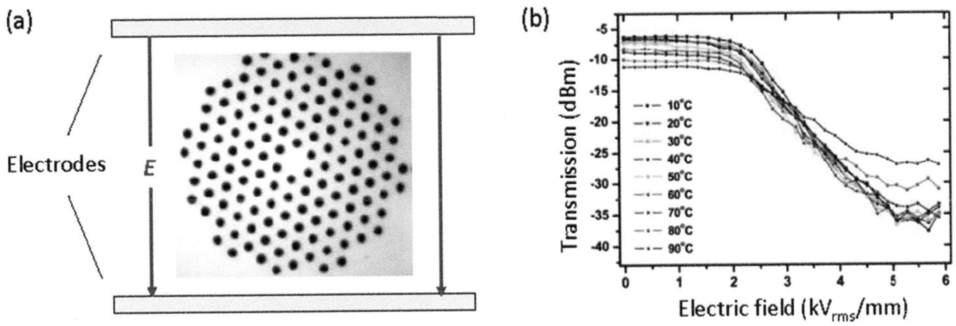

FIGURE 17.1 (a) PCF with NLC infiltrated air holes; (b) transmission at 1550 nm vs. ac electric field strength at various temperatures between 10°C and 90°C. (Reprinted with permission from [347] © The Optical Society.)

Further research has been focused on micro-structured optical fibers (MSOF), in particular photonic crystal fibers (PCF), with LC infiltrated air holes. LC infiltrated PCF have found significant interest, because their spectral and polarization properties are tunable by temperature or external electric fields; see, for example, [343] for a review. Not surprisingly, LC infiltrated PCF also have been explored as electric field and voltage sensors, for example, in [343–352]. Typically, the devices translate an applied field into a polarimetric optical phase shift (in case of a polarization maintaining fiber) [344, 345], a shift of the optical loss spectrum [348, 350–352], or optical attenuation at a fixed wavelength [346, 347]. As an example, Figure 17.1 depicts data reported by Mathews et al. [347]. The sensor comprised a PCF with solid 8.5-mm diameter core surrounded by 7 rings of 2.5-mm diameter air holes (Figure 17.1a). The air holes of the fiber were infiltrated with nematic liquid crystal over a length of about 5 mm by capillary action. AC electric fields of positive polarity up to 6 kV/mm and frequencies of 50 Hz and 1 kHz were applied. Above a threshold field of about 2.5 kV/mm, the transmission decreased by about 15–30 dB depending on temperature (Figure 17.1b). The threshold field depended on the size of the air holes and should decrease with increasing hole size. Electric field sensors based on fiber gratings embedded in liquid crystals were investigated in [353, 354], and Ko et al. presented a NLC Fabry-Pérot electric field sensor interrogated by a wavelength-swept laser [355].

17.3 MICROELECTROMECHANICAL SENSORS

Microelectromechanical sensors (MEMS) for electric fields and voltages make use of the deflection of mechanical micro-structures due to field-induced charge displacement and electrostatic forces. Such structures include diaphragms, interleaved combs, or cantilever beams. Mostly, the sensors are electrically interrogated, e.g., based on capacitance changes or piezo-resistivity [356, 357], but optically interrogated sensors are also known. Here, the deflection typically tunes a Fabry-Pérot cavity or is measured by optical position detection means [358–364], or the device acts as a field-controlled variable shutter for transmitted light [365]. The sensors are of interest, for example, to monitoring atmospheric electrostatic fields or fields under high voltage power lines.

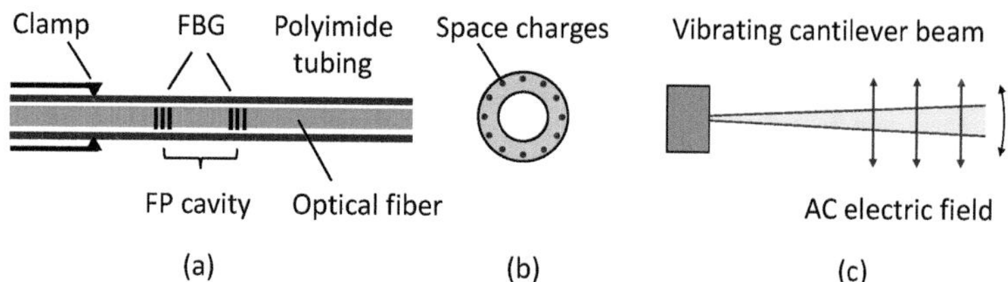

FIGURE 17.2 (a) Fiber cantilever electric field sensor with in-fiber Fabry-Pérot cavity and electrically charged polyimide (PI) tubing; (b) sketch of space charges in PI tubing; (c) scheme of vibrating sensor in an ac electric field. (Adapted from [364].)

Already in 1993, Mendez et al. reported a fiber tip Fabry-Pérot diaphragm microsensor for HV fields up to 300 kV/m, but still with a modest resolution of 40 kV/m [358]. Priest et al. measured electric fields between 135 kV/m and 650 kV/m near the sphere of a Van de Graaff voltage generator by means of a thin aluminum cantilever and low coherence Fabry-Pérot interferometry [360]. Javernik et al. reported a micro-machined fiber Fabry-Pérot sensor for dc and ac voltages up to 1 kV with resolutions between 0.1 V and 5 V [363], and Kainz et al. presented a field-controlled MEMS optical shutter with a resolution of 100 (V/m)/$\sqrt{\text{Hz}}$ up to fields of tens of kV/m and frequencies up to 300 Hz [365]. Wang et al. presented an electric field sensor consisting of an optical fiber cantilever with a thin pre-charged polyimide jacket (Figure 17.2) [364]. By appropriately choosing the cantilever length, the sensor resonance frequency was adjusted to electric power frequencies of 50 Hz or 60 Hz. Two fiber Bragg gratings formed an in-fiber Fabry-Pérot cavity to measure the cantilever vibration. The sensor had a linear response for fields up to 16 kV/m and saturated above 30 kV/m. The minimum detectable field was 0.16 V/m.

17.4 ELECTRO-LUMINOPHORES AND ELECTRO-ABSORBERS

Electro-luminophores emit light when exposed to an alternating electric field. The light intensity is proportional to $\exp(-b/E)$, where E is the electric field strength and b a material constant. Typical electro-luminophores are doped II–VI semiconductors such as ZnS:Cu emitting between 500 nm and 600 nm [366]. Electro-luminophores can be utilized to visualize surface electric fields, as demonstrated by Yang et al. [367]. They spray-coated an insulator disk with epoxy containing ZnS:Cu powder. The disk with inner and outer diameters of 20 mm and 60 mm, respectively, separated two coaxial cylindric electrodes. The arrangement resembled the geometry of gas-insulated switchgear. At ac voltages between 20 kV and 48 kV, the recorded emission images yielded the electric field strength with errors within 6%. Pustelny measured electric fields up to 1.3 kV/mm based on the emission of a fiber-coupled ZnS:Cu cell. The accuracy was within ±1% above 1 kV/mm and a few percent below [366]. An InP multiple quantum well waveguide acting as an electro-absorber with electric field dependent optical attenuation was reported by Heinzelmann et al. [368]. The device was designed for broadband operation up to the microwave regime.

17.5 STARK EFFECT-BASED ELECTRIC FIELD MEASUREMENT

The Stark effect, named after the German physicist Johannes Stark, describes the shifting and splitting of atomic and molecular energy levels in an applied electric field, which leads to modifications of the dispersion and absorption spectra of the medium. Reports on electric field measurements based on the Stark effect are rare. Höjer et al. reported spatially resolved field measurement in SF_6-gas-insulated high voltage components based on the Stark effect [369]. In their experiments, they added ammonia (NH_3) as a trace gas to the insulating SF_6-gas and probed with a CO_2 laser the strong field-induced NH_3 absorption changes around 9.5 mm, thus achieving a lateral spatial resolution of 1 mm and a field resolution of 0.25 kV/cm at fields up to 45 kV/cm.

17.6 ELECTRIC FIELD-INDUCED SECOND HARMONIC GENERATION

Electric field-induced second harmonic generation (EFISHG) is a third-order optical nonlinearity, which generates in the presence of a dc (or low frequency) electric field $E^{(dc)}$ second harmonic light of frequency 2ϖ from optical fields at a fundamental frequency ϖ. According to eq. (13.2), the induced polarization of the material at frequency 2ϖ is

$$P_i(2\varpi) \propto \chi^{(3)}_{ijkl}(2\varpi;0,\varpi,\varpi)E^{(dc)}_j E_k(\varpi)E_l(\varpi), \tag{17.2}$$

where $E^{(dc)}_j$ is the external electric field, and $E_{k,l}(\varpi)$ are optical fields at frequency ϖ. The intensity of the second harmonic is proportional to the square of the polarization:

$$I(2\varpi) \propto |P_i(2\varpi)|^2. \tag{17.3}$$

Efficient EFISHG requires high electric field strengths, typically above 1 kV/mm, and short optical pulses and has mostly been used for scientific studies, e.g., to measure fields in semiconductor quantum wells [370], piezoelectric fields in semiconductors [371], fields in plasma engineering [372], or 3D fields in liquid crystals [373], to name a few examples.

17.7 HYBRID VOLTAGE TRANSDUCERS

Hybrid voltage transformers utilize, like the hybrid current transformers of Chapter 9, conventional techniques to first convert a high voltage into an electric signal. Low power electronics then encode that signal into an optical signal and transmit it via an optical fiber ground. Werneck et al. have developed combined hybrid voltage and current transducers for 13.8 kV medium voltage distribution lines with currents up to 800 A [374]. The clamp-on type transducers can be mounted on live power lines. Resistors on a printed circuited board inside an insulator convert the voltage between two electric phases into a small electrical current. A battery-powered transmitter converts that current together with the signal from an inductive current probe into a pulse-frequency-encoded signal.

The encoded signal controls an 860-nm LED for transmission to ground via a multimode fiber cable. The reported accuracy is within ±2.5% for the voltage transducer and ±1.5% for the current transducer.

17.8 NON-OPTICAL NON-CONVENTIONAL SENSORS

Besides optical electric field and voltage sensors, a variety of non-conventional electric sensors have been developed for MV and HV power line monitoring. Typically, the devices are not meant for metering or relaying but are economic means for grid supervision. Besides voltage, they also often monitor current, temperature, and other parameters. Line-mounted sensors capacitively extract a signal representing the electric field near the line and transmit the signal wirelessly to a base station [375], while harvesting the required energy from the power line. Some sensor types utilize a surface acoustic wave (SAW) tag and radio frequency readout (RFID) of a pulse position modulated (PPM) signal [376, 377]. Others have developed ground-based field sensor arrays that estimate power line voltages from the electric fields beneath the lines; see, for example, [378]. These topics are beyond the scope of this book, however, and interested readers are referred to the extensive literature on the matter.

FURTHER READING

Selected topics of optical electric field and voltage sensing have been reviewed in [343, 379–385].

References

1. Y. R. Shen, *The Principles of Nonlinear Optics* (John Wiley and Sons, 1984), Chap. 4, pp. 53–57.
2. J. N. Nye, *Physical Properties of Crystals* (Oxford University Press, 1957, 1985).
3. A. Yariv, *Quantum Electronics* (John Wiley and Sons, 1975), Chap. 14.
4. L. Duvillaret, S. Rialland, and J.-L. Coutaz, "Electro-optic sensors for electric field measurements. II. Choice of the crystals and complete optimization of their orientation," *J. Opt. Soc. Am. B* 19, 2704–2715 (2002).
5. N. W. Ashroft and N. D. Mermin, *Solid State Physics* (Holt, Rinehart and Winston, 1976), Chap. 5.
6. K. Bohnert and J. Nehring, "Method and device for the optical determination of a physical quantity," U.S. Patent 5,715,058 (1998).
7. S. Namba, "Electro-Optical Effect of Zincblende," *J. Opt. Soc. Am.* 51, 76–79 (1961).
8. C. Li and T. Yoshino, "Simultaneous measurement of current and voltage by use of one bismuth germanate crystal," *Appl. Opt.* 41, 5391–5397 (2002).
9. Q. Chen, M. Tani, Z. Jiang, and X.-C. Zhang, "Electro-optic transceivers for terahertz-wave applications," *J. Opt. Soc. Am. B* 18, 823–831 (2001).
10. L. Duvillaret, S. Rialland, and J. L. Coutaz, "Electro-optic sensors for electric field measurements I. Theoretical comparison among different modulation techniques," *J. Opt. Soc. Am. B* 19, 2692–2703 (2002).
11. A. J. Vickers, R. Tesser, R. Dudley, and M. A. Hassan, "Fabry-Perot enhancement electro-optic sampling," *Opt. Quantum Electron.* 29, 661–669 (1997).
12. C. H. Bulmer, W. K. Burns, and R. P. Moeller, "Linear interferometric waveguide modulator for electromagnetic-field detection," *Opt. Lett.* 5, 176–178 (1980).
13. K. S. Lee, "Electrooptic voltage sensor: birefringence effects and compensation methods," *Appl. Opt.* 29, 4453–4461 (1990).
14. G. A. Massey, D. C. Erickson, and R. A. Kadlec, "Electromagnetic field components: their measurement using linear electrooptic and magnetooptic effects," *Appl. Opt.* 14, 2712–2719 (1975).
15. J. A. Osborn, "Demagnetizing factors of the general ellipsoid," *Phys. Rev.* 67(11–12), 351–357 (1945).
16. A. Garzarella and D. H. Wu, "Optimal crystal geometry and orientation in electric field sensing using electro-optic sensors," *Opt. Lett.* 37, 2124–2126 (2012).
17. C. P. Yakymyshyn, M. A. Brubaker, P. M. Johnston, and C. Reinbold, "Manufacturing challenges of optical current and voltage sensors for utility applications," *Proc. SPIE 3201, Sensors and Controls for Advanced Manufacturing*, 1998, doi: 10.1117/12.298003
18. G. Montemezzani, S. Pfändler, and P. Günter, "Electro-optic and photorefractive properties of Bi4Ge3O12 crystals in the ultraviolet spectral range," *J. Opt. Soc. Am. B* 9, 1110–1117 (1992).
19. M. Kanoi, G. Takahashi, T. Sato, et al., "Optical voltage and current measuring system for electric power systems," *IEEE Trans. Power Deliv.* PWRD-1, 91–97 (1986).

20. I. J. Laurensse, C. G. A. Koreman, W. R. Rutgers, and A. H. van der Wey, "Applications for optical current and voltage sensors," *Sens. Actuat. A* 17, 181–186 (1989).
21. A. H. Rose and G. W. Day, "Optical fiber voltage sensors for broad temperature ranges," *SPIE* 1580, 95–103 (1992).
22. T. W. Cease, J. G. Driggans, and S. J. Weikel, "Optical voltage and current sensors used in a revenue metering system," *IEEE Trans. Power Deliv.* 6(4), 1374–1379 (1991).
23. K. Kurosawa, S. Yoshida, E. Mori, G. Takahashi, and S. Saito, "Development of an optical instrument transformer for DC voltage measurement," *IEEE Trans. Power Deliv.* 8(4), 1721–1726 (1993), doi: 10.1109/61.248279
24. O. Kamada and K. Kakishita, "Electro-optical effect of $Bi_4Ge_3O_{12}$ crystals for optical voltage sensors," *Jpn. J. Appl. Phys.* 32, 4288–4291 (1993).
25. L. H. Christensen, "Design, construction, and test of a passive optical prototype high voltage instrument transformer," *IEEE Trans. Power Deliv.* 10(3), 1332–1337 (1995), doi: 10.1109/61.400913
26. J. C. Santos and K. Hidaka, "Optical high voltage measurement technique using Pockels device," *Jpn. J. Appl. Phys.* 1997, 36(Part 1, No. 4A), 2394–2398 (1997).
27. J. C. Santos, M. C. Taplamacioglu, and K. Hidaka, "Pockels high-voltage measurement system," *IEEE Trans. Power Deliv.* 15(1), 8–13 (2000), doi: 10.1109/61.847221
28. C. Li, X. Cui, and T. Yoshino, "Optical electric-power sensor by use of one bismuth germanate crystal," *J. Lightwave Technol.* 21(5), 1328–1333 (2003).
29. F. Pan, X. Xiao, Y. Xu, and S. Ren, "An optical AC voltage sensor based on the transverse Pockels effect," *Sensors* 11, 6593–6602 (2011).
30. A. Kumada and K. Hidaka, "Directly high-voltage measuring system based on Pockels effect," *IEEE Trans. Power Deliv.* 28(3), 1306–1313 (2013), doi: 10.1109/TPWRD.2013.2250315
31. K. Kyuma, S. Tai, M. Nunoshita, N. Mikami, and Y. Ida, "Fiber optic current and voltage sensors using a $Bi_{12}GeO_{20}$ single crystal," *IEEE/OSA J. Lightwave Technol.* 1(1), 93–97 (1983).
32. Y. Kuroda, Y. Abe, H. Kuwahara, and K. Yoshinaga, "Field test of fiber-optic voltage and current sensors applied to gas insulated substation," *Proc. SPIE* 586, 30–37 (1986).
33. M. Norimatsu and M. Shirasaki, "$Bi_{12}SiO_{20}$ crystal application for voltage sensor in optical fibers," *Ferroelectrics* 75, 189–196 (1987).
34. P. Lemaire and M. Georges, "Electro-optic coefficient measurements: correction of electric-field inhomogeneities in the transverse configuration," *Opt. Lett.* 17, 1411–1413 (1992).
35. P. Mihailovic, S. Petricevic, S. Stankovic, and J. Radunovic, "Temperature dependence of the $Bi_{12}GeO_{20}$ optical activity," *Opt. Mat.* 30, 1079–1082 (2008).
36. S. Petricevic, P. Mihailovic, and J. Radunovic, "A miniature Pockels cell with novel electrode geometry," *Sensors* 9, 5298–5307 (2009).
37. P. P. Chavez, F. Rahmatian, and N. A. F. Jaeger, "Accurate voltage measurement with electric field sampling using permittivity-shielding," *IEEE Trans. Power Deliv.* 17(2), 362–368 (2002), doi: 10.1109/61.997900
38. F. Rahmatian, P. P. Chavez, and N. A. F. Jaeger, "230 kV optical voltage transducers using multiple electric field sensors," *IEEE Trans. Power Deliv.* 17(2), 417–422 (2002), doi: 10.1109/61.997909
39. Y. Hamasaki, H. Gotoh, M. Katoh, and S. Takeuchi, "OPSEF: an optical sensor for measurement of high electric field intensity," *Electron. Lett.* 16, 406–407 (1980).
40. Y. Hamasaki, T. Miyamoto, Y. Kuhara, et al., "Optical fiber sensor for the measurement of electric field intensity and voltage (OPSEF)," *Fiber Integr. Opt.* 3(4), 383–389 (1981).
41. Y. Kuhara, Y. Hamasaki, A. Kawakami, et al., "BSO/fibre-optic voltmeter with excellent temperature stability," *Electron. Lett.* 18, 1055–1056 (1982).
42. A. Cruden, Z. J. Richardson, J. R. McDonald, and I. Andonovic, "Optical crystal based devices for current and voltage measurement," *IEEE Trans. Power Deliv.* 10(3), 1217–1223 (1995), doi: 10.1109/61.400899

43. J. Niewisch, P. Menke, P. Krammer, and T. Bosselmann, "Temperature drift compensation of a potential transformer using a BSO Pockels cell," in *Optical Fiber Sensors* (Optica Publishing Group, 1996), paper We15.
44. A. Cruden, Z. J. Richardson, J. R. McDonald, I. Andonovic, W. Laycock, and A. Bennett, "Compact 132 kV combined optical voltage and current measurement system," *IEEE Trans. Instrum. Meas.* 47(1) 219–223 (1998).
45. T. Mitsui, K. Hosoe, H. Usami, and S. Miyamoto, "Development of fiber-optic voltage sensors and magnetic-field sensors," *IEEE Trans. Power Deliv.* 2(1), 87–93 (1987), doi: 10.1109/TPWRD.1987.4308077
46. P. A. Williams, A. H. Rose, K. S. Lee, D. C. Conrad, G. W. Day, and P. D. Hale, "Optical, thermo-optic, electro-optic, and photoelastic properties of bismuth germanate (Bi4Ge3O12)," *Appl. Opt.* 35(19), 3562–3569 (1996).
47. J. C. Brice, *The Growth of Crystals from the Melt* (Wiley, 1965).
48. Y. A. Borovlev, N. V. Ivannikova, V. N. Shlegel, Y. V. Vasiliev, and V. A. Gusev, "Progress in growth of large sized BGO crystals by the low-thermal-gradient Czochralski technique," *J. Cryst. Growth* 229(1–4), 305–311 (2001).
49. V. N. Filippov, A. N. Starodumov, Y. O. Barmenkov, and V. V. Makarov, "Fiber-optic voltage sensor based on a $Bi_{12}TiO_{20}$ crystal," *Appl. Opt.* 39(9), 1389–1393 (2000).
50. V. N. Filippov, A. N. Starodumov, V. P. Minkovich, and F. G. P. Lecona, "Fiber sensor for simultaneous measurement of voltage and temperature," *IEEE Photonics Technol. Lett.* 12(11), 1543–1545 (2000), doi: 10.1109/68.887742
51. F. Peña-Lecona, J. F. Mosiño, V. N. Filippov, et al., "An electro-optic fibre sensor with double pass configuration for measuring high ac voltage," *Meas. Sci. Technol.* 15(6), 1129–1132 (2004).
52. S. R. M. Robertson and A. J. Rogers, "Measurement of DC electric fields using the electro-optic effect," *IEE Proc.* 132, 195–198 (1985).
53. Y. J. Rao, H. Gnewuch, C. N. Pannell, and D. A. Jackson, "Electro-optic electric field sensor based on periodically poled LiNbO3," *Electron. Lett.* 35(7), 596–597 (1999).
54. H. Hamada, "Investigation of the beam dependence of a fiber-optic voltage sensor by use of a $LiNbO_3$ crystal," *JOSA B* 18(4), 444–456 (2001).
55. M. Passard, C. Barthod, M. Fortin, C. Galez, and J. Bouillot, "Design and optimization of a low-frequency electric field sensor using Pockels effect," *IEEE Trans. Instrum. Meas.* 50(5), 1053–1058, (2001), doi: 10.1109/19.963157
56. C. Li, X. Cui, and T. Yoshino, "Measurement of AC electric power based on dual transverse Pockels effect," *IEEE Trans. Instrum. Meas.* 50(5), 1375–1380 (2001).
57. Y. Li, C. Li, and T. Yoshino, "Optical electric-power-sensing system using Faraday and Pockels cells," *Appl. Opt.* 40, 5738–5741 (2001).
58. C. Li and T. Yoshino, "Optical voltage sensor based on electrooptic crystal multiplier," *J. Lightwave Technol.* 20(5), 843–849 (2002), doi: 10.1109/JLT.2002.1007938
59. A. Garzarella, S. B. Qadri, D. H. Wu, and R. J. Hinton, "Responsivity optimization and stabilization in electro-optic field sensors," *Appl. Opt.* 46, 6636–6640 (2007).
60. T. Sawa, K. Kurosawa, T. Kaminishi, and T. Yokota, "Development of optical instrument transformers," *IEEE Trans. Power Deliv.* 5(2), 884–891 (1990), doi: 10.1109/61.53098
61. H. M. Hertz and P. Thomsen, "Optical wideband high-voltage measurement system," *Rev. Scient. Instrum.* 58(9), 1660 (1987).
62. T. S. Robinson, F. Consoli, S. Giltrap, et al., "Low-noise time-resolved optical sensing of electromagnetic pulses from petawatt laser-matter interactions," *Scient. Rep.* 7, 983 (2017), doi: 10.1038/s41598-017-01063-1
63. A. J. Rogers, "The electrogyration effect in crystalline quartz," *Proc. R. Soc. Lond. Ser. A* 353, 177–192 (1977).
64. E. Ballik and D. Liu. "Measurement of high-voltage pulses employing a quartz Pockels cell," *IEEE J. Quantum Electron.* 19(7), 1166–1168 (1983).

65. W. Epping, A. Kuchler, and A. Schwab, "Elektrische Feldstärkemessung mit doppelbrechenden und optisch aktiven Kristallen," *Arch. Elektrotech.* 67, 329–339 (1984).
66. A. Garzarella, S. B. Qadri, T. J. Wieting, and D. H. Wu, "Piezo-induced sensitivity enhancements in electro-optic field sensors," *J. Appl. Phys.* 98, 043113 (2005).
67. J. Kobayashi, T. Asahi, and S. Takahashi, "Simultaneous measurements of electrogyration and electrooptic effects of α-quartz," *Ferroelectrics* 75, 139–152 (1987).
68. V. Ivanov, M. Levichev, Y. Nozdrin, and M. Novikov, "Temperature dependence of electro-optic effect and natural linear birefringence in quartz measured by low-coherence interferometry," *Appl. Opt.* 54(33), 9911–9918 (2015).
69. V. Ivanov, A. Stepanov, V. Alenkov, and O. Buzanov, "Langasites as electro-optic materials for high-voltage optical sensors," *Opt. Mater. Express* 7, 3366–3376 (2017).
70. S. Chadderdon, R. Gibson, R. H. Selfridge, et al., "Electric-field sensors utilizing coupling between a D-fiber and an electro-optic polymer slab," *Appl. Opt.* 50(20), 3505–3512 (2011).
71. M. J. Weber, *Handbook of Optical Materials* (CRC Press, 2003).
72. R. E. Aldrich, S. L. Hou, and M. L. Harvill, "Electrical and optical properties of $Bi_{12}SiO_{20}$," *J. Appl. Phys.* 42(1) 493–494 (1971).
73. J. C. Brice, "Crystals for quartz resonators," *Rev. Mod. Phys.* 57(1), 105–146 (1985).
74. K. Bohnert, A. Frank, G. M. Müller, L. Yang, M. Lenner, P. Gabus, X. Gu, and S. V. Marchese, "Fiber optic current and voltage sensors for electric power transmission systems," in *Fiber Optic Sensors and Applications XV*, A. Mendez, C. S. Baldwin, H. H. Du, Eds., *Proc. of SPIE*, vol. 10654, 1065402, 2018, doi: 10.1117/12.2303945
75. G. A. Massey, J. C. Johnson, and D. C. Erickson, "Laser sensing of electric and magnetic fields for power transmission applications," *SPIE* 88, 91–96 (1976).
76. D. C. Erickson, "The use of fiber optics for communications, measurement and control within high voltage substations," *IEEE Trans. Power Appar. Syst.*, PAS-99 (3), 1057–1065 (1980), doi: 10.1109/TPAS.1980.319736
77. O. Tonnesen, F. Birck, T. Hansen, et al., "A new 132 kV SF_6 insulated electrooptic instrument transformer," in *Proc. EPRI Conference on Optical Sensing in Utility Applications*, 1991.
78. N. A. F. Jaeger and L. Young, "High-voltage sensor employing an integrated optics Mach-Zehnder interferometer in conjunction with a capacitive divider," *J. Lightwave Technol.* 7(2), 229–235 (1989), doi: 10.1109/50.17760
79. H. Y. Yang, "Construction principle and application test of 220kV self-healing optical voltage sensor," *Appl. Mech. Mater.* 511–512, 116–119 (2014), doi: 10.4028/www.scientific.net/amm.511-512.116
80. S. Kobayashi, et al., "Development and field test evaluation of optical current and voltage transformers for gas insulated switchgear," *IEEE Trans. Power Deliv.* 7(2), 815–821 (1992), doi: 10.1109/61.127085
81. K. Bohnert, H. Brändle, and G. Frosio, "Field test of interferometric optical fiber high-voltage and current sensors," in *10th Int. Conf. on Optical Fibre Sensors*, B. Culshaw, J. D. C. Jones, Eds., *Proc. SPIE 2360*, pp. 16–19, 1994.
82. K. Bohnert and J. Nehring, "Fiber-optic sensing of voltages by line integration of the electric field," *Opt. Lett.* 14, 290–292 (1989).
83. K. Bohnert, M. Ingold, and J. Kostovic, "Fiber-optic voltage sensor for SF6 gas-insulated high-voltage switchgear," *Appl. Opt.* 38, 1926–1933 (1999).
84. M. Ohtsuka and S. Teruo, "Influence of optical rotary power on optical voltage sensor using BGO crystal," *Electron. Commun. Jpn. (Part II: Electronics)* 77(5), 21–31 (1994).
85. A. Garzarella, S. B. Qadri, T. J. Wieting, and D. H. Wu, "The effects of photorefraction on electro-optic field sensors," *J. Appl. Phys.* 97, 113108 (2005).
86. K. S. Lee, "New compensation method for bulk optical sensors with multiple birefringences," *Appl. Opt.* 28, 2001–2011 (1989).

87. L. Li, W. Zhang, H. Li, and R. Pan, "Linear birefringence-free optical voltage sensor based on dual-crystal structure," *Appl. Opt.* 52, 8706–8713 (2013).
88. T. Sasano, "Laser CT and laser PD for EHV power transmission lines," *Electr. Eng. Jpn.* 93, 91–98 (1973).
89. A. J. Rogers, "Optical measurement of current and voltage on optical power systems," *Electr. Power Appl.* 2(4), 120–124 (1979).
90. H. D. Schlemper, D. Füchsle, G. Ramm, and J. Widmer, "Test and application of non-conventional multi-purpose voltage and current transducers," in *CIGRE Conference* 2004, paper A3-108.
91. K. Bohnert, P. Gabus, H. Brändle, and A. Khan, "Fiber-optic current and voltage sensors for high-voltage substations," in *Proc. 16th Int. Conference on Optical Fiber Sensors*, 2003, pp. 752–755.
92. R. C. Miller, "Method of deriving an AC waveform from two phase shifted electrical signals," U.S. Patent 5,001,419, Mar., 1991.
93. O. Steiger, S. V. Marchese, J. Pascal, K. Bohnert, and S. Wildermuth, "Signal processing for electro-optic voltage sensor," in *Proceedings IEEE Sensors*, 2013, pp. 1932–1936, doi: 10.1109/ICSENS.2013.6688615
94. Standard of the International Electrotechnical Commission (IEC), IEC 60044-7: Instrument transformers, Part 7: Electronic voltage transformers (1999); IEC 61869; Instrument transformers—Part 11: Additional requirements for low-power passive voltage transformers. IEC: Geneva, Switzerland, 2017.
95. S. V. Marchese, K. Bohnert, S. Wildermuth, J. L. M. van Mechelen, O. Steiger, L. C. Rodoni, G. Eriksson, J. Czyzewski, "Electro optic voltage sensor based on BGO for air insulated high voltage substations," in *Proceedings IEEE Photonics Conference 2013*, paper ThA3.2, doi: 10.1109/IPCon.2013.6656441
96. S. Wildermuth, K. Bohnert, N. Koch, et al., "High voltage sensor with axially overlapping electrodes," U.S. patent 9291650B2, priority date Apr. 25, 2013.
97. J. S. Graham, "High voltage bushings," in *High Voltage Engineering and Testing*, H. M. Ryan, Ed. (Institution of Electrical Engineers, London, 2001), pp. 467–492.
98. Q. Tan, Q. Xu, L. Chen, and Y. Huang, "A new method to improve internal electric field distributions of Pockels OVS," *IEEE Sens. J.* 17(13), 4115–4121 (2017), doi: 10.1109/JSEN.2017.2703600
99. Z. Fu, Q. Xu, X. Cai, and W. He, "Defects and improvements of Pockels transducer voltage segmentation," *IEEE Sens. J.* 19(1), 138–143 (2019), doi: 10.1109/JSEN.2018.2876500
100. N. A. F. Jaeger, F. Rahmatian, P. P. Chavez, et al., "Method and apparatus for measuring voltage using electric field sensors," U.S. patent 6,252, 388 B1, priority date Dec. 4, 1998.
101. P. P. Chavez, N. A. F. Jaeger, and F. Rahmatian, "Accurate voltage measurement by the quadrature method," *IEEE Trans. Power Deliv.* 18(1), 14–19, (2003), doi: 10.1109/TPWRD.2002.801428
102. P. P. Chavez, F. Rahmatian, N. A. F. Jaeger, and C. P. Yakymyshyn, "Voltage sensor," U.S. patent 6380725B1, Apr. 30, 2002.
103. F. Rahmatian, P. P. Chavez, and N. A. F. Jaeger, "Wide-band 138 kV distributed-sensor optical voltage transducer: study of accuracy under pollution and other field disturbances," in *2001 IEEE PES Power Summer Meeting*, 2001, vol. 1, pp. 156–161.
104. F. Rahmatian, P. P. Chavez, N. A. F. Jaeger, "Resistively shielded optical voltage transducer," in *2001 IEEE/PES Transmission and Distribution Conference and Exposition*, 2001, vol. 1, pp. 117–121.
105. P. P. Chavez, F. Rahmatian, N. A. F. Jaeger, "230 kV optical voltage transducer using a distributed optical electric field sensor system," in *2001 IEEE/PES Transmission and Distribution Conference and Exposition*, 2001, vol. 1, pp. 131–135.
106. F. Rahmatian, P. P. Chavez, N. A. F. Jaeger, "138 kV and 345 kV wide-band SF_6-free optical voltage transducers," in *2002 IEEE Power Engineering Society Winter Meeting*, 2002, vol. 2, pp. 1472–1477.

107. F. Rahmatian and N. A. F. Jaeger, "High accuracy optical electric field and voltage sensors," in *Technical Digest of 15th Optical Fiber Sensors Conference* 2002, pp. 411–414 (paper ThD3).
108. F. Rahmatian and P. P. Chavez, "SF_6-free 550 kV combined optical voltage and current transducer system," in *2003 IEEE PES Transmission and Distribution Conference and Exposition (IEEE Cat. No. 03CH37495)*, 2003, vol. 1, pp. 379–382.
109. F. Rahmatian, G. Polovick, B. Hughes, and V. Aresteanu, "Field experience with high-voltage combined optical voltage and current transducers," in *CIGRE Conference* 2004, paper A3-111.
110. X. Gu, S. Marchese, K. Bohnert, and U. Riechert, "Development of an electro-optic voltage sensor for high voltage direct current applications," in *26th International Conference on Optical Fiber Sensors*, OSA Technical Digest (Optica Publishing Group, 2018), paper TuE35.
111. M. Bordovsky, F. Cecelja, and W. Balachandran, "Comparative study of cubic crystals performance in bulk electro-optic sensor for DC and extra-low-frequency measurements," *Proc. SPIE* 2389, 166–173 (1999).
112. F. Cecelja, M. Bordovsky, and W. Balachandran, "Lithium niobate sensor for measurement of DC electric fields," *IEEE Trans. Instrum. Meas.* 50(2), 465–469 (2001), doi: 10.1109/19.918167
113. F. Cecelja, M. Bordovsky, and W. Balachandran, "Electro-optic sensor for measurement of DC fields in the presence of space charge," *IEEE Trans. Instrum. Meas.* 51(2), 282–286 (2002), doi: 10.1109/19.997825
114. F. Cecelja, W. Balachandran, and M. Bordowski, "Validation of electro-optic sensors for measurement of DC fields in the presence of space charge," *Measurement* 40, 450–458 (2007).
115. J. O. Grasdijk, X. F. Bai, I. Engin, et al., "Electro-optic sensor for static fields," *Appl. Phys. B* 125, 212 (2019), doi: 10.1007/s00340-019-7326-5
116. K. Hidaka and H. Fujita, "A new method of electric field measurements in corona discharge using Pockels device," *J. Appl. Phys.* 53(9), 5999–6003 (1982).
117. Y. Li, L. Gao, J. Wan, and J. Liu, "Optical DC electric field sensing based on the Pockels effect using bismuth germanate crystals," *Appl. Opt.* 59, 6237–6244 (2020).
118. J. Zhang and F. Chen, "An electro-optic sensor for measurement of intensive DC electric field," *IEEE Photonics J.* 14(2), 1–7 (2022).
119. C. H. Bulmer, "Sensitive, highly linear lithium niobate interferometers for electromagnetic field sensing," *Appl. Phys. Lett.* 53, 2368–2370 (1988), doi: 10.1063/1.100232
120. C. H. Bulmer, W. K. Burns, and A. S. Greenblatt, "Phase tuning by laser ablation of $LiNbO_3$ interferometric modulators to optimum linearity," *IEEE Photonics Technol. Lett.* 3(6), 510–512 (1991), doi: 10.1109/68.91017
121. A. S. Greenblatt, C. H. Bulmer, R. P. Moeller, and W. K. Burns, "Thermal stability of bias point of packaged linear modulators in lithium niobate," *J. Lightwave Technol.* 13(12), 2314–2319 (1995), doi: 10.1109/50.475569
122. N. Kuwabara, K. Tajima, R. Kobayashi, and F. Amemiya, "Development and analysis of electric field sensor using $LiNbO_3$ optical modulator," *IEEE Trans. Electromagn. Compat.* 34(4), 391–396 (1992), doi: 10.1109/15.179271
123. K. Tajima, R. Kobayyashi, N. Kuwabara, and M. Tokuda, "Development of optical isotropic E-field sensor operating more than 10 GHz using Mach-Zehnder interferometer," *IEICE Trans. Electron.* E85-C, 961–967 (2002).
124. S. A. Kingsley and S. Sriram. "Parallel-plate integrated optic high-voltage sensor," *Electron. Lett.* 31(13), 1096–1097 (1995).
125. M. Yamada, M. Saitoh, and H. Ooki, "Electric-field induced cylindrical lens, switching and deflection devices composed of the inverted domains in LiNbO3 crystals," *Appl. Phys. Lett.* 69(24), 3659–3661 (1996).
126. D. H. Naghski, J. T. Boyd, H. E. Jackson, et al., "An integrated photonic Mach-Zehnder interferometer with no electrodes for sensing electric fields," *J. Lightwave Technol.* 12(6), 1092–1098 (1994), doi: 10.1109/50.296204

127. N. A. F. Jaeger and L. Huang, "Push-pull integrated-optics Mach-Zehnder interferometer with domain inversion in one branch," *Opt. Lett.* 20, 288–290 (1995).
128. N. A. F. Jaeger and L. Young, "Asymmetric slab and strip-loaded integrated optic devices for the measurement of large electric fields," *J. Lightwave Technol.* 5(6), 745–750 (1987).
129. E. Bonek, D. Hornbachner, E. Riedl-Bratengeyer, W. Hadrian, and R. Jobst, "Electro-optic ELF field sensor," *Electron. Lett.* 21(28), 1994–1995 (1992).
130. N. A. F. Jaeger and F. Rahmatian, "Integrated optics Pockels cell high-voltage sensor," *IEEE Trans. Power Deliv.* 10(1), 127–134 (1995).
131. T. Takahashi, K. Hidaka, and T. Kouno, "New optical-waveguide pockels sensor for measuring electric fields," *Jpn. J. Appl. Phys.* 35, 767–771 (1996).
132. O. Ogawa, T. Sowa, and S. Ichizono, "A guided-wave optical electric field sensor with improved temperature stability," *J. Lightwave Technol.* 17(5), 823–830 (1999), doi: 10.1109/50.762899
133. Z. Li, H. Yuan, Y. Cui, Z. Ding, and L. Zhao, "Measurement of distorted power-frequency electric field with integrated optical sensor," *IEEE Trans. Instrum. Meas.* 68(4), 1132–1139 (2019), doi: 10.1109/TIM.2018.2864019
134. C. Gutierrez-Martinez and J. Santos-Aguilar, "Electric field sensing scheme based on matched $LiNbO_3$ electro-optic retarders," *IEEE Trans. Instrum. Meas.* 57(7), 1362–1368 (2008), doi: 10.1109/TIM.2008.917174
135. T. Meier, C. Kostrzewa, K. Petermann, and B. Schuppert, "Integrated optical E-field probes with segmented modulator electrodes," *J. Lightwave Technol.* 12(8), 1497–1503 (1994), doi: 10.1109/50.317540
136. B. Sun, F. Chen, K. Chen, Z. Hu, and Y. Cao, "Integrated optical electric field sensor from 10 kHz to 18 GHz," *IEEE Photonics Technol. Lett.* 24(13), 1106–1108 (2012), doi: 10.1109/LPT.2012.2195780
137. M. W. Howerton, C. H. Bulmer, and W. K. Burns, "Linear 1×2 directional coupler for electromagnetic field detection," *Appl. Phys. Lett.* 52(22), 1850–1852 (1988).
138. H. Jung, "Electro-optic electric-field sensors utilizing Ti:$LiNbO_3$ 1×2 directional coupler with dipole antennas," *Opt. Eng.* 52(6), 064402 (2013).
139. C. Gutierrez-Martinez and R. Ricardez-Trejo, "Remotely biasing the electro-optic response of an electric field sensing-detection system using $LiNbO_3$ asymmetric Mach-Zehnder optical retarders," *Appl. Opt.* 57(32), 9677–9682 (2018).
140. C. Gutierrez-Martinez, G. Trinidad-Garcia, and J. Rodriguez-Asomoza, "Electric field sensing system using coherence modulation of light," *IEEE Trans. Instrum. Meas.* 51(5), 985–989 (2002), doi: 10.1109/TIM.2002.806017
141. R. Zeng, B. Wang, Z. Yu, and W. Chen, "Design and application of an integrated electro-optic sensor for intensive electric field measurement," *IEEE Trans. Dielectr. Electr. Insul.* 18(1), 312–319 (2011), doi: 10.1109/TDEI.2011.5704523
142. J. Zhang, F. Chen, B. Sun, K. Chen, and C. Li, "3D integrated optical E-field sensor for lightning electromagnetic impulse measurement," *IEEE Photonics Technol. Lett.* 26(23), 2353–2356 (2014), doi: 10.1109/LPT.2014.2355209
143. J. Zhang, C. Luo, and Z. Zhao, "Design and application of integrated optics sensor for measurement of intense pulsed electric field," *J. Lightwave Technol.* 37(4), 1440–1448 (2019).
144. H. Wang, C. Zhuang, R. Zeng, S. Xie, and J. He, "Transient voltage measurements for overhead transmission lines and substations by metal-free and contactless integrated electro-optic field sensors," *IEEE Trans. Ind. Electron.* 66(1), 571–579 (2019), doi: 10.1109/TIE.2018.2826455
145. L. Chen and R. M. Reano, "Compact electric field sensors based on indirect bonding of lithium niobate to silicon microrings," *Opt. Express* 20, 4032–4038 (2012).
146. V. Calero, M.-A. Suarez, R. Salut, et al., "An ultra wideband-high spatial resolution-compact electric field sensor based on Lab-on-Fiber technology," *Scient. Rep.* 9, 8058 (2019), doi: 10.1038/s41598-019-44644-y

147. D.-J. Lee and J. F. Whitaker, "An optical-fiber-scale electro-optic probe for minimally invasive high-frequency field sensing," *Opt. Express* 16(26), 21587–21597 (2008).
148. D.-J. Lee and J. F. Whitaker, "Optimization of sideband modulation in optical-heterodyne-downmixed electro-optic sensing," *Appl. Opt.* 48(8), 1583–1590 (2009).
149. K. Sasagawa and M. Tsuchiya, "Low-noise and high-frequency resolution electrooptic sensing of RF near-fields using an external optical modulator," *J. Lightwave Technol.* 26(10), 1242–1248 (2008).
150. M. Bernier, G. Gaborit, L. Duvillaret, A. Paupert, and J.-L. Lasserre, "Electric field and temperature measurement using ultra-wide bandwidth pigtailed electro-optic probes," *Appl. Opt.* 47(13), 2470–2476 (2008).
151. H. Togo, N. Kukutsu, N. Shimizu, and T. Nagatsuma, "Sensitivity-stabilized fiber-mounted electrooptic probe for electric field mapping," *J. Lightwave Technol.* 26(15), 2700–2705 (2008).
152. S. Hisatake and T. Nagatsuma, "Continuous-wave terahertz field imaging based on photonics-based self-heterodyne electro-optic detection," *Opt. Lett.* 38(13), 2307–2310 (2013).
153. S. M. Chandani, "Fiber-based probe for electrooptic sampling," *IEEE Photonics Technol. Lett.* 18(12), 1290–1292 (2006).
154. D.-J. Lee, J.-Y. Kwon, N.-W. Kang, and J. F. Whitaker, "Calibrated 100-dB-dynamic-range electro-optic probe for high-power microwave applications," *Opt. Express* 19(15), 14437–14450 (2011).
155. W.-B. Lee, S.-U. Baek, and S.-S. Lee, "Highly sensitive electro-optic probe incorporating an ultra-high Q-factor $LiNbO_3$ etalon," *Appl. Opt.* 57(28), 8343–8349 (2018).
156. S. Kim, Y.-P. Hong, and D.-J. Lee, "Polarization insensitive electro-optic probe using birefringence-free interferometric modulation," *Opt. Lett.* 44(11), 2895–2898 (2019).
157. V. Calero, et al., "Towards highly reliable, precise and reproducible fabrication of photonic crystal slabs on lithium niobate," *J. Lightwave Technol.* 37, 698–703 (2019).
158. F. Consoli, R. De Angelis, L. Duvillaret, et al., "Time-resolved absolute measurements by electro-optic effect of giant electromagnetic pulses due to laser-plasma interaction in nanosecond regime," *Scient. Rep.* 6, 27889 (2016), doi: 10.1038/srep27889
159. K. Yang, L. P. B. Katehi, and J. F. Whitaker, "Electro-optic field mapping system utilizing external gallium arsenide probes," *Appl. Phys. Lett.* 77(4), 486–488 (2000).
160. W.-K. Kuo, Y.-T. Huang, and S.-L. Huang, "Three-dimensional electric-field vector measurement with an electro-optic sensing technique," *Opt. Lett.* 24(22), 1546–1548 (1999).
161. G. Gaborit, J.-L. Coutaz, and L. Duvillaret, "Vectorial electric field measurement using isotropic electro-optic crystals," *Appl. Phys. Lett.* 90, 241118 (2007), doi: https://doi.org/10.1063/1.2748364
162. G. Gaborit et al., "Single shot and vectorial characterization of intense electric field in various environments with pigtailed electrooptic probe," *IEEE Trans. Plasma Sci.* 42(5), 1265–1273 (2014), doi: 10.1109/TPS.2014.2301023
163. E. Suzuki, S. Arakawa, M. Takahashi, et al., "Visualization of Poynting vectors by using electro-optic probes for electromagnetic fields," *IEEE Trans. Instrum. Meas.* 57(5), 1014–1022 (2008).
164. C.-C. Chen and J. F. Whitaker, "An optically-interrogated microwave-Poynting-vector sensor using cadmium manganese telluride," *Opt. Express* 18(12), 12239–12248 (2010).
165. D. C. Wunsch and A. Erteza, "Kerr cell measuring system for high voltage pulses," *Rev. Sci. Instrum.* 35(7), 816–820 (1964).
166. E. C. Cassidy, H. N. Cones, and S. R. Booker, "Development and evaluation of electro-optical high-voltage pulse measurement techniques," *IEEE Trans. Instrum. Meas.* 19(4), 395–402 (1970).
167. R. E. Hebner and E. C. Cassidy, "Measurement of 60 Hz voltages using the Kerr effect," *Rev. Sci. Instrum.* 43(12), 1839–1841 (1972).
168. R. E. Hebner, E. C. Cassidy, and J. E. Jones, "Improved techniques for the measurement of high-voltage impulses using the electrooptic Kerr effect," *IEEE Trans. Instrum. Meas.* 24(4), 361–366 (1975), doi: 10.1109/TIM.1975.4314460

169. R. E. Hebner, R. A. Malewski, and E. C. Cassidy, "Optical methods of electrical measurements at high voltage levels," *Proc. IEEE* 65(11), 1524–1548 (1977).
170. A. Törne and U. Gäfvert, "Measurement of the electric field in transformer oil using Kerr technique with optical and electrical modulation," *Proc. ICPADM* 1, 61–64 (1985).
171. R. Shimizu, M. Matsuoka, K. Kato, et al., "Development of Kerr electro-optic 3-d electric field measuring technique and its experimental verification," *IEEE Trans. Dielect. Electr. Insul.* 3(2), 191–196 (1996), doi: 10.1109/94.486769
172. A. Helgeson and M. Zahn, "Kerr electro-optic measurements of space charge effects in high voltage pulsed propylene carbonate using parallel-plate electrodes," *IEEE Trans. Dielect. Electr. Insul.* 9(5), 838–844 (2002).
173. T. Kamiya, S. Matsuoka, A. Kumada, and K. Hidaka, "High voltage measuring apparatus based on Kerr effect in gas," *IEEE Trans. Dielect. Electr. Insul.* 22(2), 760–765 (2015), doi: 10.1109/TDEI.2015.7076773
174. C. Gao, B. Qi, Y. Gao, Z. Zhu, and C. Li, "Kerr electro-optic sensor for electric field in large-scale oil–pressboard insulation structure," *IEEE Trans. Instrum. Meas.* 68(10), 3626–3634 (2019), doi: 10.1109/TIM.2018.2881803
175. A. J. Rogers, "Polarization-optical time domain reflectometry: a technique for the measurement of field distributions," *Appl. Opt.* 20(6), 1060–1074 (1981).
176. C. Han, F. Lv, C. Sun, and H. Ding, "Silica microwire-based interferometric electric field sensor," *Opt. Lett.* 40, 3683–3686 (2015).
177. J. Hou, H. Ding, B. Wei, et al., "Microfiber knot resonator based electric field sensor," *Instrum. Sci. Technol.* 45(3), 259–267 (2017).
178. E. Hecht and A. Zajac, *Optics* (Addison-Wesley Publishing Company, 1974).
179. N. F. Borrelli, B. G. Aitken, M. A. Newhouse, and D. W. Hall, "Electric-field-induced birefringence properties of high-refractive-index glasses exhibiting large Kerr nonlinearities," *J. Appl. Phys.* 70(5), 2774–2779 (1991).
180. M. C. Farries and A. J. Rogers, "Temperature dependence of the Kerr effect in a silica optical fibre," *Electron. Lett.* 19(21), 890–891 (1983).
181. K. Bohnert, "Optical voltage sensor," European patent EP2095135B1, priority date Dec. 22, 2006.
182. S. Wildermuth, K. Bohnert, and H. Brändle, "Interrogation of birefringent fiber sensors using non-reciprocal phase modulation," in *20th International Conference on Optical Fibre Sensors, Proc. SPIE 7503*, 2009, pp. 598–601.
183. S. Wildermuth, K. Bohnert, and H. Brändle, "Interrogation of a birefringent fiber sensor by nonreciprocal phase modulation," *IEEE Photonics Technol. Lett.* 22(18), 1388–1390 (2010), doi: 10.1109/LPT.2010.2060321
184. C. Zhang, X. Feng, S. Liang, C. Zhang, and C. Li, "Quasi-reciprocal reflective optical voltage sensor based on Pockels effect with digital closed-loop detection technique," *Opt. Comm.* 283(20), 3878–3883 (2010).
185. X. Feng, L. Li, X. Wang, C. Zhang, J. Yu, and C. Li, "Birefringence elimination of bismuth germanate crystal in quasi-reciprocal reflective optical voltage sensor," *Appl. Opt.* 52(8), 1676–1681 (2013).
186. X. Feng, X. Wang, L. Li, et al., "Influences of imperfect polarization induced effects to the quasi-reciprocal reflective optical voltage sensor," *J. Lightwave Technol.* 31(16), 2777–2784 (2013), doi: 10.1109/JLT.2013.2272875
187. H. Li, L. Cui, Z. Lin, L. Li, R. Wang, and C. Zhang, "Signal detection for optical AC and DC voltage sensors based on Pockels effect," *IEEE Sens. J.* 13(6), 2245–2252 (2013), doi: 10.1109/JSEN.2013.2249581
188. H. Li, L. Bi, R. Wang, L. Li, and C. Zhang, "Design of closed-loop detection system for optical voltage sensors based on Pockels effect," *J. Lightwave Technol.* 31(12), 1921–1928 (2013).

189. H. Li, L. Cui, Z. Lin, L. Li, and C. Zhang, "An analysis on the optimization of closed-loop detection method for optical voltage sensor based on Pockels effect," *J. Lightwave Technol.* 32(5), 1006–1013 (2014), doi: 10.1109/JLT.2013.2296518
190. L. Liu, H. Li, Z. Fu, and L. Feng, "Analysis on the optimization of high-frequency performance for optical voltage sensors based on Pockels effect," *IEEE Sens. J.* 17(15), 4826–4833 (2017), doi: 10.1109/JSEN.2017.2700998
191. W. Deng, H. Li, C. Zhang, and P. Wang, "Optimization of detection accuracy of closed-loop optical voltage sensors based on Pockels effect," *Sensors* 17, 1723 (2017), doi: 10.3390/s17081723
192. W. Chu, S. Heo, and M. Oh, "Polymeric integrated-optic bias chip for optical voltage transducers," *J. Lightwave Technol.* 32(24), 4730–4733 (2014).
193. W. Chu, S. Kim, X. Wu, L. Wen, and M. Oh, "Optical voltage sensors based on integrated optical polarization-rotated reflection interferometry," *J. Lightwave Technol.* 34(9), 2170–2174 (2016), doi: 10.1109/JLT.2016.2527685
194. Y. Huang, Q. F. Xu, Q. Tan, and Z. Xu, "An OVT based on conoscopic interference and position sensitive detector," *IEEE Sens. J.* 17(2), 340–346 (2017), doi: 10.1109/JSEN.2016.2629493
195. F. da Cruz Pereira, J. H. Galeti, R. T. Higuti, et al., "Real-time polarimetric optical high-voltage sensor using phase-controlled demodulation," *J. Lightwave Technol.* 36, 3275–3283 (2018).
196. D. Wang and N. Xie, "An optical voltage sensor based on wedge interference," *IEEE Trans. Instrum. Meas.* 67(1), 57–64 (2018), doi: 10.1109/TIM.2017.2756798
197. R. A. Myers, N. Mukherjee, and S. R. J. Brueck, "Large second-order nonlinearities in poled fused silica," *Opt. Lett.* 16(22), 1732–1734 (1991).
198. T. Fujiwara, D. Wong, Y. Zhao, S. Fleming, S. Poole, and M. Sceats, "Electro-optic modulation in germanosilicate fibre with UV-excited poling," *Electron. Lett.* 31(7), 573–575 (1995).
199. J. M. B. Pereira, A. R. Camara, F. Laurell, O. Tarasenko, and W. Margulis, "Linear electro-optical effect in silica fibers poled with ultraviolet lamp," *Opt. Express* 27, 14893–14902 (2019).
200. P. G. Kazansky, A. Kamal, and P. St. J. Russell, "High second-order nonlinearities induced in lead silicate glass by electron-beam irradiation," *Opt. Lett.* 18, 693–695 (1993).
201. J. M. B. Pereira, O. Tarasenko, Å. Claesson, F. Laurell, and W. Margulis, "Optical poling by means of electrical corona discharge," *Opt. Express* 30, 20605–20613 (2022).
202. F. De Lucia, D. Huang, C. Corbari, N. Healy, and P. J. A. Sazio, "Optical fiber poling by induction: analysis by 2D numerical modeling," *Opt. Lett.* 41, 1700–1703 (2016).
203. F. De Lucia and P. J. A. Sazio, "Optimized optical fiber poling configurations," *Opt. Mater.: X*, 1, 100016 (2019), doi: 10.1016/j.omx.2019.100016
204. S. C. Fleming and H. An, "Poled glasses and poled fibre devices," *J. Ceram. Soc. Jpn.* 116(1358), 1007–1023 (2008).
205. X.-C. Long, R. A. Myers, and S. R. J. Brueck, "The electro-optic effect in poled amorphous silica," *Opt. Lett.* 19(22), 1819–1821 (1994).
206. P. G. Kazansky, L. Dong, and P. St. J. Russell, "High second-order nonlinearities in poled silicate fibers," *Opt. Lett.* 19(10), 701–703 (1994).
207. T. Fujiwara, D. Wong, and S. Fleming, "Large electrooptic modulation in a thermally-poled germanosilicate fiber," *IEEE Photonics Technol. Lett.* 7(10), 1177–1179 (1995), doi: 10.1109/68.466582
208. X.-C. Long, R. A. Myers, and S. R. J. Brueck, "Measurement of linear electro-optic effect in temperature/electric-field poled optical fibers," *Electron. Lett.* 30(25), 2162–2163 (1994).
209. P. G. Kazansky, P. St. J. Russell, L. Dong, and C. N. Pannell, "Pockels effect in thermally poled silica optical fibres," *Electron. Lett.* 31(1), 62–63 (1995).
210. X.-C. Long, R. A. Myers, and S. R. J. Brueck, "A poled electrooptic fiber," *IEEE Photonics. Technol. Lett.* 8(2), 227–229 (1996).
211. W. Xu, P. Blazkiewicz, D. Wong, S. Fleming, and T. Ryan, "Specialty optical fibre for stabilising and enhancing electro-optic effect induced by poling," *Electron. Lett.* 36(15), 1265–1266 (2000).

212. A. Michie, I. Bassett, and J. Haywood, "Electric field and voltage sensing using thermally poled silica fibre with a simple low coherence interferometer," *Meas. Sci. Technol.* 17(5), 1229–1233 (2006).
213. S. Bauer, "Poled polymers for sensors and photonic applications," *J. Appl. Phys.* 80, 5531 (1996), doi: 10.1063/1.363604
214. D. J. Welker, J. Tostenrude, D. W. Garvey, et al., "Fabrication and characterization of single-mode electro-optic polymer optical fiber," *Opt. Lett.* 23, 1826–1828 (1998).
215. P. G. Kazansky and P. St. J. Russell, "Thermally poled glass: frozen-in electric field or oriented dipoles?" *Opt. Commun.* 110(5–6), 611–614 (1994).
216. T. G. Alley and S. R. J. Brueck, "Visualization of the nonlinear optical space-charge region of bulk thermally poled fused-silica glass," *Opt. Lett.* 23, 1170–1172 (1998).
217. V. Pruneri, F. Samoggia, G. Bonfrate, P. G. Kazansky, and G. M. Yang, "Thermal poling of silica in air and under vacuum: the influence of charge transport on second harmonic generation," *Appl. Phys. Lett.* 74(17), 2423–2425 (1999).
218. P. Blazkiewicz, W. Xu, D. Wong, and S. Fleming, "Mechanism for thermal poling in twin-hole silicate fibers," *JOSA B* 19(4), 870–874 (2002).
219. P. G. Kazansky, A. R. Smith, P. S. J. Russell, G. M. Yang, and G. M. Sessler, "Thermally poled silica glass: laser induced pressure pulse probe of charge distribution," *Appl. Phys. Lett.* 68(2), 269–271 (1996).
220. D. Wong, W. Xu, S. Fleming, M. Janos, and K-M. Lo, "Frozen-in electrical field in thermally poled fibers," *Opt. Fiber Technol.* 5(2), 235–241 (1999).
221. Y. Quiquempois, N. Godbout, and S. Lacroix, "Model of charge migration during thermal poling in silica glasses: evidence of a voltage threshold for the onset of a second-order nonlinearity," *Phys. Rev. A* 65, 043816-1–043816-14 (2002).
222. N. Myrén and W. Margulis, "Time evolution of frozen-in field during poling of fiber with alloy electrodes," *Opt. Express* 13(9), 3438–3444 (2005).
223. H. An and S. Fleming, "Hindering effect of the core-cladding interface on the progression of the second-order nonlinearity layer in thermally poled optical fibers," *Appl. Phys. Lett.* 87(10), 101108 (2005).
224. W. Margulis, O. Tarasenko, and N. Myrén, "Who needs a cathode? Creating a second-order nonlinearity by charging glass fiber with two anodes," *Opt. Express* 17, 15534–15540 (2009).
225. C. Corbari, P. G. Kazansky, S. A. Slattery, and D. N. Nikogosyan, "Ultraviolet poling of pure fused silica by high-intensity femtosecond radiation," *Appl. Phys. Lett.* 86(7), 071106 (2005).
226. H. An and S. Fleming, "Creating second-order nonlinearity in pure synthetic silica optical fibers by thermal poling," *Opt. Lett.* 32(7), 832–834 (2007).
227. P. Blazkiewicz, W. Xu, D. Wong, S. Fleming, and T. Ryan, "Modification of thermal poling evolution using novel twin-hole fibers," *J. Lightwave Technol.* 19(8), 1149–1154 (2001).
228. K. Lee, P. Hu, J. L. Blows, D. Thorncraft, and J. Baxter, "200-m optical fiber with an integrated electrode and its poling," *Opt. Lett.* 29(18), 2124–2126 (2004).
229. M. Fokine, L. E. Nilsson, Å. Claesson, D. Berlemont, L. Kjellberg, L. Krummenacher, and W. Margulis, "Integrated fiber Mach-Zehnder interferometer for electro-optic switching," *Opt. Lett.* 27(18), 1643–1645 (2002).
230. N. Myrén, H. Olsson, L. Norin, N. Sjödin, P. Helander, J. Svennebrink, and W. Margulis, "Wide wedge-shaped depletion region in thermally poled fiber with alloy electrodes," *Opt. Express* 12, 6093 (2004).
231. N. Mukherjee, R. A. Myers, and S. R. J. Brueck, "Dynamics of second-harmonic generation in fused silica," *J. Opt. Soc. Am. B* 11, 665–669 (1994).
232. O. Sugihara, T. Hirama, H. Fujimura, and N. Okamoto, "Second-order nonlinear optical properties from poled silicate channel-waveguide," *Opt. Rev.* 3(3), 150–152 (1996).

233. O. Sugihara, M. Nakanishi, H. Fujimura, C. Egami, and N. Okamoto, "Thermally poled silicate thin films with large second-harmonic generation," *J. Opt. Soc. Am. B* 15, 421–425 (1998).
234. M. Janos, W. Xu, D. Wong, H. Inglis, and S. Fleming, "Growth and decay of the electrooptic effect in thermally poled B/Ge codoped fiber," *J. Lightwave Technol.* 17(6), 1037 (1999).
235. D. W. Wong, W. Xu, and S. C. Fleming, "Charge dynamics and distributions in thermally poled silica fiber," *Proc. SPIE* 3847, 88–93 (1999).
236. O. Deparis, C. Corbari, P. G. Kazansky, and K. Sakaguchi, "Enhanced stability of the second-order optical nonlinearity in poled glasses," *Appl. Phys. Lett.* 84(24), 4857–4859 (2004).
237. R. Magalhães, J. Pereira, O. Tarasenko, et al., "Towards distributed measurements of electric fields using optical fibers: proposal and proof-of-concept experiment," *Sensors* 20(16), 4461 (2020).
238. P. G. Kazansky, A. Kamal, and P. St. J. Russell, "High second-order nonlinearities induced in lead silicate glass by electron-beam irradiation," *Opt. Lett.* 18(9), 693–695 (1993).
239. A. Michie, I. M. Bassett, J. H. Haywood, and J. Ingram, "Sensing electric fields using thermally poled silica optical fibre at 50Hz," in *Optical Fiber Sensors*, OSA Technical Digest (CD) (Optica Publishing Group, 2006), paper TuC3.
240. A. Michie, I. Bassett, J. Haywood, and J. Ingram, "Electric field and voltage sensing at 50 Hz using a thermally poled silica optical fibre," *Meas. Sci. Technol.* 18(10), 3219–3222 (2007).
241. A. Michie, J. Ingram, I. M. Bassett, J. H. Haywood, P. Hambley, and P. Henry, "Lab tests of an all fibre voltage sensor system," *Proc. SPIE* 7004, 700431 (2008), doi: 10.1117/12.786914
242. P. Hambley, A. Michie, I. Bassett, P. Henry, and J. Ingram, "Self aligning fibre for a fibre optic voltage sensor," *Proc. SPIE* 7004, 70042W (2008), doi: 10.1117/12.786031
243. E. K. Johnson, J. M. Kvavle, R. H. Selfridge, et al., "Electric field sensing with a hybrid polymer/glass fiber," *Appl. Opt.* 46(28), 6953–6958 (2007).
244. J. M. Kvavle, J. Young, E. Gutierrez, et al., "Robust in-fiber electric field sensors using AJL8/APC electro-optic polymer," *IEEE Sens. J.* 11(9), 2057–2064 (2011).
245. K. Bohnert, S. Wildermuth, and H. Brändle, "High voltage measuring device using poled fiber," European patent EP2,274,569B1, priority May 14, 2008.
246. C. A. Burrus and J. Stone, "Single-crystal fiber optical devices: a Nd:YAG fiber laser," *Appl. Phys. Lett.* 26(6), 318–320 (1975).
247. K. Nagashio, A. Watcharapasorn, R. C. DeMattei, and R. S. Feigelson, "Fiber growth of near stoichiometric LiNbO3 single crystals by the laser-heated pedestal growth method," *J. Cryst. Growth* 265, 190–197 (2004).
248. S. Bera, C. D. Nie, M. G. Soskind, and J. A. Harrington, "Optimizing alignment and growth of low-loss YAG single crystal fibers using laser heated pedestal growth technique," *Appl. Opt.* 56, 9649–9655 (2017).
249. D.-H. Yoon, I. Yonenaga, T. Fukuda, and N. Ohnishi, "Crystal growth of dislocation-free $LiNbO_3$ single crystals by micro pulling down method," *J. Cryst. Growth* 142(3–4), 339–343 (1994).
250. V. I. Chani, A. Yoshikawa, Y. Kuwano, K. Hasegawa, and T. Fukuda, "Growth of $Y_3Al_5O_{12}$:Nd fiber crystals by micro-pulling-down technique," *J. Cryst.Growth* 204(1–2), 155–162 (1999).
251. J. W. Shur, W. S. Wang, S. J. Suh, et al., "Growth and characterization of Er-doped stoichiometric $LiNbO_3$ single crystal fibers by the micro-pulling down method," *J. Cryst. Growth* 229(1–4), 223–227 (2001).
252. J. B. Shim, J. H. Lee, A. Yoshikawa, M. Nikl, D. H. Yoon, and T. Fukuda, "Growth of $Bi_4Ge_3O_{12}$ single crystal by the micro-pulling-down method from bismuth rich composition," *J. Cryst. Growth* 243(1), 157–163 (2002).
253. J. B. Shim, A. Yoshikawa, A. Bensalah, et al., "Luminescence, radiation damage, and color center creation in Eu 3+-doped $Bi_4Ge_3O_{12}$ fiber single crystals," *J. Appl. Phys.* 93(9), 5131–5135 (2003).
254. V. Chani, K. Lebbou, B. Hautefeuille, O. Tillement, and J.-M. Fourmigue, "Evaporation induced diameter control in fiber crystal growth by micro-pulling-down technique: $Bi_4Ge_3O_{12}$," *Cryst. Res. Technol.* 41(10), 972–978 (2006).

255. M. Zhuravleva, V. I. Chani, T. Yanagida, and A. Yoshikawa, "The micro-pulling-down growth of $Bi_4Si_3O_{12}$ (BSO) and $Bi_4Ge_3O_{12}$ (BGO) fiber crystals and their scintillation efficiency," *J. Cryst. Growth* 310(7–9), 2152–2156 (2008).
256. S. Wildermuth, K. Bohnert, H. Brändle, J.-M. Fourmigue, and D. Perrodin, "Growth and characterization of single crystalline $Bi_4Ge_3O_{12}$ fibers for electrooptic high voltage sensors," *J. Sens.*, article ID 650572 (2013), doi: 10.1155/2013/650572
257. V. D. Golyshev, M. A. Gonik, and V. B. Tsvetovsky, "Spectral absorptivity and thermal conductivity of BGO and BSO melts and single crystals," *Int. J. Thermophys.* 29(4), 1480–1490 (2008).
258. B. M. Epelbaum, K. Inaba, S. Uda, et al., "A double-die modification of micro-pulling-down method for in situ clad/core doping of fiber crystal," *J. Cryst. Growth* 179(3–4), 559–566 (1997).
259. J. Ballato, T. Hawkins, P. Foy, et al., "Silicon optical fiber," *Opt. Express* 16, 18675–18683 (2008).
260. J. Ballato, T. Hawkins, P. Foy, et al., "Binary III-V semiconductor core optical fiber," *Opt. Express* 18, 4972–4979 (2010).
261. J. Ballato, T. Hawkins, P. Foy, et al., "Advancements in semiconductor core optical fiber," *Opt. Fiber Technol.* 16(6), 399–408 (2010).
262. J. Ballato, C. McMillen, T. Hawkins, et al., "Reactive molten core fabrication of glass-clad amorphous and crystalline oxide optical fibers," *Opt. Mater. Express* 2, 153–160 (2012).
263. B. Faugas, T. Hawkins, C. Kucera, K. Bohnert, and J. Ballato, "Molten core fabrication of bismuth germanium oxide $Bi_4Ge_3O_{12}$ crystalline core fibers," *J. Am. Ceram. Soc.* 101(9), 4340–4349 (2018).
264. H. Masai, T. Fujiwara, Y. Benino, and T. Komatsu, "Large second-order optical nonlinearity in 30BaO–15 TiO_2–55GeO_2 surface crystallized glass with strong orientation," *J. Appl. Phys.* 100(2), 023526 (2006).
265. H. Masai, N. Iwafuchi, Y. Takahashi, T. Fujiwara, S. Ohara, Y. Kondo, and N. Sugimoto, "Preparation of crystallized glass for application in fiber-type devices," *J. Mater. Res.* 24(1), 288–294 (2009).
266. N. Iwafuchi, M. Hirokazu, Y. Takahashi, and T. Fujiwara, "Electro-optic measurement in glass ceramics with highly oriented crystalline layers," *Electron. Lett.* 46(1), 69–71 (2010).
267. N. Iwafuchi, H. Masai, Y. Takahashi, T. Fujiwara, S. Ohara, Y. Kondo, and N. Sugimoto, "Crystallization behavior of optical fibers with multi layered structure for nonlinear optical devices," *J. Ceram. Soc. Jpn.* 116(1358), 1115–1120 (2008).
268. S. Ohara, H. Masai, Y. Takahashi, T. Fujiwara, Y. Kondo, and N. Sugimoto, "Fabrication of BaO-TiO_2-GeO_2-SiO_2 based glass fiber," *J. Ceram. Soc. Jpn.* 116(1358), 1083–1086 (2008).
269. S. Ohara, H. Masai, Y. Takahashi, T. Fujiwara, Y. Kondo, and N. Sugimoto, "Space-selectively crystallized fiber with second-order optical nonlinearity for variable optical attenuation," *Opt. Lett.* 34(7), 1027–1029 (2009).
270. R. Gibson, R. Selfridge, S. Schultz, W. Wang, and R. Forber, "Electro-optic sensor from high Q resonance between optical D-fiber and slab waveguide," *Appl. Opt.* 47(13), 2234–2240 (2008).
271. R. Gibson, R. Selfridge, and S. Schultz, "Electric field sensor array from cavity resonance between optical D-fiber and multiple slab waveguides," *Appl. Opt.* 48(19), 3695–3701 (2009).
272. J. Kvavle, S. Schultz, and R. Selfridge, "Ink-jetting AJL8/APC for D-fiber electric field sensors," *Appl. Opt.* 48(28), 5280–5286 (2009).
273. D. Perry, S. Chadderdon, R. Forber, et al., "Multiaxis electric field sensing using slab coupled optical sensor," *Appl. Opt.* 52(9), 1968–1977 (2013).
274. N. Stan, F. Seng, L. Shumway, R. King, R. Selfridge, and S. Schultz, "High electric field measurement using slab-coupled optical sensors," *Appl. Opt.* 55, 603–610 (2016).
275. F. Seng, N. Stan, R. King, et al., "Optical sensing of electric fields in harsh environments," *J. Lightwave Technol.* 35(4), 669–676 (2017).
276. N. Stan, F. Seng, L. Shumway, R. King, and S. Schultz, "Non-perturbing voltage measurement in a coaxial cable with slab-coupled optical sensor," *Appl. Opt.* 56(24), 6814–6821 (2017).

277. H. Sun, A. Pyajt, J. Luo, et al., "All-dielectric electrooptic sensor based on a polymer microresonator coupled side-polished optical fiber," *IEEE Sens. J.* 7(4), 1515–1524 (2007).
278. Z.-Y. Cheng, X.-D. Liu, L.-C. Ma, and M. Chen, "Developing highly efficient and electric-field-sensitive fiber-waveguide evanescent couplers," *Appl. Opt.* 60(17), 5087–5093 (2021).
279. W. Cady, *Piezoelectricity* (Dover Publications, New York, 1964).
280. B. Jaffe, W. Cook Jr., and H. Jafffe, *Piezoelectric Ceramics* (Academic Press, New York, 1971).
281. A. Barzegar, R. Bagheri, and A. Karimi Taheri, "Aging of piezoelectric composite transducers," *J. Appl. Phys.* 89, 2322–2326 (2001), doi: 10.1063/1.1339853
282. Landoldt-Börnstein, *Elastic, Piezoelectric, Piezooptic, and Electrooptic Constants of Crystals*, K.-H. Hellwege and A. M. Hellwege, Eds., New Series III 1, 2 (Springer-Verlag, 1966) and references therein.
283. M. W. Hooker, "Properties of PZT-based piezoelectric ceramics between-150 and 250 C," NASA Tech. Rep. 1.26: 208708 (Langley Research Center 1998).
284. M. Pacheco, F. Mendoza Santoyo, A. Méndez, and L. A. Zenteno, "Piezoelectric-modulated optical fibre Bragg grating high-voltage sensor," *Meas. Sci. Technol.* 10(9), 777–782 (1999).
285. G. M. Sessler, "Piezoelectricity in polyvinylidene fluoride," *J. Acoust. Soc. Am.* 70(6), 1596–1608 (1981).
286. R. Bechmann, "Elastic and piezoelectric constants of alpha-quartz," *Phys. Rev.* 110(5), 1060–1061 (1958).
287. R. T. Smith and F. S. Welsh, "Temperature dependence of the elastic, piezoelectric, and dielectric constants of lithium tantalate and lithium niobate," *J. Appl. Phys.* 42(6), 2219–2230 (1971).
288. S. Bouchy, R. J. Zednik, and P. Bélanger, "Characterization of the elastic, piezoelectric, and dielectric properties of lithium niobate from 25 °C to 900 °C using electrochemical impedance spectroscopy resonance method," *Materials* 15(13), 4716 (2022).
289. I. B. Kobiakov, "Elastic, piezoelectric and dielectric properties of ZnO and CdS single crystals in a wide range of temperatures," *Solid State Comm.* 35(3), 305–310 (1980).
290. J. C. Brice, "Crystals for quartz resonators," *Rev. Mod. Phys.* 57(1), 105–146 (1985).
291. K. M. Bohnert and J. Nehring, "Fiber-optic sensing of electric field components," *Appl. Opt.* 27(23), 4814–4818 (1988).
292. K. Bohnert and J. Nehring, "Fiber-optic sensing of voltages by line integration of the electric field," *Opt. Lett.* 14(5), 290–292 (1989).
293. B. Y. Kim, J. N. Blake, S. Y. Huang, and H. J. Shaw, "Use of highly elliptical core fibers for two-mode fiber devices," *Opt. Lett.* 12(9), 729–731 (1987).
294. J. N. Blake, S. Y. Huang, B. Y. Kim, and H. J. Shaw, "Strain effects on highly elliptical core two-mode fibers," *Opt. Lett.* 12(9), 732–734 (1987).
295. J. Brooks, R. Wentworth, R. Youngquist, et al., "Coherence multiplexing of fiber-optic interferometric sensors," *J. Lightwave Technol.* 3(5), 1062–1072 (1985).
296. K. Bohnert, C. C. de Wit, and J. Nehring, "Coherence-tuned interrogation of a remote elliptical-core, dual-mode fiber strain sensor," *J. Lightwave Technol.* 13(1), 94–103 (1995), doi: 10.1109/50.350640
297. A. S. Gerges, T. P. Newson, and D. A. Jackson, "Coherence tuned fiber optic sensing system, with self-initialization, based on a multimode laser diode," *Appl. Opt.* 29(30), 4473–4480 (1990).
298. Y. Ning, K. T. Grattan, A. W. Palmer, and B. T. Meggitt, "Characteristics of a multimode laser diode in a dual-interferometer configuration," *J. Lightwave Technol.* 8(12), 1773–1778 (1990).
299. D. A. Jackson, R. Priest, A. Dandridge, and A. B. Tveten, "Elimination of drift in a single-mode optical fiber interferometer using a piezoelectrically stretched coiled fiber," *Appl. Opt.* 19(17), 2926–2929 (1980).
300. H. Yang, Y. Guo, S. Xu, et al., "Vibration characteristics of quartz crystal and analysis of system error in fiber optic voltage sensor," *Front. Phys.* 8, 607724 (2020).
301. K. Bohnert, "Fiber-optical voltage sensor," European patent application, published as WO2022/029046A1, 2022-02-10.

302. K. Bohnert and P. Pequignot, "Inherent temperature compensation of a dual-mode fiber voltage sensor with coherence-tuned interrogation," *J. Lightwave Technol.* 16(4), 598–604 (1998), doi: 10.1109/50.664069
303. K. M. Bohnert, J. Kostovic, and P. Pequignot, "Fiber optic voltage sensor for 420 kV electric power systems," *Opt. Eng.* 39(11), 3060–3067 (2000), doi; 10.1117/1.1315023
304. T. Yoshino, K. Kurosawa, K. Itoh, and T. Ose, "Fiber-optic Fabry-Perot interferometer and its sensor applications," *IEEE J. Quantum Electron.* 18(10), 1624–1633 (1982).
305. L. Martinez-Leon, A. Diez, J. L. Cruz, and M. V. Andres, "A frequency-output fiber optic voltage sensor with temperature compensation for power systems," *Sens. Actuator. A Phys.* 102(3), 210–215 (2003).
306. S. Kim, J. Park, and W.-T. Han, "Optical fiber AC voltage sensor," *Microw. Opt. Technol. Lett.* 51(7), 1689–1691 (2009).
307. P. Niewczas, L. Dziuda, G. Fusiek, and J. R. McDonald, "Design and evaluation of a preprototype hybrid fiber-optic voltage sensor for a remotely interrogated condition monitoring system," *IEEE Trans. Instrum. Meas.* 54(4), 1560–1564 (2005), doi: 10.1109/TIM.2004.851072
308. G. Fusiek, P. Niewczas, and J. R. McDonald, "Feasibility study of the application of optical voltage and current sensors and an arrayed waveguide grating for aero-electrical systems," *Sens. Actuator. A Phys.* 147(1), 177–182 (2008).
309. G. Fusiek, P. Niewczas, and M. Judd, "Towards the development of a downhole optical voltage sensor for electrical submersible pumps," *Sens. Actuator. A Phys.* 184, 173–181 (2012).
310. P. Orr, G. Fusiek, P. Niewczas, et al., "Distributed photonic instrumentation for power system protection and control," *IEEE Trans. Instrum. Meas.* 64(1), 19–26 (2015), doi: 10.1109/TIM.2014.2329740
311. G. Fusiek, J. Nelson, P. Niewczas, et al., "Optical voltage sensor for MV networks," *IEEE Sens. 2017*, 1–3 (2017), doi: 10.1109/ICSENS.2017.8234104
312. G. Fusiek and P. Niewczas, "Photonic voltage transducer with lightning impulse protection for distributed monitoring of MV networks," *Sensors* 20(17), 4830 (2020).
313. G. Fusiek and P. Niewczas, "Construction and evaluation of an optical medium voltage transducer module aimed at a 132 kV optical voltage sensor for WAMPAC systems," *Sensors* 22(14), 5307 (2022).
314. R. C. Allil and M. M. Werneck, "Optical high-voltage sensor based on fiber Bragg grating and PZT piezoelectric ceramics," *IEEE Trans. Instrum. Meas.* 60(6), 2118–2125 (2011), doi: 10.1109/TIM.2011.2115470
315. A. Dante, R. M. Bacurau, C. C. Carvalho, et al., "Optical high-voltage sensor based on fiber Bragg gratings and stacked piezoelectric actuators for a.c. measurements," *Appl. Opt.* 58(30), 8322–8330 (2019).
316. M. N. Gonçalves and M. M. Werneck, "Optical voltage transformer based on FBG-PZT for power quality measurement," *Sensors* 21(8), 2699 (2021).
317. Q. Yang, Y. He, S. Sun, M. Luo, and R. Han, "An optical fiber Bragg grating and piezoelectric ceramic voltage sensor," *Rev. Scient. Instrum.* 88(10), 105005 (2017), doi: 10.1063/1.4986046
318. Y. He, Q. Yang, S. Sun, et al., "A multi-point voltage sensing system based on PZT and FBG," *Int. J. Electr. Power Energy Syst.* 117, 105607 (2020).
319. Y. He, Q. Yang, L. Huang, et al., "Frequency optimization of PZT-FBG voltage sensor based on temperature-independent demodulation method," *IEEE Sens. J.* 21(23), 26821–26829 (2021).
320. C. T. Law, K. Bhattarai, and D. C. Yu, "Fiber-optics-based fault detection in power systems," *IEEE Trans. Power Deliv.* 23(3), 1271–1279, (2008), doi: 10.1109/TPWRD.2008.919233
321. J. M. Lukens, N. Lagakos, V. Kaybulkin, C. J. Vizas, and D. J. King, "Intensity-modulated fiber-optic voltage sensors for power distribution systems," arXiv:2001.05412v1 [eess.SY] (2020).
322. F. Sun, G. Xiao, Z. Zhang, and C. P. Grover, "Piezoelectric bimorph optical-fiber sensor," *Appl. Opt.* 43(4), 1922–1925 (2004).

323. M. Ndiaye, "A fiber optic voltage sensor based on intensity modulation," *Proc. SPIE* 9098, 183–191 (2014).
324. K. P. Koo and G. H. Sigel, "An electric field sensor utilizing a piezoelectric polyvinylidene fluoride (PVF2) film in a singlemode fiber interferometer," *IEEE J. Quantum Electron.* QE18(4), 670 (1982).
325. L. J. Donalds, W. G. French, W. C. Mitchell, et al., "Electric field sensitive optical fibre using piezoelectric polymer coating," *Electron. Lett.* 18(8), 327–328 (1982).
326. P. D. DeSouza and M. D. Mermelstein, "Electric field detection with a piezoelectric polymer-jacketed single-mode optical fiber," *Appl. Opt.* 21(23), 4214–4218 (1982).
327. M. D. Mermelstein, "Optical-fiber copolymer-film electric field sensor," *Appl. Opt.* 22(7), 1006–1009 (1983).
328. M. Imai, H. Tanizawa, Y. Ohtsuka, Y. Takase, and A. Odajima, "Piezoelectric copolymer jacketed single-mode-fibers for electric-field sensor applications," *J. Appl. Phys.* 60(6), 1916–1918 (1986).
329. L. M. Levinson, Ed., *Electronic Ceramics, Properties, Devices and Applications* (Dekker, New York, 1988).
330. S. T. Vohra, F. Bucholtz, and A. D. Kersey, "Fiber-optic dc and low-frequency electric-field sensor," *Opt. Lett.* 16(18), 1445–1447 (1991).
331. S. T. Vohra and F. Bucholtz, "Fiber-optic ac electric-field sensor based on the electrostrictive effect," *Opt. Lett.* 17(5), 372–374 (1992).
332. L. Fabiny, S. T. Vohra, and F. Bucholtz, "High-resolution fiber-optic low-frequency voltage sensor based on the electrostrictive effect," *IEEE Photonics Technol. Lett.* 5(8), 952–953 (1993), doi: 10.1109/68.238266
333. L. Fabiny, S. T. Vohra, F. Bucholtz, and A. Dandridge, "Three-channel low-frequency fiber-optic voltage sensor," *Opt. Lett.* 19(3), 228–230 (1994).
334. M. C. Tomic, J. M. Elazar, and Z. V. Djinovic, "Voltage measurement based on the electrostrictive effect with simultaneous temperature measurement using a 3 × 3 fiber-optic coupler and low coherence interferometric interrogation," *Sens. Actuat. A Phys.* 115 (2–3), 462–469 (2004).
335. J. Zhao, H. Zhang, Y. Wang, and H. Liu, "Fiber-optic electric field sensor based on electrostriction effect," *Appl. Mech. Mater.* 187, 235–240 (2012).
336. F. Marignetti, E. de Santis, S. Avino, et al., "Fiber Bragg grating sensor for electric field measurement in the end windings of high-voltage electric machines," *IEEE Trans. Ind. Electron.* 63(5), 2796–2802 (2016), doi: 10.1109/TIE.2016.2516500
337. P. G. de Gennes and J. Prost, *The Physics of Liquid Crystal* (Clarendon, Oxford, UK, 1995).
338. S. Sato and T. Hara, "Applications of a ferroelectric liquid-crystal cell to an electric field sensor," *Jpn. J. Appl. Phys.* 32, 3664–3665 (1993).
339. F. Anagni, C. Bartoletti, U. Marchetti, et al., "Optical sensors for electric substations: a voltage presence detector using a liquid crystal cell," *IEEE Trans. Instrum. Meas.* 43(3), 475–480 (1994), doi: 10.1109/19.293470
340. B. M. Lacquet, P. L. Swart, and S. J. Spammer, "Polymer dispersed liquid crystal fiber optic electric field probe," *IEEE Trans. Instrum. Meas.* 46(1), 31–35 (1997), doi: 10.1109/19.552153
341. M. Tabib-Azar, B. Sutapun, T. Srikhirin, et al., "Fiber optic electric field sensors using polymer-dispersed liquid crystal coatings and evanescent field interactions," *Sens. Actuator. A Phys.* 84(1–2), 134–139 (2000).
342. E. Scherschener, C. D. Perciante, E. A. Dalchiele, et al., "Polymer-dispersed liquid-crystal voltage sensor," *Appl. Opt.* 45(15), 3482–3488 (2006).
343. S. Ertman, K. Rutkowska, and T. R. Woliński, "Recent progress in liquid-crystal optical fibers and their applications in photonics," *J. Lightwave Technol.* 37(11), 2516–2526 (2019).
344. L. Wei, L. Eskildsen, J. Weirich, et al., "Continuously tunable all-in-fiber devices based on thermal and electrical control of negative dielectric anisotropy liquid crystal photonic bandgap fibers," *Appl. Opt.* 48, 497–503 (2009).

345. S. Ertman, T. R. Wolinski, D. Pysz, et al., "Low-loss propagation and continuously tunable birefringence in high-index photonic crystal fibers filled with nematic liquid crystals," *Opt. Express* 17(21), 19298–19310 (2009).
346. S. Mathews, G. Farrell, and Y. Semenova, "Directional electric field sensitivity of a liquid crystal infiltrated photonic crystal fiber," *IEEE Photonics. Technol. Lett.* 23(7), 408–410 (2011), doi: 10.1109/LPT.2011.2107319
347. S. Mathews, G. Farrell, and Y. Semenova, "Liquid crystal infiltrated photonic crystal fibers for electric field intensity measurements," *Appl. Opt.* 50, 2628–2635 (2011).
348. Y. Huang, Y. Wang, C. Mao, et al., "Liquid-crystal-filled side-hole fiber for high-sensitivity temperature and electric field measurement," *Micromachines* 10(11), 761 (2019), doi: 10.3390/mi10110761
349. L. Silvestri, Y. Chen, Z. Brodzeli, et al., "A novel optical sensing technology for monitoring voltage and current of overhead power lines," *IEEE Sens. J.* 21(23), 26699–26707 (2021), doi: 10.1109/JSEN.2021.3119580
350. Q. Liu, P. Xue, Q. Wu, et al., "Electrically sensing characteristics of the Sagnac interferometer embedded with a liquid crystal-infiltrated photonic crystal fiber," *IEEE Trans. Instrum. Meas.* 70, art no. 9511209 (2021), doi: 10.1109/TIM.2021.3097402
351. Md. S. Islam, Md. A. Mollah, A. F. Alkhateeb, et al, "Surface plasmon resonance voltage sensor based on a liquid crystal-infiltrated hollow fiber," *Opt. Mater. Express* 12, 4630–4642 (2022).
352. W. L. Wang, Q. Liu, Z. Liu, et al., "Simulation of a temperature-compensated voltage sensor based on photonic crystal fiber infiltrated with liquid crystal and ethanol," *Sensors* 22(17), 6374 (2022).
353. A. Czapla, W. J. Bock, T. R. Woliński, et al., "Improving the electric field sensing capabilities of the long-period fiber grating coated with a liquid crystal layer," *Opt. Express* 24, 5662–5673 (2016).
354. X. Chen, F. Du, T. Guo, et al., "Liquid crystal-embedded tilted fiber grating electric field intensity sensor," *J. Lightwave Technol.* 35, 3347–3353 (2017).
355. M. O. Ko, S. J. Kim, J. H. Kim, et al., "Dynamic measurement for electric field sensor based on wavelength-swept laser," *Opt. Express* 22(13), 16139–16147 (2014).
356. P. Yang, X. Wen, Z. Chu, et al., "AC/DC fields demodulation methods of resonant electric field microsensor," *Micromachines* 11(5), 511 (2020).
357. J. Li, J. Liu, C. Peng, et al., "Design and testing of a non-contact mems voltage sensor based on single-crystal silicon piezoresistive effect," *Micromachines* 13, 619 (2022), doi: 10.3390/mi13040619
358. A. Mendez, T. F. Morse, and K. A. Ramsey, "Fiber optic electric-field microsensor," *Proc. SPIE* 1795, 153–164 (1993).
359. A. Roncin, C. Shafai, and D. R. Swatek, "Electric field sensor using electrostatic force deflection of a micro-spring supported membrane," *Sens. Actuator. A Phys.* 123–124, 179–184 (2005).
360. T. S. Priest, G. B. Scelsi, and G. A. Woolsey, "Optical fiber sensor for electric field and electric charge using low-coherence, Fabry-Perot interferometry," *Appl. Opt.* 36(19), 4505–4508 (1997).
361. L. M. Zhou, W. Huang, T. Zhu, and M. Liu, "High voltage sensing based on fiber Fabry-Perot interferometer driven by electric field forces," *J. Lightwave Technol.* 32(19), 3337–3343 (2014).
362. T. Chen, C. Shafai, A. Rajapakse, et al., "Micromachined ac/dc electric field sensor with modulated sensitivity," *Sensor. Actuator. A Phys.* 245, 76–84 (2016).
363. A. Javernik and D. Donlagic, "Miniature micro-machined fiber-optic Fabry-Perot voltage sensor," *Opt. Express* 27(9), 13280–13291 (2019).
364. L. Wang and N. Fang, "Power-frequency electric field sensing utilizing a twin-FBG Fabry-Perot interferometer and polyimide tubing with space charge as field sensing element," *Sensors*, 19(6), 1456 (2019).
365. A. Kainz, H. Steiner, J. Schalko, et al., "Distortion-free measurement of electric field strength with a MEMS sensor," *Nat. Electron.* 1(1), 68–73 (2018).

366. T. P. Pustelny, "Electroluminescent optical fiber sensor for detection of a high intensity electric field," *Photonics Lett. Pol.* 12(1), 19–21 (2020).
367. X. Yang, Y. Jia, L. Gao, S. Ji, and Z. Li, "A method for measuring surface electric field intensity of insulators based on electroluminescent effect," *Energy Rep.* 6, 1537–1543 (2020).
368. R. Heinzelmann, A. Stohr, M. Groz, et al., "Optically powered remote optical field sensor system using an electroabsorption-modulator," in *Proc. IEEE MTT-S Int. Microwave Symposium* 1998, pp. 1225–1228.
369. S. Höjer and H. Ahlberg, "Spatially resolved measurements of electric field distributions in gas-insulated high voltage components using a CO2 laser probe beam," *Appl. Opt.* 27(18), 3908–3913 (1988).
370. A. Fiore, F. Rosencher, V. Berger, and J. Nagle, "Electric field induced interband second harmonic generation in GaAs/AlGaAs quantum wells," *Appl. Phys. Lett.* 67(25), 3765–3767 (1995).
371. C. K. Sun, S. W. Chu, S. P. Tai, et al., "Piezoelectric-field distribution in gallium nitride with scanning second-harmonic generation microscopy," *Scanning* 23(3), 182–192 (2001).
372. M. S. Simeni, Y. Tang, Y.-C. Hung, et al., "Electric field in Ns pulse and AC electric discharges in a hydrogen diffusion flame," *Combust. Flame* 197, 254–264 (2018).
373. I.-H. Chen, S.-W. Chu, F. Bresson, et al., "Three-dimensional electric field visualization utilizing electric-field-induced second-harmonic generation in nematic liquid crystals," *Opt. Lett.* 28, 1338–1340 (2003).
374. M. M. Werneck and A. C. S. Abrantes, "Fiber-optic-based current and voltage measuring system for high-voltage distribution lines," *IEEE Trans. Power Deliv.* 19(3), 947–951 (2004).
375. R. Moghe, A. R. Iyer, F. C. Lambert, and D. M. Divan, "A low-cost wireless voltage sensor for monitoring MV/HV utility assets," *IEEE Trans. Smart Grid* 5(4), 2002–2009 (2014), doi: 10.1109/TSG.2014.2304533
376. A. Fransen, G. W. Lubking, and M. J. Vellekoop, "High-resolution high-voltage sensor based on SAW," *Sens. Actuator. A Phys.* 60(1–3), 49–53 (1997).
377. M. Yazdani, D. J. Thomson, and B. Kordi, "Passive wireless sensor for measuring AC electric field in the vicinity of high-voltage apparatus," *IEEE Trans. Ind. Electron.* 63(7), 4432–4441 (2016), doi: 10.1109/TIE.2016.2546845
378. K. Zhu, W. K. Lee, and P. W. T. Pong, "Non-contact voltage monitoring of HVDC transmission lines based on electromagnetic fields," *IEEE Sens. J.* 19(8), 3121–3129 (2019), doi: 10.1109/JSEN.2019.2892498
379. V. M. N. Passaro, F. Dell'Olio, F. De Leonardis, "Electromagnetic field photonic sensors," *Prog. Quantum Electron.* 30(2–3), 45–73 (2006).
380. D.-J. Lee, N.-W. Kang, J.-H. Choi, J. Kim, and J. F. Whitaker, "Recent advances in the design of electro-optic sensors for minimally destructive microwave field probing," *Sensors* 11, 806–824 (2011).
381. R. Zeng, B. Wang, B. Niu, and Z. Yu, "Development and application of integrated optical sensors for intense E-field measurement," *Sensors* 12, 11406–11434 (2012).
382. L. J. Li, W. H. Zhang, H. Li, and P. Pan, "An overview of optical voltage sensor based on Pockels effect," *Adv. Mat. Res.* 694–697, 987–991 (2013).
383. J. Peng, S. Jia, J. Bian, et al., "Recent progress on electromagnetic field measurement based on optical sensors," *Sensors* 19, 2860 (2019).
384. Y. Zhao, L. Wang, and Y. Zhou, "Research status of optical fiber voltage transformer," *J. Phys. Conf. Ser.* 1920(1), 012003 (2021).
385. F. Behague, C. Calero, A. Coste, et al., "Minimally invasive optical sensors for microwave-electric-field exposure measurements," *J. Opt. Microsyst.* 1(2), 020902 (2021).

Final Remarks

In the course of this book, we have seen that optical current and voltage sensors have made their way into the electric power world and other industries. Nevertheless, conventional current and voltage transformers still dominate electric power transmission and distribution and will probably continue to do so for a long time to come. CT and VT are well established, have been cost-optimized over decades, and the addition of digital interfaces and optical communication makes them compatible with modern digital substations. Besides, the conceptional simplicity of inductive transformers is hard to beat.

However, the benefits of optical sensors have been recognized. Many power companies have installed optical sensors to gain experience with the new technology. Especially in the case of non-standard requirement specifications that may include special physical constraints, remote measurement locations like in mixed line protection, or the wish to measure frequency contents as high as several 100 kHz, optical sensors are often the solution of choice. Optical sensors also have unique advantages on the DC side of HVDC transmission systems, and in electro-winning of metals and chlorine, FOCS have been well established for more than a decade already.

Another application field is medium voltage grids. The distributed integration of renewable energy increases the need of current and voltage measurement for maintaining grid stability and power quality. This includes the monitoring of voltage volatility, unbalanced loads between phases, and high harmonics. These tasks can be met with relatively simple local optical field sensors with moderate accuracy that can be easily retrofit and clamp-on mounted to power lines, attached to underground power cables, or integrated into switching cabinets. Wireless electronic sensors are a competing option, though.

Still, challenges remain. Further proof of long-term reliability and cost reduction, especially at lower-rated voltages, continue to remain priorities. Continued simplification of the sensors without loss of performance, modular designs that ease installation and exchange, sensors dedicated to specific tasks instead of all-in-one solutions, and optimized and automated production methods may help to make "light work of current and voltage measurement."

Index

Pages in *italics* refer to figures and pages in **bold** refer to tables.

E

F

G

H

I

J

K

L

M

N

O

P

Q

R

S

X

Y

Z